WAR AND REMEMBRANCE

Human Dimensions in Foreign Policy, Military Studies, and Security Studies

Series editors: Stéphanie A.H. Bélanger, Pierre Jolicoeur, and Stéfanie von Hlatky

Books in this series illuminate thorny issues in national and international security, analyzing both military and foreign policy. They highlight the human dimensions of war, such as the health and well-being of military members, the factors that influence military cooperation and operational effectiveness, civil-military relations and decisions regarding the use of force, and the challenges of violence and terrorism, as well as human security and conflict resolution. Some authors focus on the ethical, moral, and legal ramifications of ongoing conflicts and wars, while others, through the lens of policy analysis, explore the impact of military and political strife on human rights and the role the public plays in shaping international policy.

Published in collaboration with Queen's University and the Royal Military College of Canada, with the Centre for International and Defence Policy, the Canadian Institute for Military and Veteran Health Research, and the Centre for Security, Armed Forces, and Society, the series plays a pivotal role in reconceptualizing contemporary security challenges – both in the academic realm and for broader publics.

1 Going to War?
Trends in Military Interventions
Edited by Stéfanie von Hlatky and H. Christian Breede

2 Bombs, Bullets, and Politicians
France's Response to Terrorism
Christophe Chowanietz

3 War Memories
Commemoration, Recollections, and Writings on War
Edited by Stéphanie A.H. Bélanger and Renée Dickason

4 Disarmament under International Law
John Kierulf

5 Contract Workers, Risk, and the War in Iraq
Sierra Leonean Labor Migrants at US Military Bases
Kevin J.A. Thomas

6 Violence and Militants
From Ottoman Rebellions to Jihadist Organizations
Baris Cayli

7 Frontline Justice
The Evolution and Reform of Summary Trials in the Canadian Armed Forces
Pascal Lévesque

8 Countering Violent Extremism and Terrorism
Assessing Domestic and International Strategies
Edited by Stéfanie von Hlatky

9 Transhumanizing War
Performance Enhancement and the Implications for Policy, Society, and the Soldier
Edited by H. Christian Breede, Stéphanie A.H. Bélanger, and Stéfanie von Hlatky

10 Coping with Geopolitical Decline
The United States in European Perspective
Edited by Frédéric Mérand

11 Rivals in Arms
The Rise of UK-France Defence Relations in the Twenty-First Century
Alice Pannier

12 Outsourcing Control
The Politics of International Migration Cooperation
Katherine H. Tennis

13 Why We Fight
New Approaches to the Human Dimensions of Warfare
Edited by Robert C. Engen, H. Christian Breede, and Allan English

14 Canada as Statebuilder?
Development and Reconstruction Efforts in Afghanistan
Laura Grant and Benjamin Zyla

15 Women, Peace, and Security
Feminist Perspectives on International Security
Edited by Caroline Leprince and Cassandra Steer

16 The Ones We Let Down
Toxic Leadership Culture and Gender Integration in the Canadian Forces
Charlotte Duval-Lantoine

17 Postcards from the Western Front
Pilgrims, Veterans, and Tourists after the Great War
Mark Connelly

18 War and Remembrance
Recollecting and Representing War
Edited by Rénee Dickason, Delphine Letort, Michel Prum, and Stéphanie A.H. Bélanger

War and Remembrance

Recollecting and Representing War

EDITED BY

Renée Dickason, Delphine Letort,

Michel Prum, and Stéphanie A.H. Bélanger

McGill-Queen's University Press
Montreal & Kingston • London • Chicago

ISBN 978-0-2280-1068-5 (cloth)
ISBN 978-0-2280-1267-2 (ePDF)
ISBN 978-0-2280-1268-9 (ePUB)

Legal deposit second quarter 2022
Bibliothèque nationale du Québec

Printed in Canada on acid-free paper that is 100% ancient forest free (100% post-consumer recycled), processed chlorine free

Funded by the Government of Canada | Financé par le gouvernement du Canada | Canada

We acknowledge the support of the Canada Council for the Arts.
Nous remercions le Conseil des arts du Canada de son soutien.

Library and Archives Canada Cataloguing in Publication

Title: War and remembrance : recollecting and representing war / edited by Renée Dickason, Delphine Letort, Michel Prum, and Stéphanie A.H. Belanger.
Names: Dickason, Renée, editor. | Letort, Delphine, editor. | Prum, Michel, editor. | Bélanger, Stéphanie A. H., 1972– editor.
Series: Human dimensions in foreign policy, military studies, and security studies ; 18.
Description: Series statement: Human dimensions in foreign policy, military studies, and security studies ; 18 | Includes bibliographical references and index.
Identifiers: Canadiana (print) 20220135266 | Canadiana (ebook) 20220135398 | ISBN 9780228010685 (cloth) | ISBN 9780228012672 (ePDF) | ISBN 9780228012689 (ePUB)
Subjects: LCSH: War and society. | LCSH: Memory—Social aspects. | LCSH: Collective memory. | LCSH: Memorialization. | LCSH: War in mass media.
Classification: LCC HM554 .W35 2022 | DDC 303.6/6—dc23

This book was typeset in 10.5/13 Sabon.

Contents

In homage to Pr Dr Denis Mukwege
(Nobel Peace Prize Laureate, 2018)

Ayons le courage de dire la vérité et d'effectuer
le travail de mémoire.

"Let's all have the courage to tell the truth and carry out
the task of constructing memory" (our translation)

Denis Mukwege

Figures

Acknowledgments

Institutional support for this publication has come from Rennes University (France), in particular the ACE (Anglophonie: Communautés, Ecritures) research unit, and from the Royal Military College of Canada (Kingston) and the Canadian Institute for Military and Veteran Health Research Team, to whom we are most grateful.

We owe particular thanks to our very patient contributors, whose enduring commitment has enabled this collective to come to fruition and whom it was a pleasure to meet in Paris (France) in June 2016. The dedication of each one of them in reworking their book chapters in an integrative way brought meaning and harmony to the otherwise complex and multifaceted topic of war commemoration.

WAR AND REMEMBRANCE

Introduction

Renée Dickason, Delphine Letort,
Michel Prum, and Stéphanie Bélanger

For those who have not lived through war, there is no authentic memory to be shared, but there is knowledge to be gained, from listening to first-hand testimonies, observing original artefacts, reading about war narratives, and watching war films. More often than not, history books and education provide a first encounter with the wars of the past while the media and their journalistic reports offer a window onto contemporary conflicts. Memories of war can hardly be shared territory in this context, and yet the legacies of past conflicts often haunt the present, leaving, in the geographical and politico-social landscape, (in)visible marks that one has to learn to decipher.

Alison Landsberg contends that the mass media have the power to create collective memories and to share heritage that transcends divisions (of ethnicity, religion, age, gender, and geography) "by encouraging people to feel connected to, while recognizing the alterity of, the 'other.'"[1] While the construction of public monuments (e.g., statues) was the most common way of narrating the past in nineteenth- and twentieth-century societies, the development of mass media (including social media) has sustained the explosion of the "phenomenon of memory," which is largely due to the multiplication of memory claims by specific actors and activist groups, aiming for the recognition of their participation in past conflicts. The development of social media has favoured access to a diversity of means of facilitating the public expression of these memories, a phenomenon that Pierre Nora dubs the "democratization of history." In Nora's terms, "this takes the form of a marked emancipatory trend among peoples, ethnic groups and even certain classes of individual in the world today; in short, the emergence, over a very short period of time, of all those forms of memory bound up with minority groups for whom rehabilitating their past is part and parcel of reaffirming their identity."[2] The explosion of memory is also related to a collective urge to decolonize the past and to invent new ways of narrating collective history – especially wars, which are often used to feed patriotic feelings by disregarding the shameful deeds they may have entailed.

The acts of remembrance and of waging war are in a constant state of transformation. While the idea or concept of war has become (over the twentieth century and beyond) highly unstable (including through the introduction of remote killing via the development of drone warfare), the approach to memory has transformed how war is remembered. In *War beyond Words: Languages of Remembrance from the Great War* to the Present, Jay Winter calls attention to the "lenses" that "make understanding war possible at the same time as they limit what we see."[3] Winter is concerned with how our "imaginings of war," notably through the concepts of martyrdom and silence ("the suppression of troubling images and events"),[4] develop through a language of memory that is ingrained in all commemorative works. The pages that follow offer a singular intervention in the debates about war memory by examining a variety of media – speech, painting, sculpture, photography, and film – that not only carry a subjective experience of war but also frame perceptions of war.

War and Remembrance: Recollecting and Representing War focuses on the cultural memory of war through interdisciplinary approaches that illuminate how to disentangle war narratives from ideology. In approaching war memories through the lenses of history and literature, heritage and memory studies, film and media studies, museums and art history studies, the authors participating in this collective project adopt different methodologies to understand the influence of the past over the present. Some of them study fictional forms in order to explore the psychological and aesthetic imprint of wars on the creative impulse of artists and writers who attempt to bring attention to repressed parts of national and local histories. Others focus on tangible traces of the past (in the forms of photographs, architecture, human remains, etc.), analyzing the information they reveal about the past and how they are incorporated into memorial processes through museology or by other means. Most strive to enhance the repressed voices of those war actors whose deaths were dismissed by the patriotic rhetoric of martial powers.

Memory is a polysemic notion that encourages interdisciplinary dialogue. Maurice Halbwachs's founding reflections on the social frameworks of collective memory explore the interaction between memories of experience lived or transmitted (particularly within families) and the interpenetration of the individual and the collective, or of the psychic and the social, in the expression of memory.[5] Pierre Nora's *Lieux de Mémoire* evokes traces and effects of the past in the present, giving way to a politics of memory that denotes institutional (and symbolic) forms of (national) memory as history.[6] Astrid Erll takes these limited, national notions and elaborates on travelling, transcultural memory.[7] Paul Ricœur's civic and political preoccupation and concern for "good forgetting" leads him to posit that the ultimate purpose of memory lies in reconciliation.[8] There are various ways of remembering the past, especially when wars are concerned. While national narratives of war pave the way for common memories, laying stress on specific battles and spotlighting heroic names as signifiers of patriotic pride, individuals with a direct experience of armed conflict recall events that are often not even mentioned in history books. Memories of war reveal conflicts between different versions: the subjective recollections of individual soldiers and civilian witnesses often challenge the dominant narrative shaped by policy-makers who decide which memorials should be built or taken down.

At this stage, it would be very risky to attempt to establish a reasoned assessment of the question of memory, given the extent to which reflection and research have developed, both qualitatively and quantitatively, since the mid-1970s. The phenomenon of memory itself takes extremely diverse forms, sometimes driven by the decomposition or recomposition of national myths; sometimes nourished by specific, individual, and shared political or social experiences; sometimes stimulated by the circulation of paradigms. The importance of wars in collective memory is therefore dependent on the circulation of information about them either through individual endeavours, especially creative works by artists and personal interventions by those involved, or through publicly commissioned projects such as museums and monuments. The pulling down or destruction of sculptures and monuments asserting or reinforcing a dominant narrative is a testimony to the enduring traumas pervading collective memories. A case in point is the movement to destroy Confederate statues that emerged in the wake of the Charleston Church shooting in the United States in June 2015. The destructions and displacements of these statues accompany the social and political vicissitudes of each society and are not in themselves new. Yet their massive character over a short period of time must be emphasized. If some statues have simply been diverted from their original meaning, such as the statue of Lenin in Odessa remodelled into Darth Vader by the artist Alexander Milov and put in place in October 2015, many other installations or creations express a society's relationship to its past and the divisive readings that structure its cultural representations. These displaced and destroyed statues, signs of an ideological rupture, bring to light how memories are superimposed and reappropriated to reveal scarcely healed psychological wounds, corpses that were once hastily buried.[9]

The purpose of this volume is to question the contemporary forms of expression of war memory, from the most traditional to the most innovative, from the political culture inherited from the ideologies of the nineteenth century to the digital era. War memory artefacts symbolize the muddied waters of public and private memory, illuminating the different ways in which memory and war are understood, experienced, and imagined. This book offers a variety of case studies, challenging current concepts and approaches to memory in order to question the means of remembrance of war. It highlights the cultural, political, and social stakes of memory processes as sources of conflict and exchange, of resistance and opposition, of negotiation and reconciliation. This collective work helps to give other perspectives and viewpoints in the gigantic puzzle of war representations. It invites readers to revisit wars or conflicts past, some of them forgotten, other still vividly commemorated, and to taste the atmosphere of the times and situations depicted, at home and/or abroad.

The corpuses examined are diversified, purposely eclectic, and heterogeneous, though they all belong to the fathomless interpretations and representations of wars, be they written narratives or visual documents: from autobiographies to postcards, from trench journals to satirical magazines, from cinematic narratives to television comedy series, from documentaries to fictions, from sonic to visual and filmic cultures. They all attempt to ponder on how wars are presented in sometimes discrete, discreet commemorations, inexorably reduced to clichés and stereotypes. They propagate myths and mythologies in the Barthesian sense.[10] Interpretations are variously channelled through a subjectivity that is constantly revived by a multitude of phenomena enriched by the benefit of hindsight, the (re)discovery of deliberately (or not) forgotten facts and events, fallen victims to historical amnesia, often the fruit not only of official censorship or even self-censorship but also of arrangements with the truth at a given time for security or ideological reasons. All these considerations help towards the understanding of the mechanisms that build and shape not only national but also more global collective memory. They contribute to evolution in the understanding, representing, or fashioning of wars and conflicts as well as to the shaping of national identities.[11] Memories of war use a mosaic of visual, sound, and filmic narratives and interpretations, intertwined among the meanderings and intricacies of war stories and histories, whether observed through the prisms of individual or collective perspectives, comparing local or national realities with a more global understanding. Whatever the motivations, the use of texts, images and sounds – whether informative, commemorative, re-emerging, reconstituted, or fictional – is emotionally strong and central to the comprehension of events and, more broadly, of history.

The uncovering of past events can emerge through multifarious experiences and opportunities. For one reason or another, researchers sometimes chance upon unexpected findings, sometimes touching and/or mysterious, that are evocative of something strangely close to personal recollections, direct or indirect. Intuition, inspired curiosity, or serendipity bring their own rewards, which arise not from reasoned or "scientific" guidance but from a smack of the freshness of the discovery or of the excitement of the (re)discovery of hidden treasures, such as those on archeological sites. Browsing through albums or leafing through a journal can bring palpable emotion and delicate vibrancy: it can lead to a kind of forensic inner examination or personal introspection, simply stimulate the imagination to linger and wander, or, like Proust's madeleine, recall something forgotten or lost, apparently gone, yet still present and surely intimately part of life and of the self.

Each area, locally centred or more globally apprehended, is a mine of information. The originality of what follows lies in the association of reflections grounded in different methodologies, research fields, and approaches, and from the pens of authors from diverse areas of the world working on the history of their native lands or of foreign parts.

Part 1, "Remembering War from Indigenous Perspectives," revisits various war narratives through the prism of national and individual memories, using a critical historiographical approach to show how master war narratives are being decolonized by the writings of researchers. It uses various examples to demonstrate that narratives of national and international conflicts are not fixed, but subject to change as historians adopt new perspectives and viewpoints from which to examine the past. The memories of war are particularly apt for revisionism when testimonies bring up forgotten events that ideologies of national unity repress.

A case in point concerns the experience of the Australian Aboriginal Progressive Association (AAPA), which provided Aboriginal groups with a voice for their discontent in the wake of the First World War. As John Maynard retraces the activist struggles of the AAPA in the first chapter of this section, he also underlines the commitment of the AAPA's leader, his grandfather, Fred Maynard. Family ties prompt him to retrieve hidden Aboriginal stories and to fathom the legacy of the battles fought by Aboriginal people ninety years ago. He thus points out the achievements of activists who dedicated their lives to fighting against oppressive policies – including government child removal policies.

Corinne David-Ives pursues a similar investigation by focusing on the place of Indigenous memories in the national war narrative of New Zealand, a former British colony in which the Maori people were reduced to racist stereotypes. The image of the "gallant Maori" arose during the colonial wars of the nineteenth century, celebrating both the elaborate strategies of Maori resistance and their defeat by the British army. David-Ives investigates the construction of a mythology around the figure of the Maori warrior, whose pride and bravery elicited admiration from British politician William Pember Reeves (1890) and New Zealand settlers. The perceived superiority of the Maoris was built into a national myth of identity and unity, which only served to conceal the brutal nature of government policies as regards native tribes. The historiographical approach to New Zealand's national narrative fails to completely alter this narrative as Maoris themselves have appropriated the myth of the Maori warrior.

Examining the national landscape of war memorials in Australia, Sheila Collingwood-Whittick furthers reflection on selective remembering and motivated forgetting in the context of Australia's construction of national identity. She observes a striking opposition between the lack of monuments devoted to Aboriginal warriors and victims and the numerous efforts made to galvanize collective memory around the Australian and New Zealand Army Corps (Anzac) that took part in the 1915 Gallipoli campaign, marking Australia's commitment to entering the First World War on the side of the Allies. Collingwood-Whittick's chapter sheds light on all aspects of the memorialization process of this event, including the various touristic visits to the Gallipoli battlefield (between 1995 and 2005) that illustrated a new emotional relationship to the past. Anzac has thus become a stepping stone in the construction of Australian national identity, overshadowing if not silencing the memories of Aboriginal struggles.

Also working on an Australian theme, Elizabeth Rechniewski and Matthew Graves clearly link remembering and representing war to nation building. War becomes the narrative upon which national identity is constructed. Gallipoli, though a defeat for Anzac forces, has become the basis upon which the nation has been built as it sheds light on (white) Australian prowess and sacrifice. Rechniewski and Graves address the question of Aboriginal people in the war memorials. The monuments dedicated to Aboriginal and Torres Strait Islander soldiers were "few, modest and late," as historian Ken Inglis puts it. Yet the inclusion of Black Diggers in national commemorations of imperial combat, problematic though it was, was easier to secure than was the recognition of the battles waged by Aboriginal people to defend their own land against white settlers. Memorials to the Black Wars (or Frontier Wars, as Aboriginals prefer to call them) are an issue that will have to be addressed. Like memorials, war museums are physical and architectural representations of past conflicts.

Fiction may help towards a better understanding of the Indigenous experience of war. Anna Branach-Kallas calls attention to the experience of black soldiers who were drawn into the First World War from the colonies. Examining French novelist Didier Daeninckx's "Un petit air mutin" (2004), British author Andrea Levy's "Uriah's

War" (2014), and Canadian writer Dionne Brand's "Tamarindus Indica" (1999), she underscores the degraded status granted to soldiers from the colonies recruited or conscripted to fight on the front. Non-white soldiers were not treated as equals, and their contribution was a dehumanizing experience. These works of fiction nonetheless challenge the dominant gaze by giving a voice to these soldier victims of discrimination, striving to call the collective memories of war into question by building a more inclusive narrative.

Part 2, "Memories of Colonial Involvement and Civil Wars," uses a diversity of angles to point out the reifying gaze that colonists cast upon Indigenous populations. The chapters in this part analyze discriminatory practices as expressed in various visual and written documents and, once again, by adopting the colonized perspective, suggest how such a previous ideological stance can be challenged.

Through the magic of touching and sensing artefacts of the past, Gilles Teulié investigates the lavish patchworks of sights and scenes of First World War Indian colonial combat troops exhibited in a collection of 250 postcards produced by the British Imperial First World War troops. The subject is not devoid of emotions and is amply worthy of the fascination and passion of the researcher. Teulié's study demonstrates that the postcards contributed to spreading stereotypical representations of the Gurkhas (Nepalese soldiers fighting alongside the British troops). As products of colonial propaganda, they give an insight into the enduring distorting and racist images that were embedded in European minds, conveying a portrait of the martial Gurkhas as warriors to be feared.

Such depiction was misleading, something that Florence Cabaret also demonstrates through the analysis of a novel that adopts the conventions of autobiography to relate the life of Lalu Singh, a fictional Indian character who joined the British forces during the First World War. Mulk Raj Anand's *Across the Black Waters* (1939) eschews the Orientalist gaze by exploiting humour as a narrative strategy that humanizes Lalu Singh and allows the protagonist to level criticism at the military codes with which he must comply while bonding across the ethnic line. The novel reverses standpoints and invites readers to see the war from the subaltern's perspective, aiming to undercut the dominant narrative concerning colonial and Indigenous troops.

Art historian Keith Bell relates, with vibrant and somewhat nostalgic emotion, the confrontation between his remembrance of things past and official and family albums dating back to his colonial childhood in Kenya at the time of the Mau Mau uprising. This is a very private approach to original research artefacts, a personal and intimate mise en abyme, which goes so far as to suggest a quasi-anthropological perception of the object of study. Interestingly, in this context, the perception and interpretation of archives, private or official, is tinged with subjectivity, thus posing a challenge to researchers faced with the delicate task of (re) writing "History."

Remembering past discussions of the US Civil War, Stephen J. Whitfield also questions how narratives of war may reflect power struggles. As a historian aiming for a truthful narrative, Whitfield questions the regional roots of the "myth of the Lost Cause," which has long prevailed in the South after Reconstruction thanks to influential, yet biased, literature and textbooks. An inquiring look at election results in Mississippi demonstrates that the colour line continues to operate as a divisive force and that the past endures in the present. The narrative of the past thus becomes a colonized narrative, which the researcher aims to decolonize.

Dominique Otigbah calls attention to an often neglected slice of history in a chapter that focuses on the experiences of ethnic minorities in the face of the civil war that tore through Nigeria between 1967 and 1970. She examines how the experience of the Biafrans and the Igbos has dominated the memories of that war, explaining that this is because of historical accounts and propaganda campaigns that downplayed ethnic diversity in the eastern regions, where members of non-Igbo communities (Efik, Ijaw, Ibibio, Ekoi, Ikwerre, and Ogoni, among others) were massacred along with 7 million Igbos. Otigbah aims for a more nuanced narrative that pleads for a duty of memory for all victims.

Part 3, "Recollections of World Wars," uses both primary and secondary materials as the authors concentrate on visible and aural/audible representations of war so as to apprehend and comprehend highlights and fragments of official, public, private, personal, fully or partly imagined pieces of the gigantic puzzle of local and/or global memories of war. Juxtaposing thoughts and ideas, new or revisited, and dismantling approved, created, or suggested interpretations of bygone days, helps to reset the subjectivities nurturing the creation and elaboration of memorial processes and devices.

Renee Dickason, who has spent over thirty-five years of her academic life reflecting on how cultures are understood through the lenses of visual media in contexts of war and peace, here chooses to take on two missions. The first is to offer an intimate construction of war memories through the prism of a trench journal, the *Wipers Times*, conceived in the worst days of the First World War in one of the most deadly sectors of the Western Front and revealing the essence of human emotions and cultural adherence in the face of potential immediate destruction – a narrative intended for immediate, on-the-spot consumption. The second is to offer a "Cartooning Mission" by focusing on a guided reading of cartoons from the renowned and satirical *Punch or The London Charivari* during the Second World War – a gem per se, illustrating the writing of war memories that are at once serious and "entertaining." The style, form, and atmosphere of *Punch* suggest that the publication also envisaged that it might serve to prompt later somewhat pleasing recollections. In the event and against all the odds, the *Wipers Times* survived the war as a fascinating piece of archive, both stimulating memories and revealing new insights into combined individual and collective preoccupations.

The atmosphere of a war can be revived through various stratagems and devices: visual images, oral discourses, musical repertoires, fictional or factual filmic reconstitutions, and verbal testimonies. In this regard, film historian Iván Villarmea Álvarez reminds us that "the free combination of visual materials from different sources, such as private home movies, archival footage and imaginary re-enactments, conveys a 'subjective truth' that reveals the filmmaker's perception of historical events."[12] This is what film historian Anita Jorge develops in her analysis of aural perception of the Blitz, as expressed in personal writings, interacting with the emergence of an official discourse on sound, a central consideration in cinematographic works devoted to the Second World War in Britain. Her chapter examines Richard Johnson's concept of a "cultural circuit," grounded in the assumption that public discourses are appropriated in personal narratives that are themselves incorporated into collective discourses. Anita Jorge studies the power of sound and the development of the sense of hearing in London underground shelters during air-raids along the lines established by mass-observation anthropologist Tom

Harrisson and war memoirist Harvey Klemmer, who pondered on oral and aural testimonies of war memories found in the literature of the Blitz. The sounds of war, which induce direct immersion in the suffering endured during bombardments, with civilians panting for breath as they anxiously await the all clear, give a different perception of the traumatic experience of wars and constitute an indelible source of war memories. Muffled resonance and echoes help to complete the recollection of facts and events.

Along with sounds, the spoken word (in the form of voices, speeches, vocal records, or broadcasts) is also part of what people might remember of an event, of a war. In Britain, one of the most notable examples in the Second World War was the Sunday evening *Postscripts*, written by established novelist J.B. Priestley and broadcast by the BBC in the summer and autumn of 1940, which attracted a peak audience of 16 million. Delivered in an intimate manner and in a recognizably northern English accent, Priestley produced something close to an unfinished spoken symphony, each movement of which included variations on familiar national themes – popular pre-war memories, courage and resolution, duty and sacrifice – while favourably comparing the British way of doing things with the dehumanized behaviour of the enemy. The carefully selected words of the speaker's skilfully crafted "personal, independent comments" made it easy for listeners to imagine for themselves the visual details of the scenes he was describing. The first of these broadcasts, 5 June 1940, quoted below, is widely acknowledged as having been instrumental in the shaping of the narration of the British evacuation of Dunkirk only a few days before, which is arguably the most enduring and frequently evoked memory of the whole war:

> I wonder how many of you feel as I do about this great Battle and evacuation of Dunkirk. The news of it came as a series of surprises and shocks, followed by equally astonishing new waves of hope ... What strikes me about it is how typically English it is. Nothing, I feel, could be more English ... both in its beginning and its end, its folly and its grandeur ... We have gone sadly wrong like this before; and here and now we must resolve never, never to do it again ... What began as a miserable blunder, a catalogue of ... misfortunes ended as an epic of gallantry. We have a queer habit – and you can see it running through our history – of conjuring up such transformations ... And to my mind what was

> most characteristically English about it ... was the part played not by the warships ... but by the little pleasure-steamers. We've known them and laughed at them, these fussy little steamers, all our lives ... These "Brighton Belles" and "Brighton Queens" left that innocent foolish world of theirs – to sail into the inferno, to defy bombs, shells, magnetic mines, torpedoes, machine-gun fire – to rescue our soldiers.[13]

But war can also be remembered in other ways. Accepting the premise that, retrospectively, once the horror of war is removed, its absurdity becomes even more apparent, humour, jokes, and comedy in general help to shed a gentler light on serious matters. British television series have (had) their say in the matter. *'Allo, 'Allo* (BBC, 1982–92) and, more particularly, *Dad's Army* (BBC, 1968–77) are, in their way, quite instructive sitcoms about the narration of the Second World War. Predictably enough, they present the common lot of citizens in the Second World War in a burlesque or even grotesque way, with an abundance of clichés, stereotypes, and catch phrases. Nevertheless, as the sometimes dubious and often absurd activities depicted were part and parcel of daily life in wartime, a certain impression of plausibility, if not of authenticity is created, stimulating happy or nostalgic memories.

This construction of an alternative vision of the war also pervades what Tatiana Prorokova-Konrad identifies as the "literature of intention." A series of American war novels written between 1948 and 1982 narrate the conflict of the Second World War through the eyes of soldiers. Shunning the heroic deeds and the brilliant leadership of commanders to be found in the adventures of the war as sold by the propaganda of American exceptionalism, the characters in these novels are but foot soldiers facing the "horrors of bureaucracy" and the egotistical ambition of officers willing to sacrifice lives for their personal advancement. The retrospective assessment of the war challenges the ideological slant that prevails around national memories of "the Good War."

Paul Stocker examines Second World War commemoration emanating from members of the far-right British National Party (BNP). Here recollecting the past is clearly an attempt to manipulate it in order to make it say the contrary of what it actually stood for. Thus BNP leader Nick Griffin unabashedly adopted the V sign with his fingers in front of TV cameras after his party's historic score at the

2009 European elections, thus reappropriating Churchill's famous gesture in his fight against Nazism during the Second World War. The heroic past – Churchill's commitment to resistance against the far-right Axis countries – is not *represented* but, rather, blatantly misrepresented, or deflected, to celebrate the very ideas that the British war cabinet kept opposing. Such examples of turning the tables are comparatively few, but representations of war are never mere photographs of facts: they always imply some kind of interpretation and often follow a hidden or unmasked agenda. To *re-present* something, through art or in other ways, is to make it present again, but this does not mean restoring it as it actually was. To summon up war memories is to take sides, to support some belligerents against others, to glorify or vilify their leaders or war itself, and it often tells more about the narrators or artists than about the actual warfare.

Finally, Victoria Young considers the power of war museums that are not there just to commemorate and celebrate but also to educate future generations. In this volume, she focuses on the United States's National World War II Museum in New Orleans, Louisiana. The choice of Louisiana, one of the poorest states in the United States, for a national museum dedicated to such an international event, rather than Washington, DC, is quite remarkable. What is also striking is the emphasis on peace, hope, and liberation, not on loss and devastation, laid by architect Bartholomew Voorsanger. The very architecture of the museum is a hymn to peace, including the "iconic" huge "Canopy of Peace," metaphorically shielding American troops and marking in the most visible way the cityscape of New Orleans.

Part 4 plumbs the depths of the dialectic implied in its title: "Remembering and Forgetting War." Jay Winter contends that representations of war take part in the "struggle against forgetting, mediated in a host of ways in social practices, in literature and the arts. It is not only that much of the violent history of the twentieth century is intrinsically worth remembering, but rather that those who die or who were injured can so easily be forgotten."[14]

In his chapter, historian Jeffrey Demsky focuses precisely on American sitcoms that contribute to (mis)remembering the destruction of European Jewry seventy-five years after the end of the Second World War. Demsky ponders on the historical distancing of a "bygone foreign memory" by studying the evolution of Holocaust-themed joke work on sitcoms skits, following the decline of the so-called rescuer lore through which the US identified itself with such values as tolerance, democracy, and pluralism. Demsky argues that current comic Holocaust remembrance may be either constructive or destructive or both. This is due either to its potential to revive this history in contemporary discourses, thereby helping to ensure that people continue to remember, or to its propensity to "destabilize facts by fabricating alternative scenarios that supplant historical realities, an especially insidious threat among youthful viewers" (328).

Mapping war history in memories leads us to examine the notions of the absurdity of war and troubled inheritance, a key notion in the study of the filmic representations of a war largely ignored in the Western world – the Sri Lankan ethnic conflict (1983–2010) involving one of the first terrorist organizations of modern times, the Liberation Tigers of Tamil Eelam (LTTE). Due to censorship, film historian Vilasnee Tampoe-Hautin battled with a paucity of material when researching evidence of the ongoing conflicts in Sri Lanka. She acknowledges the help given by the work of scholars and filmmakers, who pursued research or filming in an unsafe militarized environment, in the hope of capturing "the psychological and physical dilemmas of people caught in the throes of the conflict" (349).This chapter demonstrates how films operate as custodians of popular memory by interweaving various narrative strands from personal to social and cultural memory. They thus provide tools of remembrance where silence prevails.

Historian András Lénárt, whose research is based on Hispanic wars, takes here a special interest in the field of international, global film culture regarding war depiction, concentrating on the Spanish-American War (1898) and its exploitation in the United States of America, Cuba, and Spain. The first war filmed on the spot, this conflict pinpoints the complexity of representing war memories in various types of movies (actuality films, romantic fiction, action movies) and with the divergent perspectives on the same historical event adopted by different national film industries. It shows what historian Robert A. Rosenstone calls a revision of the past. It is indeed a kind of new highlight, a combination of recycling of thoughts and ideas that triggers new interpretations, the elaboration of a new rhetoric and a new conception based on the fact that the past cannot be separated from the present. All this, therefore, illustrates another perspective on the shaping of memories of war and the construction, the distortion, of the memorialization process.

The next chapter, by film scholar Raphaelle Costa de Beauregard, examines post–1918 films, notably William Wellman's *Wings* (1927) and Ernst Lubitsch's *Broken Lullaby* (1932), through the prisms of the psychological drama of memory and modern anti-hero narrative as they explore the complexities of remembering, forgetting, and coping with trauma. The films under study evoke peace as being contingent on oblivion when the trauma of war is too painful to live with. While the films broach memories through the angle of the individual, they participate in the construction of First World War collective memories.

Last but not least, Georges Fournier focuses on *Warriors* (1999) and *The Promise* (2011), Peter Kominsky's two documentary representations of soldiers in peacekeeping forces (in Bosnia in the 1990s and in Palestine in the late 1940s). The films articulate another vision of ongoing traumas fomented by wars, which contributes to a more complete understanding of conflict and the role of memory in the writing of history. Both conflicts are forgotten wars – "a blind spot in modern history" (394) – that brought the troops involved neither victory, nor recognition, nor satisfaction. Interestingly, as Fournier points out, the making of *Warriors* reawakened the memories of those who had served in Palestine and who subsequently provided the first-hand accounts of their own uncannily similar experiences of peacekeeping, which formed the basis for *The Promise*.

Finally, Part 5, "Intimate Memories of War," assesses a diversity of material – both fiction and nonfiction, filmed and written. In an interview for *Le Temps des médias* (2005),[15] Marc Ferro commented on the fact that fictional films may be more real than reality, and he pondered on the effective power of their images of war as opposed to those found in documentaries or current affairs programs. Such plausibility and authenticity is not necessarily to be found in a blockbuster or box-office success, as film historian Nicole Cloarec points out in the first chapter of this section, in which she studies Stuart Cooper's *Overlord* (1975), a low-budget, low-key, and intimate film in black and white that contrasts with most movies shot to "show" how things were in those days. This distinction offers an original testimony on the fusing of war narratives and memories, which is all the more pertinent since, as Cloarec reminds us, the Second World War, affectionately called the "People's War," has left a strong imprint on collective memory and is central to the shaping of war films as a genre.

Nathalie Saudo-Welby opposes the two representations of the Napoleonic wars, separated by almost a century. Eaton visited the field just a few weeks after the battle while Lady Butler was born over three decades later. Yet both are decentred witnesses, being women who did not participate in the fighting. Their points of view, as Saudo-Welby clearly shows, are opposite. Eaton tries to play down her absence on the day of the confrontation by positioning herself as a first-hand witness. She emotionalizes her military descriptions to close the gender gap that separates her from her subject, whereas Butler, relying on the testimony of old survivors whom she had met as a very young woman, uses as many filters as she can, never attempting to hide them. Both women are like Stendhal's hero Fabrice who, in *The Charterhouse of Parma* (1839), saw nothing at Waterloo, but Eaton boasts of her alleged proximity while Butler takes advantage of her decentred and mediated viewpoint to get the right perspective.

Marzena Sokołowska-Paryż's contribution is based on a comparative analysis of two novels, Nigel Farndale's *The Blasphemer* (2010) and Tatiana de Rosnay's *Sarah's Key* (2007), which display similar narrative structures. While the British writer investigates the narrator's family past during the First World War, the French novelist (who wrote her book in English) considers the hidden legacies of the Vél d'Hiv round-up through her protagonist's quest for family truths. The novels portray characters for whom history is of no importance until they are confronted with traces of the past in their own lives. The national and family pasts intertwine and impose themselves on the present, making it impossible for the characters to escape the weight of events that collective memories would rather repress.

Sylvie Pomiès-Maréchal's work on Christine Morrow's posthumously published war diary, *Abominable Epoch*, emphasizes the incomparable value of personal diaries in terms of authenticity and assesses their precious contribution as complementary to, and more flesh-and-blood than, the "dispassionate" analysis given by History. History feeds on such narratives as her war diary. Quoting historian François Dosse, who draws upon philosopher Paul Ricoeur's writings, Pomiès-Maréchal underlines the links between individual memory, collective memory, and history. Through Morrow's position as a Commonwealth citizen (she was born in Australia) and PhD student stuck in Second World War France during the Occupation, her diary sheds light on how Britain was perceived among the French population. It depicts a community characterized by anglophilia and who welcome her as if she were "a messenger from Winston Churchill" or "from Heaven."

Back in Europe, Emmanuel Roudaut examines the First World War through Robert Briffault's letters sent to his daughters in New Zealand. They provide invaluable information about warfare in the trenches as well as life on the British home front. Roudaut analyzes this as yet little explored corpus, which is one of the very few sets of correspondence we possess, written by a single person throughout the entirety of the war. The French anthropologist (and surgeon) is also a precious witness: like Christine Morrow in France, he views the war scenes with the distance of foreign eyes – even if his mother was Scottish and he spent most of his life in English-speaking surroundings. He also gives us a wealth of material details about writing letters from the trenches in appalling conditions. As he disapproves of the conduct of the conflict, he develops an atypical feeling of anglophobia, which is at odds with Morrow's perception of the war. Briffault's contribution to war literature is certainly original and worth studying. The critical distance of Briffault's testimony cannot compare with the actual distance of Charlotte Eaton's writings about, and Lady Butler's paintings of, the Waterloo battlefield.

War and Remembrance: Recollecting and Representing War considers the gaps between collective and individual memories, the shifts between dominant and minority narratives, through investigations into the historiography of specific wars. Whether written by scholars striving to decipher inconsistencies in the narratives of war to highlight how ideology takes hold of memory, or conducting research into the ego documents left by soldiers themselves, the chapters composing this book shed light on neglected war memories – those that are often outshone by the construction of official memorials and monuments honouring the dead. Fictional war literature pursues the same goal, using the made-up plot to revive forgotten events of collective memory. War literature is a genre that uses the historical narrative of the war as a background to individual stories. The plots enhance the impacts of the war on individual fates, delving into the intimate responses that war decisions engender. This literature often serves as a critique of dominant narratives by giving a voice to the victims of conflicts, sometimes within the armies themselves.

War memories have fuelled various channels of communication, including a whole range of films, fictional or not, on the small screen and the large screen, varying according to the desired colouration in the interpretations. Picturing and screening wars offers constructed representations where imagination and subjectivity are prerequisites. These more or less elaborate compositions weave the threads of narratives that diversify viewpoints and contribute to somewhat pointillist commemorative portraits, and have a certain cathartic virtue. Documenting H/history is a complex task that is subject to various notions that characterize the work of researchers: the discovery of new sources, the deepening and extension of knowledge through particular snatches and snippets of facts and events belonging to every individual's history, to every nation's history, as well as to the global apprehension of history. The scientific cultural gaze adopted here aims at opening other types of scrutiny and prisms and at furthering an enduring surface-event based approach, as historian Laurent Henninger likes to present the revival of war history/histories.[16] Evoking feelings, capturing/grasping human pulsations, and seizing social vibrations through micro-analyses emanating from a wide range of archives and materials help transcend the established views and visions of war representations in the constantly evolving writing of history, a never-ending challenge in the quest for what is part and parcel of identity, society, memory, and the history of humankind. Emmanuel Le Roy Ladurie brilliantly expresses this view on the representation of wars and conflicts in the following terms: "Let's rehabilitate the event as through a magnifying glass as it enables us to decipher the inner structures of a society."[17]

NOTES

1 "Mass culture makes particular memories more widely available, so that people who have no 'natural' claim to them might nevertheless incorporate them into their own archive of experience." See Alison Landsberg, *Prosthetic Memory: The Transformation of American Remembrance in the Age of Mass Culture* (New York: Columbia University Press, 2004), 9.

2 Pierre Nora, "The Reasons for the Current Upsurge in Memory," *Tr@nsit Online* 22, 19April 2002 (viewed 18 October 2020), https://www.eurozine.com/reasons-for-the-current-upsurge-in-memory/.

3 Jay Winter, *War beyond Words: Languages of Remembrance from the Great War* to the Present (Cambridge: Cambridge University Press, 2017), 1.
4 Ibid., 2.
5 Maurice Halbwachs, *On Collective Memory* [1941], ed. and trans. Lewis A. Coser (Chicago: University of Chicago, 1992).
6 Pierre Nora, "Between Memory and History: Les Lieux de Mémoire," *Representations (Special Issue: Memory and Counter-Memory)* 26 (Spring 1989): 7–24, https://doi.org/10.2307/2928520.
7 Astrid Erll, *Memory in Culture*, trans. Sara B. Young (Basingstoke, UK: Palgrave Macmillan, 2011).
8 "The strategies of forgetting are directly grafted upon this work of configuration: one can always recount, by eliminating, by shifting the emphasis, by recasting the protagonists of the action in a different light along with the outlines of the action … The resource of narrative then becomes the trap, when higher powers take over the emplotment and impose a canonical narrative by means of intimidation or seduction, fear or flattery. A devious form of forgetting is at work here, resulting from stripping the social actors of their original power to recount their actions themselves." See Paul Ricœur, *Memory, History, Forgetting*, trans. Kathleen Blamey and David Pellauer (Chicago: University of Chicago Press, 2004), 448.
9 Renaud Bouchet, Hélène Lecossois, Delphine Letort, and Stéphane Tison, "Introduction," *Résurgences conflictuelles: La mémoire entre arts et histoire*, ed. R. Bouchet et al. (Rennes: Presses Universitaires de Rennes, 2021), 17–29, 18.
10 Roland Barthes, *Mythologies* (Paris: Seuil, 1957).
11 Christian Delporte, Denis Maréchal, Caroline Moine, Isabelle Veyrat-Masson, eds., *La Guerre après la guerre: Images et construction des imaginaires de guerre dans l'Europe du XXe siècle*, Coll. "Histoire culturelle" (Paris: Nouveau monde éditions, 2010).
12 Iván Villarmea Álvarez, "Cinema as Testimony and Discourse for History: Film Cityscapes in Autobiographical Documentaries," *Revue* LISA/*LISA e-journal* 12, no. 1 (2014); *Documentary Filmmaking Practices: From Propaganda to Dissent*, ed. Delphine Letort and Georges Fournier, https://doi.org/10.4000/lisa.5579.
13 J.B. Priestley, *Postscript*, radio broadcast, 5 June 1940.
14 Jay Winter, *Remembering War: The Great War between Memory and History in the Twentieth Century* (New York: Yale University Press, 2006), 12.

15 Anne-Claude Ambroise-Rendu and Isabelle Veyrat-Masson, "Entretien avec Marc Ferro: guerre et images de guerre," *Le Temps des médias* 1, no. 4 (2005): 239–51.

16 Laurent Henninger, "Le renouveau de l'histoire de la guerre," *L'Histoire aujourd'hui*, ed. Jean-Claude Ruano-Borbalan (Auxerre: Sciences Humaines Editions, 1999), 207–24.

17 Our translation of "Réhabilitons aussi l'événement-loupe: Il permet de déchiffrer les structures profondes d'une société," *Nouvelle Histoire-Bataille. Cahiers du centre d'études d'histoire de la Défense, Cahier 9, Ministère de la Défense, Secrétariat général pour l'Administration* (1999): 37.

I

Remembering War from Indigenous Perspectives

1

War Voices: Australian Aboriginal Political Revolt Post-First World War

John Maynard

The Australian Aboriginal Progressive Association (AAPA) was the first all Aboriginal political organization to form in Australia. The AAPA was established in 1924 and expressed a growing resentment among Aboriginal people with the oppressive and restrictive Australian state government policies that ruled over their lives. My grandfather Fred Maynard led the AAPA in a bitter five-year campaign against the New South Wales (NSW) state government's Aborigines Protection Board. The Great War had a significant impact on Aboriginal political mobilization in its aftermath. The catalyst for a growing national Aboriginal resistance movement in the wake of the First World War was tied to threats against Aboriginal land, Aboriginal children, the experiences of Aboriginal soldiers returning home, and international black political influences.

The escalation of stripping Aboriginal children away from their families during and after the First World War and placing them into institutions had a catastrophic and long-term traumatic impact on Aboriginal Australia. The children later became known as the "Stolen Generations."[1] The Aboriginal activists were adamant that government child removal policies were in fact "a quiet working scheme [that] looked like an attempt to *exterminate* the race."[2] The recognition of a genocidal government policy was clear from an Aboriginal perspective. Fred Maynard, having witnessed through personal observation the suffering of Aboriginal people, set out to mobilize support to challenge government directives. Maynard was an articulate and outspoken Aboriginal leader. He forcefully stated the obvious: "Make no mistake. No doubt, they are trying to *exterminate* the Noble and Ancient race of sunny Australia ... What a horrible conception of so-called legislation, Re any civilised laws, I say deliberately stinks of the Belgian Congo."[3] In 1925, Maynard stated that the aim of the government's policy of tearing Aboriginal children away from their families was clear. The "objectionable practice of segregating the sexes as soon as they reach a certain age should be abolished for it meant rapid *extinction*."[4]

The other major issue for these early Aboriginal activists was the fight to protect their rights to their land. The Aboriginal leadership of the AAPA was drawn from Aboriginal men and women from right across NSW. Some were returned soldiers from the Great War. Others were driven to defend their land and children. Maynard states: "the Australian [Aboriginal] people are the original owners of the land and have a prior right over all other people in this respect."[5] It is little known today that, following the original impact of British invasion, occupation, dispossession, and cultural destruction, by the later stages of the nineteenth century Aboriginal people had begun to re-establish their connection and place on traditional country. Aboriginal people had prospered on their regained land for four to five decades from the 1860s onwards, combining traditional subsistence with Western farming methods. Susan Johnstone estimates that, by the turn of the twentieth century, nearly 90 per cent of Aboriginal people in NSW were self-sufficient, combining traditional hunting and gathering food production with Western farming methods.[6] Aboriginal farmers were clearing large returns on their crops and were winning major prizes in the agricultural shows for their produce.[7]

Despite the success that Aboriginal farmers had on this regained land, beginning in 1910, the NSW Aborigines Protection Board began to revoke the land from Aboriginal control with no compensation. The revocations accelerated during the First World War, with Aboriginal people forced from their land. Aboriginal discontent over the policies targeting their children and the taking of their land witnessed rising resentment. Adding to this growing anger some Aboriginal men were beginning to form international connections that affected Aboriginal political mobilization. Ninety years ago, Maynard and other members of the AAPA membership were greatly influenced by Marcus Garvey and his organization in the United States.[8] Today, Garvey is recognized as the charismatic leader of the largest Black Nationalist movement ever mobilized in the United

States. His capacity to influence a massive grassroots following that spread rapidly across the globe before mass media technology remains unprecedented, and the transnational impact and influence of Garveyism was incredible. Garvey became the inspiration for many with a powerful platform of genuine self-determination, and he was cutting in his assessment in an interview that appeared in the *Literary Digest* USA:

> During the World War nations were vying with each other in proclaiming lofty concepts of humanity. Make the world safe for democracy. Self-determination for smaller people reverberated in the Capitals of warring nations opposed to Germany. Now that the war is over we find those same nations making every effort by word and deed to convince us that their blatant professions were just meaningless platitudes never intended to apply to earths darker millions. We find the minor part of humanity – the white people constituting themselves lords of the universe, and arrogating to themselves the power to control the destiny of the larger part of humanity.[9]

It was Garvey's call for racial pride and a strong connection to a homeland and self-determination that attracted Aboriginal activists. Aboriginal wharf labourers, including Fred Maynard, had joined Garvey's Universal Negro Improvement Association, and a branch of this association, which included black seamen from many nations, was established in Sydney in 1920. This group operated until 1924, when the Aboriginal members of the UNIA broke away to form the AAPA. The Aboriginal activists had obviously realized that they would be best served with an organization of their own; nevertheless, they carried with them many of the political ideals of Garveyism, including connection to land and a platform organized on bettering conditions economically, socially, politically, and culturally.

In Australia, there was rising and widespread Aboriginal discontent and a growing Aboriginal voice of dissatisfaction, including that of returned Aboriginal servicemen. Alfred Bews, an Aboriginal returned soldier, argued: "There have been Aborigines who have nobly responded to our Empire's 'call to arms' overseas. Many have paid the supreme sacrifice ... we feel humiliated to know that we are still looked upon as a servile cringing race."[10] Aboriginal James Harris wrote to the press in 1921, outlining the Aboriginal commitment during the First World War: "Some of our numbers took part in the Great War and made the supreme sacrifice, while others have returned to find that they are no nearer to getting a fair deal. We haven't got a vote in the country, nor a voice in the framing of the laws that govern us."[11] Another Aboriginal man argued: "Three of us went to the great war out of my family one was killed. I always thought that fighting for our King and country would make me naturalise[d] british subject and a man with freedom in the country but ... they place me under the act and put me on a settlement like a dog. It seems as if the Chief Protector thinks that a returned soldier doesn't want justice."[12]

THE 1925 LAUNCH OF THE AAPA

It was this discontent that proved the catalyst for the AAPA's holding the first ever Aboriginal political convention at St David's Church and Hall in Surrey Hills in April 1925. The organization was instantly front-page news, with headlines trumpeting "On Aborigines Aspirations – First Australians to Help Themselves – Self Determination" and "Aborigines in Conference – Self Determination Is Their Aim – To Help a People." This is an amazing revelation in the contemporary setting – self-determination as a platform being expressed by Aboriginal activists five decades before the Whitlam government's directive,[13] which is widely attributed with instigating self-determination as Aboriginal policy.

It was noted that over two hundred enthusiastic Aboriginal people were in attendance, and "they heartily supported the objectives of the association."[14] Maynard outlined the association's directives in his inaugural address: "Brothers and sisters, we have much business to transact, so we will get right down to it ... We aim at the spiritual, political, industrial and social. We want to work out our own destiny. Our people have not had the courage to stand together in the past, but now we are united, and are determined to work for the preservation for all of those interests which are near and dear to us." The crowd represented a diverse array of Aboriginal people: "The old and young were there. The well-dressed matronly woman and the shingled girl of 19. The old man of 60 and the young man of athletic build. All are fighting for the preservation of the rights of Aborigines for self-determination."[15] Maynard declared that "Aboriginal people were sufficiently advanced in the sciences to control their own affairs."

Indeed, the Aboriginal activists of the 1920s period were articulate, eloquent, and educated statesmen and stateswomen far removed from the wider misconceptions of the time, which portrayed Aboriginal people as belonging to the Stone Age, unable to be educated, and comprising a dying race. The AAPA would eventually hold four conferences in Sydney (1925), Kempsey (1925), Grafton (1926), and Lismore (1927), respectively. They attracted widespread support from Indigenous communities, eventually establishing thirteen branches and four sub-branches with a membership that exceeded six hundred. News of the AAPA spread rapidly through an active Indigenous community network, and the formation of the organization filled Aboriginal people with hope and inspiration and with the knowledge that some of their own were now speaking out against the oppressive policies that confronted them and their communities. One old man "wrote from a far back settlement, asking that someone should come and tell them about the 'Freedom Club.'"[16]

The AAPA platform was clearly expressed in a manifesto that was sent to newspapers and politicians at both state and federal government levels. They wanted around forty acres of land to be granted to each and every Aboriginal family in the country. They wanted the Protection Board policy of removing Aboriginal children from their families stopped and the board itself to be scrapped and replaced by an all-Aboriginal body to oversee Aboriginal affairs.[17] They wanted citizenship within their own country, a Royal Commission into Aboriginal affairs, and the federal government to take charge of Aboriginal affairs. And they wanted the right to protect a strong Indigenous cultural identity. Maynard highlighted the fact that Aboriginal affairs was "a national subject ... [W]e implore your generous assistance in our effort to obtain simple justice ... [O]ur requests are few and their equity cannot be denied ... to give our people and their children every reasonable opportunity in our own land."[18]

The New South Wales government, through the influence and negative advice of the New South Wales Aborigines Protection Board, completely dismissed the AAPA manifesto. It was satisfied that Aboriginal people were adequately cared for by the board and that the government should continue to ignore the demands of these "agitators." When Maynard received word of the rebuttal he penned a powerful and articulate reply:

> I wish to make it perfectly clear on behalf of our people, that we accept no condition of inferiority as compared with the European people. Two distinct civilisations are represented by the respective races. On one hand we have the civilisation of necessity and on the other the civilisation co-incident with a bounteous supply of all the requirements of the human race. That the European people by the arts of war destroyed our more ancient civilisation is freely admitted, and that by their vices and diseases our people have been decimated is also patent, but neither of these facts are evidence of superiority. Quite the contrary is the case ... The members of this Board have also noted the strenuous efforts of the Trade Union leaders to attain the conditions which existed in our country at the time of invasion by Europeans – the men only worked when necessary – we called no man "Master" and we had no king.[19]

During its period of operations, the AAPA gained widespread publicity and, on several occasions, severely embarrassed the Protection Board, police, and government over their appalling treatment of Aboriginal people. In another letter, Maynard wrote to a young Aboriginal girl informing her she was but one of many young Aboriginal girls who had suffered abuse through the government child removal policy. The girl was taken from her family at the age of fourteen and placed into domestic service, where she was raped by the manager. When she became pregnant, the Protection Board sent her to Sydney by train, where records state that she had a baby that died on birth. The girl was then placed on another train and sent back to the same place where she had been abused.

Maynard was alerted to this girl's plight and wrote to her seeking details of the perpetrator of the assaults as he would see the man in court. Sadly, the girl never received the letter as it was intercepted by the manager, who alerted the Protection Board to its inflammatory content. Maynard was highly critical of the government's policy and its intentions: "these so called civilised methods of rule, under Christianized ideals so they claim, of civilising our people under the pretence of love ... [T]hese tyrannous methods, under the so called administrative laws re the Aboriginal Act, have got to be blotted out as they are an insult to intelligent, right thinking people. We are not going to be insulted any longer than it will take to wipe it off the Statute Books. That's what our Association stands for, liberty, freedom and the right to function in our own interest."[20]

He drove it home that Aboriginal people had been pushed too far: "Are we going to stand for these things any longer? Certainly not! Away with the damnable insulting methods, which are degrading. Give us a hand, stand by your own native Aboriginal officers and fight for liberty and freedom."[21] Instead of being alarmed at the abuse of a young girl under its care, the Protection Board intensified police surveillance and its attacks against Maynard and the AAPA. In November 1927, the AAPA, in a meeting covered in the press at Nana Glen on the north coast of New South Wales, stated that the organization had formed "to give the Aboriginal an opportunity of voicing an appeal to the government from their own point of view. So far the administration and the legislation had all been fixed up from the viewpoint of the legislator and to suit the ideas of those in power." Aboriginal people "had received many promises from past politicians, but so far nothing had come to light."[22]

The question remains why the AAPA slipped from public view in 1928. It seems hard to fathom, especially when press reports that appeared only just before this abrupt end indicated exciting new initiatives that included a new newspaper to be called the *Corroboree*.[23] There are a number of most likely causes for the public demise of the AAPA, including the impact of the Great Depression (which did have a major impact on Aboriginal employment and, of course, police intimidation and force). Historian David Huggonson speculates on the level of intimidation that "officers of the Board may have made in relation to taking Maynard's children into state care if he continued his agitation."[24] This is corroborated by my father's recalling that, in the later 1930s, he and another young Aboriginal boy aged only five or six were picked up by the police and taken to the police station in Canterbury. My father has no hesitation in stating that the time he spent in that police station was the most terrifying moment of his life. The police terrorized the young boys for hours, and, when finally released, my father ran home and hid in his room. He never said a word to either his mother or father of the incident. It was only on reflection many years later that he realized the police were just getting a message through to his father: "we can pick up your kids anytime we like so shut up."

Despite these threats, the AAPA leadership continued to meet in my grandparents' kitchen. Maynard continued to talk at La Perouse and Salt Pan Creek in Sydney, and his last public talk was reported as being in the mid-1930s on the grounds of the

University of Sydney. His name was put forward to speak at the NSW government inquiry into Aboriginal affairs in 1938, but he was prevented from speaking. In the late 1930s, Aboriginal activist Bill Ferguson reflected that he was aware of the AAPA struggle the decade before, stating that they "held three annual conferences, but were hounded by the police officer acting for the Protection Board." The NSW police unquestionably played a major role in the breakup of the AAPA. A revealing 1927 newspaper interview with Maynard indicates the level of intimidation to which he was subjected: "he had been warned on many occasions that the doors of Long Bay [Gaol] were opening for him. He would cheerfully go to gaol for the remainder of his life, he declared, if, by so doing, he could make the people of Australia realize the truly frightful administration of the Aborigines Act."[25]

In 1931, the Australian Communist Party published *Rights for Aborigines: Draft Program for the Struggle Against Slavery*. It adds further weight to my grandfather's insistence that the AAPA was indeed broken up due to severe police threats and intimidation: "the conditions of the Aborigines have not been considered by workers in the revolutionary movement, and the rank and file organization set up by the [A]borigines was allowed to be broken up by the A.P.B., the missionaries, and the police."[26]

THE AFTERMATH

The AAPA did succeed for a short time in slowing the taking of Aboriginal children from their families and the revocations of Aboriginal land. But the single biggest impact of the AAPA was through the inspiration it provided by delivering a united Aboriginal political voice. Aboriginal people were encouraged from this point to step forward and voice protest and demands on behalf of their people. The members of the La Perouse Aboriginal community of Sydney refused to be moved off their land in 1928, stating forcefully: "We, the undersigned [A]borigines of the La Perouse reserve, emphatically protest against our removal to any place. This is our heritage bestowed upon us: in these circumstances, we feel justified in refusing to leave."[27]

The AAPA had ignited the fire of rebellion. Two Aboriginal men, Edward and William Anderson, appeared barefoot in Kogarah Police Court in Sydney charged with riding a train to Wollongong on the

South Coast without having paid their fare. In response to the charge, Edward Anderson informed the court that they were penniless, and he defiantly declared: "This country belongs to us. Our brother is in hospital, and we wanted to go and see him. We could not get to Wollongong by any other means."[28] An organized Aboriginal political movement would be rejuvenated again in the late 1930s, led by a new wave of Aboriginal political activists that included the likes of William Cooper, Bill Ferguson, Jack Patten, and Pearl Gibbs. These activists most notably conducted the "Day of Mourning" protest in Sydney on 26 January 1938, coinciding with the sesquicentenary of British colonization of Australia celebration. In late 1938, these Aboriginal activists marched on the German embassy in Melbourne to protest the treatment of the Jewish people in Nazi Germany. They were the first group in Australia to publicly come out and speak out against the ill-treatment of the Jewish people. The German embassy refused to meet the deputation. In the press, the Aboriginal activists revealed similarities between their experience and that the Jews: "[like] the Jews our people have suffered cruelty and exploitation as a national minority." The statement continued: "We are a poor people, and few in number but in extending our sympathy to the Jewish race we also pledge ourselves to help them by all means in our power."[29]

The Aboriginal political movement of this period would be forced into hibernation by the start of the Second World War. It would not be until the vibrant 1960s that organized Aboriginal political protest would be revived. Most notably, Charles Perkins would emulate the "Freedom Rides" in the United States led by Dr Martin Luther King, and, in 1965, through the media, he would expose the shocking Third World living conditions and segregation experienced by Aboriginal people across New South Wales. This event ignited full-scale Aboriginal political revolt. In 1966, Gurindji stockmen in the Northern Territory walked off the job, demanding equal pay and the return of their traditional land, and, in 1967, the Australian government's national referendum resulted in over 92 per cent voting to recognize Aboriginal people as citizens of their own country and to have the Commonwealth government take charge of Aboriginal affairs. In 1972, in one of the most significant moments in Aboriginal political history, the "Aboriginal Tent Embassy" protest was set up on the lawns of Parliament House in Canberra. The Aboriginal political fight and demands initially voiced by the AAPA over ninety years ago continue to this day.

This was the legacy of the AAPA. We are left today to recognize and remember these early Aboriginal freedom fighters who were prepared to bravely step forward to challenge the tight government and police control over Aboriginal lives. It is amazing that, for so long a period, this history lay forgotten and erased from Australian historical memory. The battles Aboriginal people were fighting ninety years ago are still being fought today. The sad reality is that, in the twenty-first century, Aboriginal people have the worst health statistics in the country, with a life expectancy roughly twenty years less than that of non-Indigenous people. Aboriginal people suffer the worst employment, housing, and educational statistics, and they experience the highest rates of incarceration in the country. Recognizing the early struggle for Aboriginal political justice and equality restores a significant but long forgotten moment to Australian historical memory.

NOTES

1 The Stolen Generations were the children of Australian Aboriginal people who were removed from their families by the Australian federal and state government agencies and church missions under acts of their respective parliaments. Tens of thousands of Aboriginal children were removed and placed in institutions. The 1997 *Bringing Them Home* report provided extensive details about the removal programs and their effects. It ended, in 2007, with Prime Minister Kevin Rudd apologizing to Indigenous people over the policies of the past. The eruption of the Australian "History Wars" during the late 1990s witnessed debate and denial among politicians, media commentators, and historians, some arguing that the "Stolen Generations" have been exaggerated.

2 *Northern Star,* 3 August 1927.

3 Fred Maynard, "Letter to Aboriginal girl," New South Wales State Archives (hereafter NSWSA), 1927, A27/915.

4 *Voice of the North,* 10 December 1925.

5 Fred Maynard, "Letter to Premier Jack Lang," NSWSA, 1927, A27/915.

6 Susan L. Johnstone, "The New South Wales Government Policy towards Aborigines, 1880–1904" (MA thesis, University of Sydney, 1970).

7 John Maynard, *Fight for Liberty and Freedom* (Canberra: Aboriginal Studies Press, 2007), 61–2.

8 Ibid., 55.

9 *Morning Bulletin* (Rockhampton), 11 June 1921, 10.

10 *Daylight*, 29 February 1928, 294.
11 *Sunday Times* (Perth), 30 October 1921, 12.
12 Philippa Scarlett, *Aboriginal and Torres Strait Islander Volunteers for the AIF – Indigenous Response to World War I* (Canberra: Indigenous Histories, 2014), 39–41.
13 Gough Whitlam, twenty-first Australian prime minister (1972–75). Whitlam's government is remembered for introducing radical social, political, and cultural reforms.
14 *Daily Guardian*, 24 April 1925.
15 Ibid., 7 May 1925.
16 *Macleay Chronicle*, 19 August 1925.
17 The 1915 amendment to the New South Wales Aboriginal Protection Act, 1909, gave the board and its array of bureaucrats the powers and provisions to remove any Aboriginal child from its parents for, in practice, little or no other reason than the fact that they were Aboriginal.
18 *Australian Natives Association Journal*, main reading room, Australian collection, N9 919.405 ANA, vol. 1:1–3:5 (7 December 1925 – 7 May 1928).
19 Fred Maynard, NSW Premiers Department Correspondence Files (1927), NSWSA, A27/915.
20 Maynard, "Letter to Aboriginal girl."
21 Ibid.
22 *Daily Examiner*, 16 November 1927.
23 *Voice of the North*, 10 January 1928, 18.
24 David Huggonson, "Aborigines and the Aftermath of the Great War," *Australian Aboriginal Studies* 11, no. 1 (1993): 8.
25 *Newcastle Sun*, 7 December 1927.
26 *Workers Weekly*, 24 September 1931.
27 *Sydney Morning Herald*, 4 April 1928, 24.
28 Ibid., 12 June 1931, 8.
29 *Workers Weekly*, 20 December 1938, 3.

2

War Memories and Indigenous Stereotypes: The Fabrication of the Maori Warrior

Corinne David-Ives

The previous chapter delves into the forgotten experience of Aboriginal soldiers in the Anzac, demonstrating that the place of Indigenous minorities in colonial and postcolonial narratives of national identity has been largely conditioned by episodes of warfare. Colonial wars have been an inexhaustible source of stereotypes that sought to establish a form of primal difference between settlers of European descent and a range of "savages" or "Barbarians" with which colonial states had to contend. Although it has long claimed to occupy a special place in the former British colonies of settlement in the treatment of its own Indigenous people, New Zealand was no exception to the rule. In 1840, the Treaty of Waitangi established New Zealand as a British colony on the basis of the consent of the Maori people whose rights were guaranteed and who were to enjoy the privileges – and obligations – of British subjects. The bloody New Zealand Wars, or Maori Wars, which broke out some twenty years later did not obliterate Maori rights and did not ascribe the traditional place of the arch-enemy to Indigenous people. In fact, the notion of mutual respect dominated the postwar narrative. Memories of the New Zealand Wars appropriated the existing stereotype of the fierce Maori warrior, albeit while reinforcing the dominant belief in a hierarchy of races, which justified the terms of colonization.

This chapter is concerned with the legacy of the New Zealand Wars on the national narrative of identity and, in particular, with the place of Maori within that narrative. It seeks to shed light on the variations that occurred in the perceptions of the myth of the Maori Warrior, which was also played on by Maori leaders in their successive strategies for empowerment. I rely on New Zealand historiography to review the various avatars of the stereotype and its resurgence on the occasions of the two world wars, then go on to assess its influence to this day.

THE NEW ZEALAND WARS AND THE VICTORIAN HIERARCHY OF RACES: THE GALLANT MAORI

In 1890, British politician Sir Charles Dilke, a militant imperialist and a radical, evoked "the warlike and intelligent Maori race" in the New Zealand chapter of his treatise on the British Empire. He noted about these Indigenous people: "They have shown a great aptitude for civilisation, and have won the respect of white colonists – *a most unusual thing in the case of any dark-skinned race*."[1] That appraisal came less than twenty years after the final episode that brought the New Zealand Wars to a close. It was very much in tune with the dominant Victorian ideology that sought to present the British as a benevolent master race, while recognizing certain qualities in specific colonial "races." This brief introduction to Maori was building on a positive vision that pre-dated the New Zealand Wars and had obviously not been significantly altered by them. The context, though, was important: a time period of intense nation building, when New Zealand had elected a reformist Liberal government that was going to produce a very influential narrative of identity for the young colony. To understand the conditions of the production of that narrative it is necessary to go back to the early days of the colonial project.

The official annexation of New Zealand in 1840 had been largely conditioned by a favourable vision of the "natives," who ranked rather highly on a pre-Darwinian scale that was dominant in the colony until at least the 1860s–70s.[2] A considerable body of research has examined the role of the missionary-dominated "humanitarian" lobby in London in this appraisal.[3] The positive view of Maori was mainly based on the following observations: they were a sedentary people; they practised a form of agriculture; although deprived of a proper "body politic," Maori society exercised a form of sovereignty over the land or at least had a "political capacity."[4] This had led to the recognition of native rights according to the principles of natural law that had been enunciated principally by Emmerich De Vattel in the eighteenth century.

The New Zealand frontier was therefore construed not so much as the relentless progress of civilization over savagery (as in Australia),[5] or as a pitched battle fought against hostile hordes (as in South African), but, rather, as the extension of a compromise between Maori and settlers of British descent. The basis of this compromise had been defined by the Treaty of Waitangi, which established British sovereignty over New Zealand while recognizing Maori customary property, conferring on the native population "the rights and privileges of British subjects." If one leaves aside the humanitarian concerns of the 1840s, this compromise had been rendered necessary by a very simple fact readily admitted by London: the warring tribes of New Zealand were not to be subjugated by force, a daunting and expensive endeavour not easily justifiable internationally and, therefore, to be avoided as much as possible. The Treaty thus provided a legal framework within which to proceed with "amalgamation": a form of assimilation highly favoured by the missionaries and within which the two races would happily mix.[6]

Those idealistic beginnings were to flounder against the harsh realities of colonization. Within twenty years, Maori frustration over the interpretation of the Treaty of Waitangi, the settlers' hunger for land, and the emergence of a supra-tribal resistance movement known as the King Movement resulted in a typically colonial conflict centred on land control and ownership. The New Zealand (Land) Wars, or Maori Wars, erupted in 1860. They consisted in a series of skirmishes and violent episodes involving traditional tribal and/or prophetic movements and came close to degenerating into an all-out civil war. The situation was very tense until the end of the decade. The wars were not simply waged by settlers against Maori: they involved rival tribes – namely, rebels – against loyalist, or *kupapa*, tribes. And they involved thousands of British regular forces – between fourteen thousand and eighteen thousand – which was more than were stationed in India at the time. These troops were sent to New Zealand to quell the rebellion.

The New Zealand Wars were used to justify the colonial state's onslaught on the much coveted land in the centre of the North Island as massive land confiscation (*raupatu* in Maori) was the price the Waikato tribes of the "King Country," the leaders of the rebellion, had to pay. But the Wars paradoxically contributed to reinforcing the perceived superiority of Maori over other Indigenous peoples within the British Empire and beyond. At that time, the amazing achievement of a total of probably five thousand warriors who were rather poorly equipped was admittedly "a unique feat of resistance to nineteenth century European expansion."[7] It was soon integrated into the legend of the New Zealand frontier. This was reflected in the first accounts of the New Zealand Wars, which, in the form of newspaper articles or diaries, were published in the 1860s by colonial participants in the conflict. They were written by colourful figures such as Prussian-born Major von Tempsky who fought with the Forest Rangers, an irregular colonial force better able to deal with guerrilla tactics in the New Zealand bush than was the British regular army. Von Tempsky was a war correspondent for the *Daily Southern Cross* and also wrote *Memoranda of the New Zealand Campaign*, but he is best remembered for his many watercolour illustrations of a mix of peaceful Maori village scenes and battle scenes. His works provided an imaginative backdrop for the war epic, and his death in action in 1868 in Taranaki turned him into a popular folk hero.[8]

In the aftermath of the New Zealand Wars, a number of undeniable facts had to be made to fit the Victorian assumptions about race: the drawn-out nature of the conflict, the difficulties experienced in defeating the rebels, and the de facto secession of certain tribes in the Urewera who ruled their land in complete isolation for another twenty-five years. In the face of efficient Maori resistance, the official discourse chose to emphasize heroic deeds *on both sides*, pointing out that, nevertheless, in the end the Maori rebels had lost their fight, this being the natural result of the superiority of British civilization.[9]

This defeat was then exploited through a new ideological lens. Social Darwinism hit New Zealand in the latter part of the nineteenth century, but its influence was mitigated by the view of Maori that had been prevalent until then. "Fatal impact" was a popular notion that focused on the idea that, unfortunately, Maori were doomed as a race. The New Zealand Wars had amplified a noted and supposedly natural phenomenon, whereby contact with a superior race led to the dying out of the less fit, inferior race.[10] Despondent after their defeat, Maori as a people suffered a form of collective depression or melancholy, which announced their imminent downfall: a fate to which they were now resigned.[11] Demographic figures confirmed this alleged fact as, in the 1891 census, the native population was shown to have reached an absolute low.[12]

This idea was notably expressed by Liberal intellectual and leading politician William Pember Reeves in his famous account of New

Zealand, *The Long White Cloud – Ao Tea Roa.*[13] That exhaustive presentation of the young colony depicted New Zealand as a social laboratory and as a better, antipodean Britain, focusing on its political and economic history and its progressive labour laws. It included well-documented chapters on Maori history, culture, and mythology, and, as such, it functioned as what Benedict Anderson terms "a continuous narrative of identity"[14] – namely, an attempt to establish national continuity and to connect the emerging nation to a time-immemorial tradition shrouded in myth. In doing this, Reeves keeps oscillating between the typically Victorian appraisal of an intelligent but primitive race and respect for the many virtues of a people deemed almost the equals of Britons (very much like depictions of the Celts). He systematically builds on the stereotype of the formidable Maori warrior, more of a "Barbarian" than a "Savage," whom he contrasts to North American Indians, who did not deserve respect as warriors because they practised "cold-blooded torture."[15] He devotes two whole chapters to the New Zealand Wars, emphasizing "native prowess which epithets can hardly exaggerate" and "the sturdy resistance of the natives … due to their splendid courage and skilful use of rifle-pits and earthworks."[16] He notes how "clever the Maori engineers were," how "bravely the Brown warriors defended their entrenchments." Of course these achievements were doomed, "seeing how overwhelming was the White force."[17] This being clear, Reeves expands on "the boldness, rising at moments to heroism, with which clusters of badly armed savages met again and again the finest fighting men of Europe."[18] His account of the famous siege of Orakau, a fortified native village, is given in a definitely epic vein. Expanding on the desperate courage of the besieged, who refused to surrender, *and* the humane conduct of the assailants, who offered to spare the women and children, he concludes: "The earthworks and the victory remained with us, but the glory of the engagement lay with those whose message of 'Aké, aké, aké' [We fight right on, for ever], will never be forgotten in New Zealand."[19]

Reeves had contributed to the New Zealand Wars epic and to the making of heroes on both sides. The Wars had given way to memories of the Wars, glorious rather than bitter. Reeves had also confirmed that the national pact continued to rest on the original alliance between two peoples, sealed by the Treaty of Waitangi. This was an essential statement on the part of a leading colonial politician who was keenly aware of the dangers of division. And, most significantly, he had consolidated the national myth according to which the treatment of Maori by the colonial authorities was "just and often generous."[20] This was to be the building block of the official discourse on the racial harmony that reigned in New Zealand.

The New Zealand Wars had therefore not destroyed the positive image of Maori in the colony – all the more so as the pivotal role of the *kupapa* (loyal) tribes could not be ignored. After the Wars, resentment was directed at Britain (rather than at Maori) for abandoning its loyal subjects in the midst of serious difficulties: the colonists felt that the withdrawal of all British troops for financial reasons in 1870 was a betrayal.[21] The Wars had been really the first occasion for New Zealand to graduate to nationhood, through what contemporary French observer André Siegfried terms "the rough and healthy school of adversity."[22] The Wars were eventually blurred in the overall popular tale of the New Zealand frontier, which was typically rendered in a humorous vein.[23] After the New Zealand Wars there was little official interest in memorial celebrations. Some medals were awarded to local veterans, essentially to recall that imperial troops had been supported by the settlers and that they had truly *needed* that help. Here again, war memories were useful in contrasting the strong New Zealand colonial race, well adapted to its new homeland, to the British troops and to emphasize how brave and clever the New Zealand Maori were, in contrast, notably, to Aboriginal peoples in Australia. Certainly, initial unease about the fragile peace and the effects of the massive confiscation of native land were a damper on settler satisfaction, all the more so as the heart of the Maori King Movement had not been crushed: the Maori communities of the district had even set up an *autaki*, a line forbidding European encroachment into their territory. The Boer War would soon be the source of much more popular enthusiasm.

THE NEW ZEALAND WARS AND THE OFFICIAL NARRATIVE OF HISTORY: THE MAORI WARRIOR'S LONG ROAD TOWARDS EQUALITY

The long reign of the Liberals in New Zealand (1890–1912) had permitted the reformulation of the compromise between *Pakeha* (settlers) and Maori. The New Zealand Wars had been relegated to a convenient if not very distant past. A new generation of educated Maori – the self-termed Young Maori, particularly those issuing from the *kupapa* tribes – had been permitted to take part in the political life of the nation, and some had even been accepted in government. The Liberals had finally succeeded in putting an end to the last rebels' withdrawal from colonial life, cleverly offering the Maori chief a seat in the New Zealand Senate, granting old rebel chief Te Kooti an amnesty, and at last making the Urewera native lands into protected Native Reserves.[24] In 1911, the government had a memorial in the form of an obelisk erected on the site of the battle of Orakau. It was adorned with bronze plaques, one of which reads: "On this site in an unfinished pa / about 300 Maoris with some / women and children, poorly / armed and with little food / and no water, held at bay 1,500 / better equipped British and / Colonial troops. Refusing to / surrender, on the third day / a remnant of the Maoris / escaped across the Puniu / River."[25] The memorial was inaugurated in 1914 with pomp and circumstance on the occasion of the fiftieth anniversary of the battle. Very significantly, the ceremony was attended by James Allen, the minister of defence, and prominent Maori politicians, such as government minister Maui Pomare, while the local press compared Chief Rewi's last stand to the ancient Greeks' battle at Thermopylae.

How would New Zealand historians from then on recall the New Zealand Wars? Another generation elapsed before serious academic attention was devoted to this topic. The first comprehensive account of the conflict was published in 1922 by James Cowan, initially a writer and a journalist who was commissioned by the Dominion's government between 1916 and 1920 to write a history of the New Zealand Wars. He produced what, for decades, was considered to be the most definitive account of the events.[26] Serious critical academic examination of his work would not occur until the 1960s–70s. It is interesting to point out that Cowan was the son of an Irish Waikato settler who had fought in the Wars and been awarded confiscated land in Taranaki for farming. Cowan had been brought up on the memories of the Wars, in proximity with Maori, and was himself fluent in Maori.[27] The main interest of his work lies in its careful collection of accounts from eyewitnesses on both sides of the conflict and in his use of Maori oral history as a reliable source of information.

Cowan's *New Zealand Wars* thus became a master narrative of the Wars and, beyond that, of the relationship between Maori and Pakeha. Cowan recognized the importance of the Wars for "nation-building" but went beyond that fact, proposing the "correct perspective" on the phenomenon.[28] In other words, he consciously supplied a discourse of identity that was going to be very influential in New Zealand for the next forty years and reach well beyond academic circles. He would first posit the difference between the conflicts of the New Zealand frontier and those of neighbouring Australia. Establishing a fundamental difference with Australia was a typical New Zealand exercise that, again, was rooted in facts: Aboriginal resistance to colonization in Australia followed another path altogether.[29] However, Cowan was also building on a popular racial stereotype: the superiority of Maori over "Aborigines," which implied a difference in the nature of colonization in Australia. He states: "Australia's pioneering-work was of a different quality from ours, mainly because the nation-makers of our neighbour encountered no powerful military race of indigenes to dispute the right of way."[30] Then, again, this would enable him to build on the age-old cliché for which war truly forged the national character: "The inevitable shock of battle between the tribesman of Aotearoa and the white man who coveted and needed his surplus lands is a feature of our history which has had no small influence upon our national existence and national type."[31]

And from there, Cowan enounces what was, for him, the crux of New Zealand's identity. "It was in the last and unavoidable test, when bayonet met long-handled tomahawk and when British artillery battered Maori stockades, that the two races came to gauge each other's manly calibre, and came, finally, *to respect each other* for the capital virtues that only trial of war can bring to mutual view."[32] And thus, right from the beginning, in the first chapter, conveniently entitled "The Old Race and the New," Cowan's *New Zealand Wars* reiterates the basis of the New Zealand compromise, which had been redefined under the Liberals twenty years before: Maori were worthy of the respect of their colonizers and they could eventually graduate to civilization and equality. From then on, the text abounds with multiple references to the noble qualities of Maori warriors, which appear as the cement for future national unity: "The soldiers ... marvelled at the devotion of such a race and then *came to love them for their savage chivalry*. The wars ended with a strong mutual respect, tinged with a real affection, which would never have existed but for this ordeal by battle."[33]

For the author, the shared knowledge of the episodes of the New Zealand Wars therefore served the purpose of reinforcing patriotic sentiments among the descendants of the settlers. Here again, both Maori and Pakeha could commune over the shared love of their country, and the settler "race" could understand the feelings of the natives for the land, as "the passionate affection with which the Maori clung to his tribal lands is a quality which undeniably tinges the mind and outlook of the farm-bred, country-loving, white New Zealander to-day."[34] Thus, far from estranging settlers from Maori, the Wars forged a new form of unity. This was the traditional building block of the New Zealand narrative of identity, not so much altered by the Wars as digested and reformulated.

In spite of Cowan's own sincere admiration for Maori military prowess, his language of equality and respect was mixed with racial considerations, suggesting a fundamental difference that explains the final result of the Wars. Commenting on public opinion in the 1860s, he notes "an admission that the naked Maori was a better warrior than the heavily armed British soldier, *man for man*, in the forest environment in which he had been schooled to arms and the trail from his infancy. Each admitted the other's pre-eminence *under certain conditions*, and each protagonist came to admire the primal quality of valour in his opponent."[35] The message was clear and did not fundamentally depart from the Victorian assessment of Maori on a racial scale: the "natives" shared essential qualities of courage and honour with the whites, but they were superior only in their own environment. The whites had "Civilization" on their side, and Civilization would prevail. Maori were aware of this too: "well they knew that in the end they could not hope to prevail over men of such mettle."[36] And their desperate courage in the face of doom was really the source of the whites' admiration for them: "holding heroically to nationalism and *a broken cause*."[37] *New Zealand Wars* thus took on an epic dimension very similar to that of Reeves's account mentioned above: the battle of Orakau, for instance, "imperishably remains as an inspiration to deeds of courage and fortitude."[38] The text is replete with references to "gallant resistance to overwhelming force,"[39] to the "gallant Maoris,"[40] again to "the splendid devotion and fearlessness displayed by the Maori heroes."[41]

Of course, the context within which Cowan did his research and writing played an important role. In the aftermath of the First World War, when the book was published, nostalgia for an old-fashioned confrontation in which untamed nature had the lead role was very conspicuous: "Regarding these old wars in the light of the ordeal of battle from which the civilized world has lately emerged, the *Pakeha-Maori* conflicts seem chivalrous tournaments."[42] The comparison built on the warrior stereotype even more readily as war on an industrial scale had erased the possibility of relating to the enemy through codified behaviour. Cowan also derails the "overwhelming faith in the white man's military invincibility," which characterized early British officials in New Zealand[43] – a criticism of British officers rather than of settler forces, echoing similar criticism of the British High Command during disastrous First World War episodes involving New Zealand troops. In the same way, certain passages describing the use of modern machinery by the whites against Maori fortified villages during the New Zealand Wars were reminiscent of the confrontations of the First World War. "Suffering the tortures of thirst, half-blinded with dust and powder-smoke, many bleeding from wounds which there was no time to stanch, ringed by a blaze of rifle-fire, with big-gun shells and grenades exploding among them, the grim band of heroes held their crumbling fort till this hour against six times their number of well-armed, well-fed foes."[44]

What would be the legacy of such ordeals for the young nation? Cowan devotes room in his work to memorial celebrations and, in particular, to "the chivalrous example the Maoris [set] to the *pakeha* in the care of their antagonists' graves."[45] In a closing chapter fittingly entitled "Lest we forget – The soldiers' graves at Ohaeawai," another obvious reference to the First World War, he recalls how, "as an aged chief, formerly conspicuous among our enemies, said to me, 'The brave warriors of both races, the white skin and the brown, now that all strife between them is forgotten, may sleep side by side until the end of the world."[46] He quoted the governor general of the Dominion on this: "'I question,' the Governor concluded, 'if there be a more touching episode in the annals of the warfare of even civilized nations in either ancient or modern times' (Appendices to Journal of the House of Representatives, 1871)."[47] This provides a very appropriate conclusion to Cowan's *New Zealand Wars*, insisting on forgiveness and remembrance on both sides, while viewing the colonial wars within the modern context of the First World War.

Of course, contemporary Maori participation in the world conflict had but enlarged the positive vision from which they benefitted in the national narrative of history. Maori leaders had obtained the creation of a Pioneers Battalion, mostly composed of Maori volunteers – a strategy whose aim was to reinforce their inclusion in the

New Zealand nation, while capitalizing on the indomitable warrior stereotype.[48] The 1916 photographs of Conservative prime minister Massey visiting Maori troops stationed in Flanders, where they performed a haka for him in their New Zealand army uniforms, had circulated at home, a powerful symbol of Maori loyalty to the nation and a tribute to their unique culture. The myth could then live on undisturbed, and Cowan's account of the colonial land wars was not subjected to any serious criticism for a long time, although his tendency to romanticize Maori life and culture was noted by more academically minded historians of his generation.[49]

QUESTIONING THE OFFICIAL NARRATIVE: FROM "REVISIONIST" TO "NEO-REVISIONIST" HISTORY

It was not until the 1960s–70s that this idealized vision of the nineteenth-century wars would be called into question by a new generation of historians. Cowan had provided a precise chronological account of the New Zealand Wars that relied on testimonies from all the parties involved. He had tried to estimate the number of dead, in particular Maori, as accurately as possible (about twenty-five hundred). However, he had never analyzed the colonial project as such, nor questioned the natural right of the settlers to take what he called "Maori surplus land."[50] In earlier works he had manifested his belief in the theory of fatal impact, and, in the same vein, he explained the Wars as a fatal confrontation, "the last and unavoidable test" between Maori and Pakeha, which inevitably resulted in the victory of the European side.[51] His epic rendition of the various battles blurred the reality of massacres and the brutality of the assaults of the most modern army in the world at the time.

Cowan's account had blended in perfectly with the renewed myth of the gallant Maori warrior who had now proved his loyalty to the nation through glorious participation in two world wars and had sought to gain recognition for this through "the price of blood."[52] This myth was inextricably linked to New Zealand's national identity as a paradise of race relations and of integration, which was the face that the young independent nation presented to the world in the 1950s. Although new research tackled the New Zealand Wars, the perspective was not fundamentally changed. In the preface of his first book, entitled *The Origins of the Maori Wars*, Keith Sinclair, considered the father of modern New Zealand history, thus declares in 1957 that the "Maori Wars" "formed a necessary prelude to the growth of a new nation which embraces two races."[53] Up until 1966, official history maintained the traditional narrative around the New Zealand Wars and the noble Maori warrior. In its chapter on the famous battle of Orakau, the *Encyclopedia of New Zealand* thus notes: "Sir Duncan Cameron arrived on the field and, being impressed by the defenders' courage, he offered them an opportunity to surrender."[54]

Historians started to scratch timidly at the surface of the official discourse on national history as Maori activism burst on the stage in the late 1960s, fighting for true equality and empowerment. In the late 1960s and early 1970s, the idea started to emerge that the "paradise of race relations" was an official construction that, in reality, owed a great deal to Victorian views on races and native peoples, and the inherited and continued belief in a natural hierarchy of races. The official discourse on Pakeha/Maori relations started to be examined more critically, although leading scholars could not yet bring themselves to seriously challenge it. The titles given to scholarly articles are in themselves extremely revealing. Keith Sinclair thus asked himself in 1971, in an article for the *New Zealand Journal of History*, "Why Are Race Relations in New Zealand Better Than in South Africa, South Australia or South Dakota?"[55] Examining the impact of Social Darwinism in New Zealand, he drew the conclusion that Maori were not so much considered inferior to the whites as a race than as superior to other natives. He states: "European New Zealanders ... believed in the superiority of the Anglo-Saxons, but never accepted any notion of inherent Maori inferiority."[56] There may well have been a paradox of that sort in nineteenth-century New Zealand, but Sinclair's optimistic vision falls short of questioning the official discourse. In reality, the comparisons with segregationist societies were a reiteration of the New Zealand governments' views at the time, and they justified monoculturalism in New Zealand and the domination of Maori.

Other historians are much more straightforward. As early as 1967, Alan Ward reconsidered the New Zealand Wars through the lens of colonial aggression and greed, exposing "the cultural snobbery of the educated settler elite and the brutish racial superiority of ignorant whites," noting the colonial authorities' urge "to end Maori resistance by force of arms."[57] In 1969, in his excellent *Politics of the New Zealand Maori: Protest and Cooperation, 1891–1909*,

American historian John Williams reviews Maori strategies of resistance to the colonial onslaught, from war to political action, pointing out the brutal character of native policies, even under supposedly enlightened governments such as the Liberals. In 1973, Alan Ward published *A Show of Justice: Racial "Amalgamation" in the Nineteenth Century in New-Zealand*, which also questions the supposedly enlightened colonial policies.[58] In 1975, old-timer Keith Sorrenson, who had been quietly working since the late 1950s on repressive native land policies, asked ironically, in the *New Zealand Journal of History*, "How to Civilize Savages: Some 'Answers' from Nineteenth Century New Zealand," revisiting colonial views on Maori aptitude for civilization.

Such indictments paved the way for a complete review of New Zealand history and a debate that was going to be further fuelled by the Maori Renaissance in the 1980s. The official narrative collapsed under the onslaught of a new school of historians animated by the desire to take on a postcolonial perspective.[59] In this context, the Land Wars were notably tackled by James Belich, who published the seminal *The New Zealand Wars and the Victorian Interpretation of Racial Conflict* in 1986. Belich focuses on the narrative of war and deconstructs the myth of the gallant Maori, analyzing instead how sophisticated Maori military organization had been. He insists on how important it had been for the Victorian ideologues to provide an acceptable account of the conflict and explain away the serious difficulties experienced by the British Army. Belich also published *I Shall Not Die: Tītokowaru's War, New Zealand, 1868–9*, and he later produced a television series about the Wars (1998). In his two-volume history of New Zealand, he consolidates his theory that the Wars had been an essential moment in the construction of New Zealand identity through which "the two peoples made each other – as peoples rather than tribes and provinces."[60]

All these new developments did not necessarily obliterate certain well-documented aspects of the New Zealand Wars that were part of the traditional historical narrative: for instance, in the 1992 edition of *The Oxford History of New Zealand*, Keith Sorrenson recalls: "on numerous occasions ... both sides displayed courage and chivalry of the highest order, and in the end a warm comradeship between victor and vanquished. Maori 'generals' and British officers developed a considerable respect for one another. Moreover, the wars did not become race wars."[61]

Meanwhile, the process of reconciliation with Maori that had been initiated through the Tribunal of Waitangi since 1975 through to 1985 was funding a massive body of research, including much needed work on the local effects of the New Zealand Wars, in particular land confiscation.[62] Ward, Sorrenson, Claudia Orange, and a number of others were recruited to provide scholarly studies that would be used to substantiate the claims made before the Tribunal by various tribes. Recently, this research was ironically nicknamed "Tribunal history" by historians who criticized its "revisionist agenda."[63] The tendency to oppose "victims and villains" for the purpose of compensating the tribes for the abuse they suffered may well have warped the Tribunal's approach.[64] However, the Waitangi process marked the necessary coming of age of New Zealand history, characterized by new voices and new perspectives – and some inevitable strife among New Zealand historians. Yet another generation of historians is now moving away from these "neo-revisionists" and their more positive assessment of colonial authorities. The latest research on the New Zealand Wars focuses on the effects of the conflict on specific tribes, going on to reassess the number of casualties, which appears much higher than had been previously estimated.[65]

An important legacy of the continuing Waitangi process is the official endorsement of a revised interpretation of the nineteenth-century Wars, almost a hundred years after Cowan. The New Zealand Wars now have a new "correct perspective": the latest biography, available on *NZHistory* online, of Governor George Grey, a major figure in early colonial history and in the New Zealand Wars, thus informs the reader that: "His [Governor Grey's] reputation is tarnished, however, by his later policies in Taranaki, his invasion of Waikato, and the massive confiscation (*raupatu*) of Māori land which followed. The confiscations, in particular, caused decades of bitterness and deep division."[66]

But beyond the inclusion of Maori perspective on the Wars, did the myth of the Maori warrior survive the dawning of the postcolonial era? The Maori Warrior has lately been officially revived through the celebrations around the anniversaries of both world wars. Maori veterans from the 1940 Maori Battalion were duly honoured. Some of the photographs of the Maori Battalion have become iconic, in particular that of the famous haka in Egypt.[67] Maori contribution to the war effort both in 1940 and in 1914 has been the subject of recent research.[68] Serious contemporary historical accounts still include references to the indomitable character of Maori soldiers. They bear witness to the enduring nature of the myth and also illustrate a continuing Maori strategy to re-colonize that myth. *Te Ara – The Encyclopedia of New Zealand* online features an article authored by Maori historian Soutar, who notes: "Māori who served overseas in wartime have often been known as fierce and determined fighters – a force to be reckoned with."[69]

As to the colonial land wars, which are the origin of the Maori Warrior myth, they still have not been the subject of much memorial activity. Fiction and film have appropriated the legend of the nineteenth-century wars, insisting on their grimness and the atrocities committed on both sides, and emphasizing the devastation caused by the colonial project in New Zealand: a point that New Zealand historians have made over the last thirty years and that has sipped into popular culture.[70] Very recently there have been some calls not only for memorials to be erected for the Maori dead,[71] but also for more official attention to be paid to the New Zealand Wars, these being now "largely ignored" by the New Zealand public, as noted by historian Monty Soutar.[72] Contemporary historians have lately pointed out the discrepancies between the centennial celebrations of world conflicts, in particular the focus on major events such as Gallipoli, and the paucity of official remembrance around the New Zealand Wars, a pivotal time in New Zealand history.[73] Although certain episodes have been re-enacted locally on the occasion of their 150th anniversary, little political attention has been devoted to these efforts and, more significantly, little government money was spent on them. The Maori (Land) Wars have obviously been overshadowed by the recent celebrations of the two world wars, in particular the First World War. This has given rise to a new debate around competing memories and the treatment of nineteenth-century Maori resistance, which, for historians such as Vincent O'Malley, is a "cause no less worthy of remembrance."[74]

This is certainly the agenda of a number of Maori *iwi* whose forebears were involved. In their dedicated website, *Battle of Gate Pa – The Founding of Our City*, the members of the Tauranga Tribes Trust explain their educational action: "To establish the Battle of Gate Pa as an occasion to be commemorated every year and remind ourselves of how far we have come as a community, and how far we have to go." They also state:

By understanding *our history*
We can see how far *our community* has come
And look forward to *our future* together.[75]

CONCLUSION

The early discourse on Maori and the history of the Dominion interpreted the nineteenth-century New Zealand Wars as a heroic last stand in which the natives lost their final battle against the march of "Civilization." It produced a distinctive myth around the Maori Warrior and his many admirable virtues. This was the building block of New Zealand's national identity, which was further consolidated by the official narrative of history, which was put together in the aftermath of the First World War and lasted well into the 1950s. This myth was, in turn, appropriated by Maori themselves, particularly regarding their participation in the two world wars.

In the latter part of the twentieth century, this gave way to a new approach, which exposed the underlying assumptions informing the national narrative of identity and the relationship between Maori and Pakeha. From "revisionist historians" to self-termed "neo-revisionists," the New Zealand Wars have been the focus of research and discussion relating to the colonial project in New Zealand as well as to the necessary transition to a postcolonial era. The recognition of the New Zealand Wars as a pivotal episode in the nation's history has been very much connected to the national process of reconciliation, in which New Zealand has been engaged since 1985, mainly through the Tribunal of Waitangi. Tribunal historiography has unearthed contending discourses and voices, and generally helped to produce innovative research on topics and issues that used to be considered definitively settled.

It is thus symptomatic that Conservative prime mister John Key, in an obvious reiteration of the old traditional discourse on the country's benevolent treatment of its Indigenous minority, caused a major uproar in late 2014 when he stated that New Zealand had been "settled peacefully." Historians immediately reacted, recalling the nineteenth-century wars and the devastation they caused.[76] As New Zealand again turns to introspection on the occasion of the celebration of the centenary of the First World War and of events traditionally connected to its coming of age (such as the battle of Gallipoli), it is fitting that other defining moments of its history also be recalled and given official attention. There is a growing sense that the "historical amnesia" pertaining to the New Zealand Wars is political because "they do not rouse nationalist pride." As was ironically asked by a leading scholar on the subject: "Who wants troubling introspection when we can have heart-warming patriotism instead?"[77]

As to the Maori Warrior myth, it remains an enduring vision that is perpetuated by Maori themselves. Aside from the clichés amplified by the media (essentially with regard to rugby hakas), the Maori Warrior is periodically revived in times of protest. This produced a striking effect when, in February 2016, Maori in traditional dress and facial tattoos led the demonstrations against the Pacific Trade Agreement in all of New Zealand's major cities.

NOTES

1 Charles Dilke, *The Problems of Greater Britain* (London: Macmillan and Co, 1890), 414–15 (emphasis added).
2 Keith Sorrenson, "How to Civilize Savages: Some 'Answers' from Nineteenth Century New Zealand," *New Zealand Journal of History* 9, no. 2 (1975): 97.
3 See, for instance, Peter Adams, *Fatal Necessity: British Intervention in New Zealand* (Auckland: Auckland University Press, 1977); and Claudia Orange, *The Treaty of Waitangi* (Wellington: Allen and Unwin, 1987).
4 Keith Sorrenson, "Maori and Pakeha," *The Oxford History of New Zealand*. 2nd ed., ed. Geoffrey W. Rice (Auckland: Oxford University Press 1992), 149.
5 See, for instance, Richard Broome, *Aboriginal Australians – A History since 1788*, 4th ed. (Sydney: Alllen and Unwin, 2010).
6 James Belich, *Making Peoples: A History of the New Zealanders from Polynesian Settlement to the End of the Nineteenth Century* (Auckland: Penguin Books, 1996), 183.
7 Belich, *Making Peoples*, 326.
8 N.A.C McMillan, "Tempsky, Gustavus Ferdinand von," *Dictionary of New Zealand Biography. Te Ara – The Encyclopedia of New Zealand*, 11 March 2014, http://www.TeAra.govt.nz/en/biographies/1t90/tempsky-gustavus-ferdinand-von.
9 James Belich, *The New Zealand Wars and the Victorian Interpretation of Racial Conflict* (Auckland: Auckland University Press, 1986).

10 André Siegfried, *La démocratie en Nouvelle-Zélande: Thèse présentée à la faculté des lettres de l'université de Paris* (Paris: Librairie Armand Colin, 1904), 12.
11 John Williams, *Politics of the New Zealand Maori: Protest and Cooperation, 1891–1909* (Seattle: University of Washington Press, 1969), 12.
12 E. J. Von Dadelsen, Registrar-General, *The New Zealand Official Yearbook 1892. Prepared under the instructions of the Honourable J. Ballance, Premier* (Wellington: Government Printer, 1892), 273.
13 William Pember Reeves, *The Long White Cloud – Aotearoa* (Christchurch: Golden Press, [1898] 1950).
14 Benedict Anderson, *Imagined Communities: Reflections on the Origins and the Spread of Nationalism. Revised Edition* (London: Verso [1983], 1991).
15 Reeves, *Long White Cloud*, 57.
16 Ibid., 201.
17 Ibid., 208.
18 Ibid., 223.
19 Ibid., 209.
20 Ibid., 25.
21 Keith Sinclair, *A History of New Zealand* (Auckland: Penguin Books 1983), 215–16.
22 Siegfried, *Democracy in New Zealand*, 10.
23 Most notably by Edward Maning, *Old New Zealand: A Tale of the Good Old Times, Together with a History of the War in the North of New Zealand against the Chief Heke in the Year 1845 as told by an Old Chief of the Ngapuhi Tribe, also Maori Traditions* (Christchurch: Whitcombe and Tombs Limited 1863–1930).
24 Reeves, *Long White Cloud*, 225–6.
25 "Ōrākau NZ Wars memorial" (Ministry for Culture and Heritage, 24 June 2014), http://www.nzhistory.net.nz/media/photo/orakau-nz-wars-memorial.
26 James Cowan, *The New Zealand Wars: A History of the Maori Campaigns and the Pioneering Period*, vol. 2, *The Hauhau Wars (1864–72)*, vols. 1 and 2 (Wellington: Government Printer, 1922 / Wellington: R.E. Owen, 1956), http://nzetc.victoria.ac.nz/collections.html.
27 David Colquhoun, "Cowan, James (1870–1943)," *Dictionary of New Zealand Biography. Te Ara – The Encyclopedia of New Zealand*, vol. 3 (1996), http://www.teara.govt.nz/en/biographies/3c36/cowan-james.
28 Cowan, *New Zealand Wars*, v.

29 Broome, *Aboriginal Australians*.
30 Cowan, *New Zealand Wars*, 1.
31 Ibid., 2.
32 Ibid., 2 (emphasis added).
33 Ibid., 3 (emphasis added).
34 Ibid. 3.
35 Ibid. 3 (emphasis added).
36 Ibid. 3.
37 Ibid., 366 (emphasis added).
38 Ibid., 395.
39 Ibid.
40 Ibid., 400.
41 Ibid., 401.
42 Ibid., 5.
43 Ibid.
44 Ibid., 395.
45 Ibid., 549.
46 Ibid.
47 Ibid. (emphasis added).
48 Corinne David, "Regards autochtones sur la Seconde Guerre mondiale: L'expérience unique du Bataillon Maori de Nouvelle-Zélande," in *Expériences de guerres: Regards, témoignages, récits*, ed. Renée Dickason (Paris: Éditions Mare et Martin, Collection Politéia, 2012), 4.
49 Cowan would later further build on the Maori Warrior myth by writing a history of the Maoris in the First World War. See James Cowan, *The Maoris in the Great War: A History of The New Zealand Native Contingent and Pioneer Battalion* (Auckland: The Maori Regimental Committee 1926).
50 Cowan, *New Zealand Wars*, 2.
51 Ibid.
52 Apirana Ngata, *The Price of Citizenship: Ngarimu, Victoria Cross* (Wellington: Whitcombe and Tombs, 1943).
53 Keith Sinclair, *The Origins of the Maori Wars* (Wellington: New Zealand University Press, 1957), 1.
54 Bernard John Foster, "The Battle of Orakau," in *The Encyclopedia of New Zealand*, ed. A.H. McLintock, (Wellington: Department of Internal Affairs, 1966).
55 Keith Sinclair, "Why Are Race Relations in New Zealand Better Than in South Africa, South Australia or South Dakota?," *New Zealand Journal of History* 5, no. 2 (1971): 126–7.

56 Ibid.
57 Alan Ward, "The Origins of the Anglo-Maori Wars: A Reconsideration," *New Zealand Journal of History*, 1, no. 2 (1967): 170.
58 Alan Ward, *A Show of Justice: Racial "Amalgamation" in the Nineteenth Century in New-Zealand* (Toronto: University of Toronto Press, 1973).
59 Peter Gibbons, "Cultural Colonization and National Identity," in *New Zealand Journal of History* 36, no. 1 (2002): 5–15.
60 Belich, *Making Peoples*, 441.
61 M.K.P. Sorrenson, "Maori and Pakeha," in *The Oxford History of New Zealand*, 2nd ed., ed. W. Rice (Auckland: Oxford University Press, 1992), 157.
62 "Set up by the Treaty of Waitangi Act 1975, the Waitangi Tribunal is a permanent commission of inquiry that makes recommendations on claims brought by Māori relating to Crown actions which breach the promises made in the Treaty of Waitangi." See Waitangi Tribunal, 21 June 2016, http://www.waitangitribunal.govt.nz/. Since 1985 and the Waitangi Tribunal amendment act, the tribunal has been allowed to deal with claims related to historical grievances.
63 Giselle Byrnes, *The Waitangi Tribunal and New Zealand History* (Melbourne: Oxford University Press, 2004).
64 Giselle Byrnes, "Jackals of the Crown? Historians and the Treaty Claims Process in New Zealand," *Public Historian*, 20, no. 2 (1998): 9, https://doi.org/10.2307/3379416.
65 Vincent O'Malley, *Beyond the Imperial Frontier: The Contest for Colonial New Zealand* (Wellington: Bridget Williams Books 2014), 129.
66 Ministry for Culture and Heritage, "George Grey," (2014), emphasis added, http://www.nzhistory.net.nz/media/photo/george-grey-painting. This can be contrasted with Keith Sinclair's 1990 online biography of Governor Grey, which does not dwell on the confiscations and notes that Grey showed "scholarly interest in Maori language and culture." See Keith Sinclair, "Grey, George," in *Dictionary of New Zealand Biography*, *Te Ara – the Encyclopedia of New Zealand* (1990), http://www.TeAra.govt.nz /en/biographies/1g21/grey-George; "Maori Battalion haka in Egypt, 1941," http://www.nzhistory.net.nz/media/photo/maori-battalion-haka-in-egypt-1941.
67 "Maori Battalion haka in Egypt, 1941," http://www.nzhistory.net.nz/media/photo/maori-battalion-haka-in-egypt-1941.
68 Monty Soutar, *Nga Tama Toa / The Price of Citizenship: C Company (28) Maori Battalion 1939–1945* (Auckland: David Bateman Ltd, 2011).
69 Monty Soutar, "Story: Ngā pakanga ki tāwāhi – Māori and overseas wars," in *Te Ara – The Encyclopedia of New Zealand* (updated 2

December 2015), http://www.teara.govt.nz/en/nga-pakanga-ki-tawahi-maori-and-overseas-wars.

70 Two important feature films were produced locally on the subject of the wars: Geoff Murphy, *Utu* (New Zealand 1983); and Vincent Ward, *The River Queen* (New Zealand, 2005).

71 Danny Keenan, *The New Zealand Wars/ Ngā Pākanga Whenua O Mua* (2016), http://newzealandwars.co.nz/us/.

72 Libby Wilson, "Maori History Largely Ignored, Say Top Historians," Stuff.co.nz, 17 September 2015, http://www.stuff.co.nz/national/72142660/Maori-history-largely-ignored-say-top-historians.

73 Vincent O'Malley, "Historical Amnesia over New Zealand's Own Wars," *Dominion Post*, 23 April 2015, http://www.stuff.co.nz/dominion-post/comment/67944795/Historical-amnesia-over-New-Zealands-own-wars.

74 O'Malley, *Beyond the Imperial Frontier*, 129.

75 *Pukehinahina Charitable Trust, Battle of Gate Pa – The Founding of Our City* (2014), http://www.battleofgatepa.com/.

76 Chris Bramwell, "Peaceful Settlement' View Challenged," Radio New Zealand, 25 November 2014, http://www.radionz.co.nz/news/political/260174/'peaceful-settlement'-view-challenged.

77 O'Malley, "Historical Amnesia."

3

"This Day Is Not for You": The Commemorative Displacement of Black Wars in White Australia

Elizabeth Rechniewski and Matthew Graves

All instances of commemoration, including the construction of monuments and memorials, involve a process of displacement since they seek to impose a simplified and oriented interpretation of an event that necessarily excludes or displaces other narratives. Certain actors are foregrounded, certain actions highlighted; a certain public is invoked, indeed interpellated, by being addressed as members of a group or a nation. In the case of monuments and memorials, this orientation may be achieved through material representation (the statues of the Australian Digger in repose; the horsemen on the new Boer War memorial in Canberra); through the inscriptions and the nomination of events, such as lists of battles, names of the fallen, and dates to be remembered; and through the placement in the landscape. The ceremonies associated with the memorials contribute to "filling out" the narrative through the panoply of symbols, such as flags, that can be deployed, the selection of attendees and special guests and speakers, and the speeches that are made. Owen Dwyer proposes that we understand this creation of commemorative meaning through the concept of "symbolic accretion": "commemoration can be understood as an attempt to impose a partial (in both senses of the word) interpretation of past events on the memorial landscape, in effect, trying to condense and harden – to accrete – a layer of meaning above all others."[1]

There are at least two further meanings to commemorative displacement: the original intention of a monument can be "displaced" when the memorial is appropriated to serve another function. Matthew Graves illustrates this phenomenon in his article on the controversy surrounding the construction of the monument built to commemorate the assassination of Alexandre I of Yugoslavia in 1934, noting that the memorial briefly became the focus for popular resistance to the Vichy regime and to the German occupation of France during the Second World War.[2] Such appropriation can have a greater or lesser intentional political purpose. In Australia, in the early years after the First World War, the existing Boer War memorials were sometimes used as a suitable venue to commemorate the fallen of the later war, until new monuments were built.

Commemorative displacement also, and crucially, occurs through the decisions that are made (usually at an official level) as to the events that are deemed "worthy of remembering" in the nation's memorial agenda. By a kind of sleight of hand that conceals as much as it reveals, the decision to commemorate, among the plethora of possible events, those that correspond to the dominant narrative contributes to telling an officially sanctioned story about the nation, its significant events, populations, and personalities. Commemoration as the mobilization of affect in the interests of national unity and the manufacture of patriotism supposes the exclusion of other social groups' commemoration. Such commemorative choices become more evident when a formerly excluded or repressed minority campaigns for recognition of events and people significant to its own history. In a further twist, however, the response to such demands may be to offer a form of inclusion and recognition within the dominant narrative that neither challenges nor fundamentally changes this narrative but simply extends or modifies it, a form of *récupération* of the counter-memories.[3]

A form of symbolic violence, commemorative displacement consists in the imposition of categories of thought and perception on dominated social agents.[4] If, in general terms, symbolic violence refers to everyday practices of discrimination against a particular group, infringing on its rights to live equally with others, the sociologist Fatma Göçek proposes expanding this definition to cover "the violence inherent in the production of knowledge, especially in relation to one social group's contribution to the cultural and social fabric."[5] The right to "live equally" should include the right to record and commemorate its own history, but typically minorities are displaced from the national story and the repression that has been exercised against them by dominant populations and elites is excluded from the narrative. This is particularly true of the founding stories of colonial nations, writes Göçek, where the foundational violence perpetrated against the original Indigenous populations cannot be acknowledged: "among all acts of violence committed directly or indirectly by states and their governments, those that are temporally closest to the nation's creation myth are silenced and denied the most and the longest because they constitute a foundational violence."[6]

In colonial settler societies, argues the Australian historian Ann Curthoys, national identity is built on "the exclusion of Indigenous peoples from foundational historical narratives."[7] Representations of the "struggle to tame the land," of the hardships and sacrifice of the early settlers, serve to obscure the sufferings of the displaced Indigenous peoples. In the case of Australia, the myth of the Anzacs at Gallipoli (to which the journalists Ellis Ashmead-Bartlett and Charles Bean contributed heavily) extolls the wartime sacrifice of the Australian soldiers, the "Diggers," as laying the foundation of post-Federation national identity. The commemoration of Gallipoli and the associated Anzac legend, Curthoys maintains, replaces and displaces the Black Wars from the national narrative: the foundational violence on which the nation was built is thereby identified as that deployed against an external enemy.

A concerted and extensive official policy of commemoration has over the last century highlighted the Anzac legend to the quasi exclusion of the foundational wars of Australia: the Black Wars against Aboriginal peoples. The commemorative agenda of Australia was colonized throughout the twentieth century by a militaristic interpretation of its history centred on the exploits of white soldiers in the service of the British Empire. The deeds of these men were commemorated on thousands of war memorials in towns large and small, and were memorialized in a dedicated institution – the Australian War Memorial, which is both museum and shrine – and promoted by extensive government funding of education programs.

This chapter explores how analysis through the prism of these overlapping forms of commemorative displacement contributes to an understanding of, first, the relationship between demands for recognition of the Black Wars, including at the Australian War Memorial (AWM), and the recent steep rise in commemoration of Black Diggers in many fora as well as at the AWM; and, second, the commemorative conflict that arises from the recent campaign to ramp up commemoration of the Boer War. The contested question of the recognition of the "Black Wars" offers a particularly striking illustration of the struggle over the Australian commemorative calendar as demands to recognize the previously marginalized history of Aboriginal martial response to invasion and colonization have begun to challenge the prevailing, hegemonic narratives of white Australia.

CONTESTING THE "BLACK WARS"

The existence of "Black Wars" between the British military and settlers on the one side, and Aboriginal peoples on the other, has for several decades been the object of intense academic debate and political acrimony: Was there a concerted military offensive against Aboriginal peoples as white settlement progressed across the continent? In the early 2000s, historians who decried the "black armband" view of history, led by Keith Windschuttle, denied that there had been systematic or even widespread killing of Aboriginals.[8] This claim has since been fairly comprehensively debunked by many authors, including Bain Attwood,[9] but it still occupies a largely unchallenged place among the tenets of the right. A more technical question concerns the nature of the engagements between white and black: Were they on the scale or of the nature of a "war" as Europeans understood it? Certainly they lacked the battlefield confrontations of grouped forces typical of conventional warfare, the Aboriginal fighters generally resorting, for sound tactical reasons, to guerrilla strategies of ambush and surprise attacks.[10] And yet the phrase "Black War" was already being used in the nineteenth century to describe the rounding up and removal of the blacks of Tasmania, and it is the term that we use in this chapter.[11] Although we do not have the space here to review the extensive literature that has accompanied the last two decades of debate, our previous research highlights the extent to which the evidence of widespread killing can be found, even though often euphemized under terms such as "dispersal," in the records of the time.[12] A volume of articles and letters from the *Queenslander*, collected and published by its editor Carl Feilberg in 1880, illustrates in harrowing detail the depredations inflicted on Aboriginal people by the whites and also by the native police the latter employed.[13] Terms such as "war of extermination" and "massacres" are regularly used by the letter-writers to describe the "nameless deeds of horror" they have witnessed or heard openly discussed by whites.[14] A recent paper by two Queensland researchers estimates that the numbers of Aboriginals killed in Queensland alone across a forty-year period to be over forty thousand, on the most conservative of estimates, and suggests that the total death toll may be over sixty thousand, which would bring Aboriginal deaths close to those of Australian troops in the First World War.[15]

It will be readily understood, given the controversy that has surrounded and still surrounds the very existence of "Black Wars," and the long history of concealment and denial that characterized colonial society, that the question of their place in the commemorative practices of Australia was and still is a contested one. Memory of the Black Wars was displaced and overlaid, at the time and since, by commemoration of the military engagements undertaken by the new Federation of Australia (1901), notably in the Boer War (1899–1902) and in the First World War. These engagements, although fought at the behest of the imperial master in far-flung countries, allowed the young nation to assert its military prowess, the legitimacy of its claim to join the community of nations, and the superiority of the white and, particularly, the "British race" to which its citizens consciously belonged. The institution that has been central to the promotion of this military legend is undoubtedly the Australian War Memorial in Canberra.

THE AUSTRALIAN WAR MEMORIAL: GUARDIAN OF THE NATION'S MEMORY

Originally planned by Charles Bean to house the relics of the First World War, the Australian War Memorial was only opened in 1941 in the midst of another world war, and, after some debate, its collections were extended to encompass this one too, and gradually to include the many other wars in which Australia has fought. The Memorial has played a "unique role among the world's war memorials," writes Michael McKernan, official historian and one-time associate director (1981), because of its dual purpose of "commemoration and understanding or more accurately commemoration through understanding."[16] This dual function was the inspiration of the war journalist Charles Bean, whose vision for the memorial was: (1) that it should both honour and commemorate those who had, as he had witnessed at Gallipoli, given their lives in the terrible battles of the First World War and (2) that it should preserve the relics, records, diaries, and official archives that would contribute to an understanding of the war in which they had been engaged, for the benefit of family and friends, the general public, and historians. For Bean there was no contradiction between these functions – the sacralization of memory and the preservation of relics – since the relics were themselves sacred,[17] and they were to be exposed as museum items so that they could aid in remembering and understanding the action of fighting men. In his review of the noted historian Ken Inglis's *Sacred Places: War Memorials in the Australian Landscape* in the *Australian Humanities Review* of December 1998,[18] Martin Ball refers to the "innate schizophrenia" of the Australian War Memorial – schizophrenia that at its simplest might be summarized by the question: Is it a Museum or a Memorial? This is a dilemma that gives rise to a series of complex questions about the scope, nature, and functions of the institution today.[19] What room, if any, is there in the Australian War Memorial for an exhibition about the early wars between Aboriginals and white settlers? Should there be a gallery of "colonial warfare," recognizing the skirmishes, massacres, and battles that took place following the British invasion of the continent? The arguments that have taken place around this issue, and the virulent opposition that such proposals have often encountered, are revelatory of the centrality of the Memorial in the promotion of the official narrative of nationhood.

There is little doubt that, throughout the twentieth century, the AWM offered a perspective on Australian national identity that was closely aligned with a British heritage, offering a field for what the social anthropologist Ghassan Hage calls the "rituals of White empowerment."[20] It was a perspective that excluded the martial role of Aboriginals in defence of their land and also, for a long time, their service as soldiers in the regular forces. For many years, the only presence of Aboriginals at the Memorial was to be found in the courtyard, where an Aboriginal man and woman are represented among a series of stone sculptures depicting Australian fauna. Despite controversy and protests, the sculptures are still in place.

According to Michael McKernan, the historian Geoffrey Blainey, when he was asked to provide ideas in 1979 for a "better organisation of the Memorial's displays, based on historical principles," suggested that, in his view, the Memorial should include a section on Aboriginal-European warfare "within the next decade," and added further that it could be included in an exhibition on the theme of "home ground advantage" in war.[21] It was an interesting suggestion from one who would later be accused of conservatism, even racism, but his idea was ahead of its time for the AWM. Advice sought by the board from the military historian Alec Hill argued – foreshadowing later debates – that it was inappropriate to use the term "war" to describe conflicts between white and Aboriginal Australians.[22] Macintyre and Clark argue that Blainey's proposal did not, however, provoke the heated debate, either on the council or in the media, that might have been expected because it predated the "History Wars" and the crystallization of attitudes that then occurred.[23]

The publication of Ken Inglis's *Sacred Places* nearly two decades later, in 1998, once again brought the issue to the fore. At the launch of the book at the AWM by Governor General Sir William Deane, Inglis proposed some future directions for the Memorial, including "the representations of warlike encounters between black and white."[24] The official reply from the Memorial was that "the story did not belong in the Memorial" but could be told elsewhere, for example in the new National Museum, where, indeed, there is a gallery on the "First Australians" that features various forms of Aboriginal resistance to occupation, though little on direct black/white armed confrontation.[25] This time the debate became caught up in the burgeoning "History Wars" of the turn of the century, the very notion of a "war" between black and white being described by conservatives such as Keith Windschuttle as a "left-liberal fiction."

In defence of his call for memorialization of black/white armed conflict at the AWM, Inglis, in the third edition of his book (2008), quotes from the *Atlas of Australian Wars*, published in 2001 and on display at the AWM, which refers to the "brutal, bloody and sustained confrontation that took place on every significant piece of land across the continent" until the 1930s, with the characteristics of a civil war.[26] Inglis gives further examples of the use of the term "war" in contemporary accounts of engagements with the "Aborigines of Australia," and he comments on the differences with New Zealand where the confrontation between Maori and whites is represented as fully fledged war. By 1915, he writes, there were some thirty memorials in New Zealand to men who fell in battle against the Maori: "They were inscribed to men who *fell*. Nobody was said to have *fallen* in battle against the natives of Australia."[27] Moreover, the New Zealand Army Museum at Waiouru includes part of a gallery devoted to the "New Zealand Wars."

In March 2008, ACT Labor Senate candidate and former Hawke government adviser Peter Michael Conway, on behalf of the Canberra Institute,[28] wrote to then prime minister Julia Gillard to request that the Memorial "examine the possibility of nationally recognising military style conflicts ... with Aboriginal clans, commonly referred to as the 'Aboriginal Wars.'" His letter asks for a monument to be built at the southern end of Anzac Parade, on the axis between Parliament House and the AWM, or on the same axis between the High Court and the National Library, and for a section on the wars to be included among the Colonial War dioramas.[29] The submission nominates a number of conflicts to be commemorated, including the Pemulwuy-led Hawkesbury and Nepean Wars from 1790, the Black Wars of Tasmania, the Port Phillip District Wars from 1830 to 1850, the Kalkadoon Wars of North West Queensland from 1870 to 1890, and the Western Australian Conflict from 1890 to 1898. Some historians date the continuation of armed conflict into the 1920s.[30]

The event that had triggered the request, according to the letter, was the announcement that a monument to commemorate Australian participation in the Boer War would be built on iconic Anzac Parade, Canberra, the avenue that leads up to the Memorial and that is lined with monuments to Australia's battles. This decision, writes Conway, undermines the previous reasons given for refusing to recognize the Aboriginal Wars at the Memorial. For it had been argued that the key phrase in its mission statement was "Australian wars," and Australia as a united country, with a united military force, had not of course existed before 1901. If only post-Federation conflicts were included at the AWM, then the Aboriginal wars – whether they were wars or not – were excluded. But accepting the placement of a monument to the Boer War (1899–1902) alongside those to post-Federation battles meant that this objection no longer held.[31] Moreover, Conway points out that certain colonial wars that were being waged at the same time as the Aboriginal wars are recognized in the Hall of Valour, which honours holders of the Victoria Cross and includes three VCs from the Boer War and, in certain of the dioramas, the 3D models of battles. Indeed the "Colonial Conflicts" gallery recounts Australia's early military history from European settlement to the end of the Boer War. It includes reference to Australian participation in the fight against the Maori in New Zealand but not to the Black Wars.[32]

Peter Conway's letter was reported in several press outlets, and it garnered some fairly predictable reactions. Major-General (Ret) Bill Crews of the Returned and Services League (RSL) told the ***Sunday Telegraph*** that his organization would oppose the plan. He said there was already a memorial for Aboriginal servicemen and servicewomen behind the Australian War Memorial and cited the criteria of external conflict for inclusion: "All of the memorials that have been established generally commemorate the role of Australians in conflicts outside Australia and there is no precedent for a civil-style conflict to be commemorated."[33] This is the current position of the Memorial too. In an address to the National Press club in 2013, the then director Brendan Nelson reiterated: "The Australian War Memorial ... is about Australians going overseas in peace operations and in war in our name as Australians."[34] Moreover, a December 2013 posting on its blog states that the Memorial "has found no substantial evidence that home-grown military units, whether state colonial forces or post-Federation Australian military units, ever fought against the Indigenous population of this country."[35]

The determined exclusion of the Black Wars (or even of acknowledgment that armed conflict took place) illustrates the inertia of this institution and the obstacles to recognition of internal conflict posed by its organizational structure and special status. The nation's war memory, official ceremonies, and commemorations are largely governed by an agenda set by this institution, which is in turn subject to government oversight. Its generous budget is the object of a special direct grant, it is answerable to the Department of Veterans' Affairs through annual reports, and it is subject through the military members of its board to the conservative influence of the RSL and veterans' associations.[36]

PROMOTING THE BLACK DIGGERS

Noticeable in Bill Crews's response outlined above is confusion between commemorating Aboriginal Wars and acknowledging the service of Aboriginals in the regular armed forces. On this latter issue, it is indeed true that, since 1988, a plaque dedicated to Aboriginal servicemen has been situated in bushland some distance behind the Memorial, funded with money donated by private individuals.

Since 2007, it has acquired greater prominence in the commemorative calendar as an official annual service has taken place there each Anzac Day. This illustrates the increasing attention that has been paid, at the Memorial and elsewhere, to the role of Aboriginal servicemen in the regular army, even while the Black Wars continue to be sidelined. The centenary of the First World War saw a flurry of initiatives undertaken by the Memorial to record and highlight the participation of "Black Diggers" – Aboriginal soldiers – in the regular forces, with projects that included providing input into some fifteen documentaries and programs, compiling a rollcall of Black Diggers, and collecting individual stories of Indigenous military personnel that will feature in the redeveloped galleries.

In a dramatic reversal of their previous neglect, authorities at federal, state, and city levels that, for so long, overlooked Indigenous service, have, in the last fifteen years or so, begun to promote their memory in a major and sustained way. It is interesting to speculate on whether the recent "excessive" recognition of Black Diggers may be seen as an attempt to "recuperate" Aboriginal people into the hegemonic national story: to portray Aboriginal warriors as soldiers of the Queen and King, not as defenders of their lands.[37] The monuments dedicated to Aboriginal and Torres Strait Islander soldiers that Ken Inglis described in 1998 as "few, modest and late" have greatly increased in number. While those dedicated in the 1990s were often small local memorials or plaques, the more recent ones have been much more ambitious and prominent: in Adelaide, a monument dedicated by the governor in November 2013 lays claim to be the first national memorial. Another, designed by Aboriginal artist Tony Albert, was inaugurated in April 2015 in central Sydney's Hyde Park, near the existing war memorial.

In March 2019 a monument to the service of Indigenous soldiers in all wars since the nineteenth century was dedicated in the AWM sculpture garden. Titled "For Our Country" and designed by an Aboriginal artist, this acknowledgment at the AWM of the service of Indigenous men and women offers a further displacement of focus away from the Frontier Wars, which still await recognition at the Memorial. The title echoes the phrase used by Aboriginal people when referring to their land – "For Country" – and yet the monument is dedicated to the men and women who served with the Australian forces overseas, often in imperial conflicts. Is this ambiguity a deliberate attempt to recuperate the meaning of "country" to serve a patriotic end? Noah Riseman aptly poses the question in the title of his book on Indigenous soldiers in the Pacific War: "Defending Whose Country?"[38]

BLACK WARS AND BOER WARS

Contemporaneous with the rediscovery of the role of Black Diggers and of the debate over recognition of the Black Wars is the revitalization of commemoration of the Boer War. Although it is rare to discuss side by side the Boer War and the war against Aboriginal peoples, the latter was, at the time of the Boer War, still being fought in the North and the central desert regions of Australia. The press articles of the time reveal, writes historian Henry Reynolds, that parallels were drawn between the fight against the Boers and the fight against the Aboriginals as both involved extending the imperial frontier in the interests of the British Empire. "It was common in the late nineteenth century to relate this domestic conflict to Britain's many other colonial wars. [The Australian frontier, it was often thought, was one part of the much larger, more widespread imperial frontier.] And the leaders of the time were convinced that force was legitimate both to secure the continent and then to hold it as the exclusive domain of the 'white race.' This was their implicit foreign policy, their own imperial project."[39]

Reynolds recounts the public enthusiasm that greeted the decision by the Australian colonies to send soldiers to the Boer War to aid the mother country. While in the parliaments some dissentient voices were raised, and radical nationalists such as the *Bulletin* writers were vehemently opposed to the war, once it was under way these voices tended to be drowned out or even attacked as treasonous. The celebration of the colonial soldiers as harbingers of a glorious Australian military future began in earnest before the Boer War was over, and the myth of the Diggers as troops who manifested "peculiarly Australian characteristics," as "natural horsemen, inured to hard outdoor living, capable of living off the land and able to find their way across country," was already under construction, despite the predominantly urban origin of the soldiers.[40]

The outpouring of imperial jingoism that greeted the initial engagements of the war was tempered postwar by renewed doubts as to its legitimacy, doubts fostered by the comments of the British prime minister Campbell-Bannerman who, in 1906, condemned it as "infamous and criminal, and wholly unnecessary." His comments were widely reported, though usually without elaboration, in Australian newspapers in early September 1906,[41] yet only a few years later enthusiasm for a foreign war would again reach its peak on the outbreak of the First World War. Commemoration of the Boer War, which had been modest and sometimes belated in the immediate postwar period,[42] was eclipsed by the increasingly central place that Anzac Day, 25 April, came to occupy in Australian national life. Charles Bean argued in 1918 that the sacrifice of the Diggers at Gallipoli justified the young Australian nation's claim to occupy the continent.[43] The actual battle for Australia, won against the Aboriginal peoples by means that included murder, poison, and the use of far superior arms, a battle that brought little credit to the "fighters," was concealed behind the growing mythology of the Diggers; for the former was not a fight that could be commemorated or even, in many cases, admitted.

The recent upsurge in commemoration of the Boer War points up the contrast between the recognition given to the whites' military deeds and the sidelining of the Black Wars. A century on, the nation's "selective memory" has enabled a return to the uncritical valorization of the engagement of the colonial troops in South Africa.[44] The Boer War Memorial website records a score of ceremonies held on or around National Boer War Day on 31 May 2019 (date of the signing of the Treaty of Vereeniging) in all the state capitals and some regional towns.[45] The founding of the Boer War Memorial Committee in 2008 to raise funds to erect a national memorial on Anzac Parade in Canberra culminated with the dedication in May 2017 of a striking monument featuring mounted troopers, a four-man scouting party emerging from the bush. On the low walls surrounding the sculptures are inscribed a verse by Banjo Paterson and extracts from the diary of a Boer War soldier.

The publicity for the Melbourne Boer War Memorial service in 2016 records that the soldiers were fighting and dying "during the very birth of our Nation," they were the "Fathers of the Anzacs." On the Boer War Memorial website specific reference is made to the pre-Federation engagement of troops against the Aboriginals but in a way that tends to minimize the fighting ("quite fierce"; "in some areas"; "minor"; "quickly put down") and leaves the role of Aboriginal people in defence of their lands unclear: "Up until 1899 for Australians there had been quite fierce fighting in some areas as

European settlement expanded across the lands of the Aboriginal peoples, and two minor rebellions on the Australian mainland were quickly put down by British garrison troops."[46]

Old Boer War monuments have been refurbished and, since interest has been shown in identifying Aboriginal people who participated in the Boer War,[47] new plaques have appeared dedicated to Aboriginals who served in South Africa, including to Private John Searle in Perth in 2013 and to George Madigan at Ingham, North Queensland, in 2014.[48] This is the preferred story that still garners official sanction: that of the foreign wars that "made Australia," with Aboriginal soldiers, who, like all the Australians who fought abroad, were volunteers, playing a subordinate but loyal role in support. Alissa Macoun and Elizabeth Strakosch summarize this approach: "'Exclusion and inclusion have operated as twin strategies of settler colonialism throughout the Australian experience."[49]

THE SLOW MARCH OF RECOGNITION

It has been left to local community initiatives to retrace and reinscribe commemoration of the massacre of Aboriginal people onto the landscape, often in the context of the Reconciliation movement. Among the first was the construction of a monument, at the initiative of the Uniting Church, to commemorate the massacre of approximately thirty Wirrayaraay people at Myall Creek in 1838. The inaugural ceremony in 2000 brought together descendants of the victims, of the survivors, and of the perpetrators of the massacre, and it has become an annual event. The Coniston Massacre Memorial Plaque at Arrwek was dedicated in 2003 on the seventy-fifth anniversary of the massacre of sixty to seventy Aboriginals in a series of reprisal raids led by Constable William Murray. Once again, it brought together descendants of the Aboriginal people killed and family members of Constable Murray (Liza Dale-Hallett, Murray's great niece). The Appin Massacre Monument was dedicated in 2007 at Cataract Dam, Campbelltown. Once again it was an initiative closely associated with the Reconciliation movement, sponsored by the Winga Myamly Reconciliation Group.

In the last few years, the issue has been taken up by Aboriginal activists in new ways and at the national level by Aboriginal organizations. It is no longer just white historians demanding recognition for the Black Wars (which Aboriginal activists prefer to call the Frontier

Wars), as had tended to be the case, but Aboriginal people taking charge of telling their own history and demanding that recognition be given at the highest, national level, including at the iconic institution that is the Australian War Memorial. The Aboriginal Tent Embassy in Canberra held a "Frontier Wars Story Camp" during Anzac Week 2016 that organized public events, discussion, and education, and has launched a petition to recognize and remember

> those Sovereign Tribal Original People who were slaughtered during the colonisation of Australia. We also ask for an Official National Day of Remembrance to occur every year for the Frontier Massacre of the Original people in Australia from 1788 onwards. This Day of remembrance is to occur on a significant date during colonisation, not Anzac day, as this is a separate issue.
>
> We also ask that a proper memorial be constructed on Anzac Parade in front of the Australian War Memorial in Canberra to honour our Ancestors who were slaughtered during the colonisation of Australia.[50]

Such demands mirror the practices, the material objects, and the symbols that white Australia has deployed to substantiate its own history, including a monument on Anzac Parade, the ability to join in marches and wreath-laying, and a National Day. Over the past decade a "shadow" march has followed the Anzac Day procession in Canberra to remember the thousands who died in the Frontier Wars. In 2015, those participating in this march met with police obstruction and were prevented from laying wreaths at the Anzac monument or from marching with an Aboriginal flag – this even extended to the case of an Aboriginal ex-serviceman. The title of this chapter quotes from the police response to the marchers: "This day is not for you."[51]

There has also been a recent shift in focus as Aboriginal protestors and activists demand recognition not only for the massacres of which Aboriginal people were victims but also for the role of Aboriginals as warriors in defence of their lands. This movement finds an antecedent in the limited recognition given to Indigenous leaders like Red Kangaroo (Gunnedah, NSW) in the 1980s, whose historical significance was restricted to inter-tribal warfare.[52] Aboriginal people now attach increased importance to the warrior Pemulwuy, who kept the British Army at bay for years around Parramatta and Hawkesbury in the late eighteenth century. Since 2010 there has been a campaign for the return of his skull, sent to England by Sir Joseph Banks, although so far it has proved impossible to locate it. A plaque to Pemulwuy was also unveiled at the National Gallery in 2015.

CONCLUSION

There can be little doubt that the officially promoted, extensive, and expensive commemorative events that marked the centenary of the First World War in Australia tended to overshadow and displace demands to critically revisit the narrative of the wars that "built a nation." Where Aboriginal people have attempted to inscribe their own combat in defence of country onto the memorial calendar, they have been rebuffed and reminded that "this Day is not for you" and, more broadly, that there is no place for the Frontier Wars in the hegemonic narrative that claims white Australians' military exploits overseas as the true foundation of the nation. The recent campaigns for commemoration of the Boer War add a new layer of "symbolic accretion" to that story and have pushed it further back, from the First World War into the nineteenth century, where it is joined by Australian colonial engagements in the Sudan, the Maori Wars, and the Boxer Rebellion. These interventions have been emptied of the controversies that accompanied them at the time – the highly contestable rationale for Australian involvement in them and the way they were conducted – to become mere ciphers in a narrative of white military sacrifice and prowess in the construction of the nation.

It is ironic that the memorial inaugurated by the Gaythorne RSL in 2009, to take one example, which lists these nineteenth-century engagements as well as twentieth-century wars, is dedicated to those who died "in defence of their country."[53] It can be said with certainty that of all the combats undertaken by the inhabitants of Australia, the one fought by Aboriginal peoples was the one that was most clearly in defence of country. Evans and Orsted-Jensen write that the Frontier War was "our Great War – a War for both the defence and the conquest of Australia." Though the AWM and the official Australian commemorative agenda have long avoided confronting the issue of frontier conflict, we conclude that it must eventually be faced. For only then, "armed with an encompassing integrity, can we move forward to a process of nation-building that is ethically based rather than ethnically constructed."[54]

NOTES

1 Owen J. Dwyer, "Symbolic Accretion and Commemoration," *Social and Cultural Geography* 5, no. 3 (2004): 422, https://doi.org/10.1080/1464936042000252804.
2 Matthew Graves, "Memory and Forgetting on the National Periphery: Marseilles and the Regicide of 1934," *Portal* 7, no. 1 (2010), https://doi.org/10.5130/portal.v7i1.1291.
3 Elizabeth Rechniewski, "Remembering the Black Diggers: From the 'Great Silence' to 'Conspicuous Commemoration?,'" in *War Memories. Commemorations, Recollections and Writings on War*, ed. Stéphanie Bélanger and Renée Dickason (Montreal and Kingston: McGill-Queen's University Press, 2017), 388–408.
4 Pierre Bourdieu, *Language and Symbolic Power* (Cambridge: Polity Press, 2002).
5 Fatma M. Göçek, *Denial of Violence: Ottoman Past, Turkish Present, and Collective Violence against the Armenians, 1789–2009* (Oxford: Oxford University Press, 2014), 16.
6 Göçek, *Denial of Violence*, 18.
7 Ann Curthoys, "National Narratives, War Commemoration and Racial Exclusion," in *Becoming Australia*, ed. Richard Nile and Michael Peterson (St Lucia: University of Queensland Press, 1998), 178.
8 Keith Windschuttle, *The Fabrication of History*, vol. 1, *Van Diemen's Land, 1803–1847* (Sydney: Macleay Press, 2002).
9 Bain Attwood, *Telling the Truth about Aboriginal History* (Crows Nest, NSW: Allen and Unwin, 2005).
10 The historian David Day emphasizes the limitations that their way of life placed on the ability of Aboriginal people to sustain armed conflict: guerrilla attacks were more suited both to the terrain (in Tasmania, for example) and to their need to gather food daily. See David Day, *Claiming a Continent: A New History of Australia* (Sydney: HarperCollins, 1996), chaps. 4–7.
11 For example, "The Black War," *Cornwall Chronicle* (Launceston, Tas.), 8 September 1860, 3.
12 The reality underlying the euphemism of "dispersal" can be seen in reports such as this: "News has been received that a party of natives surrounded three troopers who were attempting to arrest some of their number for cattle stealing and spearing a boundary rider. The police, however, dispersed the mob, killing nine and wounding two." See "Trouble with the Natives. Nine killed by police. Perth August 24, 1896," *Telegraph* (Brisbane), 24 August 1896, 5.

13 Carl Adolph Feilberg, *The Way We Civilise: Black and White, the Native Police* (Brisbane: G. and J. Black, 1880), http://nla.gov.au/nla.obj-52760287/view#page/n43/mode/1up.
14 See, for example, the letters from "Outis" and "A.N.," in Feilberg, *Way We Civilise*, 30–1 and 37.
15 Raymond Evans and Robert Orsted-Jensen, "'I Cannot Say the Numbers That Were Killed': Assessing Violent Mortality on the Queensland Frontier," paper presented at the Australian Historical Association's 33rd Annual Conference, University of Queensland, 7–11 July 2014.
16 Michael McKernan, *Here Is Their Spirit: A History of the Australian War Memorial, 1917–1990* (St Lucia: University of Queensland Press, 1991), xii.
17 In September 1917 Bean wrote an article titled "Australian Records Preserved as Sacred Things" for the *Commonwealth Gazette,* cited in McKernan, *Here Is Their Spirit*, 42.
18 Ken Inglis, *Sacred Places: War Memorials in the Australian Landscape*, 3rd ed. (Melbourne: Miegunyah Press, 2008).
19 Martin Bell, "Lost in the Monumental Landscape," *Australian Humanities Review*, December 1998, viewed 12 January 2020, http://australianhumanitiesreview.org/1998/12/01/lost-in-the-monumental-landscape/.
20 Ghassan Hage, *White Nation: Fantasies of White Supremacy in a Multicultural Society* (Sydney: Pluto Press 1998): 241.
21 McKernan, *Here Is Their Spirit*, 293–4.
22 Ibid.
23 Stuart Macintyre and Anna Clark, *The History Wars* (Melbourne: Melbourne University Press, 2003), 205.
24 These words were wrongly attributed to the governor general, creating a minor storm in the press. Inglis recounts this episode in the epilogue to the third edition of *Sacred Places*, 502–3.
25 A section is devoted to the Coniston massacre of 1928. A striking memorial to "commemorate the thousands of Aboriginal people who perished in the course of European settlement" is prominently displayed near the entrance to the National Gallery of Australia, Canberra.
26 Quoted by Inglis, *Sacred Places*, 504. He points out that the book was commissioned by the Australian Army and its author is Lieutenant-General John Coates.
27 Inglis, *Sacred Places*, 23, emphasis in original.
28 Conway was the managing director of the institute, a lobby group.
29 Letter from Peter Conway, Director of the Canberra Institute, to the Hon J. Gillard, Acting Prime Minister, 27 March 2008.

30 Alan Atkinson, "Conquest," in *Australia's Empire*, ed. Deryck Schreuder and Stuart Ward (Oxford: Oxford University Press 2008), 34.
31 The colonial armies were officially united as the Commonwealth Forces only by the Defence Act, 1903.
32 The website describes the gallery thus: "Late in the century colonial forces became involved in overseas conflicts in support of the British Empire in New Zealand, in the Sudan, in China and South Africa."
33 "'Aboriginal Wars' Memorial Plan Under Fire," *Sunday Telegraph*, 8 June 2008.
34 Brendan Nelson, "National Press Club Address," 18 September 2013, https://www.awm.gov.au/talks-speeches/national-press-club-address/.
35 Marylou Pooley, "Will the Australian War Memorial Tell the Story of the Colonial Conflicts?," 17 December 2013.
36 A recent grant of $500 million to expand the exhibition space has been controversial, with some critics arguing that the money could be better spent on veteran services.
37 Rechniewski, "Remembering the Black Diggers."
38 Noah Riseman, *Defending Whose Country? Indigenous Soldiers in the Pacific War* (Lincoln: University of Nebraska Press, 2012).
39 Henry Reynolds, *Unnecessary Wars* (Sydney: University of NSW Press, 2016), 6–7.
40 Ibid., 195.
41 For example, *Sunday Telegraph* (Sydney), 9 September 1906, 4.
42 When the first commemoration of the Boer War took place in Brisbane in June 1921, the mayor lamented the long delay. It is noteworthy that he found it necessary in his speech to assert that the war was a "just one," while another speaker, Brigadier-General Browne, declared that "the South African War was a gentleman's fight. Both sides strictly observed the rules of warfare and humanity." See "Heroes Honoured. South African Campaign. First Commemoration," *Telegraph* (Brisbane), 1 June 1921, 5.
43 Charles Bean, *In Your Hands Australians* (London: Cassell, 1918).
44 Reynolds, *Unnecessary Wars*, 201.
45 Boer War Memorial website, https://www.bwm.org.au/boer_war_day/Boer_War_Day2019.php.
46 Boer War Memorial website, http://bwm.org.au/site/Boer_War_Day2016.php#Melbourne.
47 John Maynard, "'Let Us Go … It's a 'Blackfellows' War': Aborigines and the Boer War," *Aboriginal History* 39 (2015).
48 Viewed 7 July 2020, https://www.bwm.org.au/soldiers/John_Searle.php.

49 Alissa Macoun and Elizabeth Strakosch, "The Ethical Demands of Settler Colonial Theory," *Settler Colonial Studies* 3, nos 3–4 (2013): 429.
50 Viewed 3 August 2020, https://www.change.org/p/australian-war-memorial-acknowledge-the-indigenous-frontier-massacres-during-colonisation.
51 "'This Day Is Not for You': Police Stop Black Digger from Marching for Frontier Wars," *New Matilda*, 27 April 2015, https://newmatilda.com/2015/04/27/day-not-you-police-stop-black-digger-marching-frontier-wars/?utm.
52 "The mighty fighter" was first remembered by a plaque to the "Red Chief" (inspired by the eponymous 1950s novel) unveiled by the Gunnedah Historical Society in the 1960s. In 1984, a new memorial inscription was agreed with local Aborigines, which restored "The Red Kangaroo's" indigenous name: Cumbo Gunnerah. The rededication sees an inflection of commemorative discourse from the earlier stereotype of exclusively white agency, albeit one of limited import: Gunnerah's battles were fought against other tribes and he died in 1745, four decades before colonization. http://monumentaustralia.org.au/themes/people/indigenous/display/21470-red-kangaroo-red-chief.
53 Queensland War Memorial Register, http://www.qldwarmemorials.com.au/memorial/?id=553.
54 Evans and Orsted-Jensen, "Assessing," 7.

4

Allies or Enemies? The Representation of Black Soldiers in Recent French, British, and Canadian Great War Fiction

Anna Branach-Kallas

Remembering the First World War a hundred years after the conflict directs attention to the neglected contributions of colonized soldiers, absent for many reasons from official memory. In his introduction to *Race, Empire and First World War Writing*, Santanu Das estimates that above 4 million non-white men, combatants and non-combatants, were mobilized into the British, French, and American armies during the First World War.[1] While Britain and the United States were cautious about sending soldiers of colour to European battlefields, and France deployed 500,000 colonial soldiers in Europe, Germany openly protested against this practice, seeing it as an offence against white prestige.[2] Unlike Western men, who left hundreds of letters and diaries, and who later immortalized their experiences in poetry and fiction, colonized soldiers often came from illiterate or semi-literate backgrounds, with the result that their experiences were hardly recorded during and after the Great War.[3] They might have survived in what Jan Assmann calls communicative memory, based on direct linguistic transmission; this, however, begins to disappear a hundred years after the events in question.[4] We find echoes of communicative memory in recent First World War fiction that attempts to reconstruct the experience of soldiers of colour, exposing racist practices in the army and situating the conflict, as well as its remembrance, in new perspectives.

Commenting on the scarce authentic documentation available about colonized subjects during the 1914–18 conflict, Das asks:

"What do we know about the daily lived war experience of these men from the former colonies and from different racial and ethnic groups? How were they perceived by the white soldiers and civilians and what was the degree of contact between them within and outside the war zone?"[5] Exploring the experience of Black soldiers in the First World War,[6] French novelist Didier Daeninckx's "Un petit air mutin" (2004), British writer Andrea Levy's story "Uriah's War" (2014), and Canadian author Dionne Brand's "Tamarindus Indica" (1999) contest the official past and reveal much, not only about the war itself but also about those involved in the process of its remembrance. According to Katharine Hodgkin and Susannah Radstone, "The idea of contest in the literal sense is apparently a straightforward one: it evokes a struggle in the terrain of truth. If what is disputed is the course of events – what really happened – new answers, particularly by groups whose knowledge has previously been discounted, may challenge dominant or privileged narratives. But to contest the past is also, of course, to pose questions about the present, and what the past means in the present."[7] As the two world wars have been increasingly commemorated all over the world in recent years, Daeninckx, Levy, and Brand stress the need to remember the war contributions of soldiers of colour.

RETRIEVING THE VOICES AND THE BODIES OF THE KANAKS

Didier Daeninckx incorporates the history of New Caledonia in "Un petit air mutin,"in which he imagines a Kanak ex-serviceman visiting Champagne fourteen years after the war.[8] Arrested by two French veterans of the Great War who suspect him of hunting for trophies on the former battlefields, Wanakaeni is incarcerated with a Frenchman who collects scrap metal for a living. He tells him about the past of New Caledonia, a penal and settler colony with a population of 58,098 inhabitants, including 28,075 Kanaks. The Indigenous peoples of the French colonies, not being French citizens, were not subject to conscription but could be recruited as volunteers. In total, the French government managed to recruit 1,078 Kanaks, out of whom 948 left for France. Three hundred and eighty-two were killed, approximately 35.4 per cent of the Kanak volunteers, the highest price any Indigenous population paid to France in the war. They served in the Bataillon mixte du Pacifique, which comprised four companies of Kanak and Tahitian ***tirailleurs***. The first Niaoulis arrived in France in April 1915. In 1918, the Kanaks took part in the bloody battles near Soissons and on the Chemin des Dames, where thirty lost their lives.[9]

Recruitment became particularly difficult in 1917 after the news of severe Kanak casualties had reached the local tribes. The French falsely promised the Kanaks they would be rewarded with medals, employment, citizenship, and land grants. The recruiters emphasized that France did not really need Kanak soldiers to win the war but wanted all its subjects to be part of the victory. The Indigenous people from French-occupied New Caledonia saw the irony of their participation in a war to defend German-occupied France. The French coerced the Kanaks to "volunteer" by pressing the chiefs of the tribe into providing a certain number of soldiers, often choosing among the youngest and the weakest individuals. During the April 1917 and January 1918 revolts, the Kanaks burned down farms and killed several *colons*. In response, the army (with the assistance of some Indigenous soldiers) violently repressed and even annihilated some tribes.[10]

The protagonist of "Un petit air mutin" has come to Champagne to pay tribute to his tribe members killed on the front and abandoned on the battlefield without proper burial. When talking to him in jail, the French prisoner is surprised to learn that the Kanak soldiers were unwilling to fight for France. Wanakaeni emphasizes that the Kanaks were frightened by the condition in which France returned its Indigenous soldiers, in particular by a *gueule cassée* who had suffered terrible pain without any medical assistance. To avoid military service, Wanakaeni fled into the bush and was caught by Tahitian soldiers; he was beaten, enchained, and imprisoned for three months before sailing to Europe. He blames the local chiefs for selling the Kanak recruits to the French. Speaking about the rebellion in the North, he stresses the devastation of the country, the cruelty of the French soldiers towards the Indigenous population, and the treacherous murder of Noël, the leader of the rebellion. Wanakaeni also comments on the terrible consequences of the Great War on the Pacific archipelago. When he came back to New Caledonia in 1919, he could not recognize his village: "La vérité, c'est que la brousse est devenue un désert, les cultures ont été laissées en jachère, les maladies comme la peste ou la lèpre se sont emparées des corps affaiblis."[11]

Daeninckx uses bitter irony to depict Wanakaeni's imprisonment fourteen years after the hostilities ended at the precise site where he fought during the Great War. The two French veterans who arrest "le noir" do not recognize him as an ex-serviceman. Ironically, the protagonist is one of the Kanaks who were exhibited as a human zoo in the Jardin d'Acclimatation in Paris during the Colonial exhibition of 1931. On display, almost naked in the middle of winter, they were asked to imitate the act of cannibalism.[12] They were offered contracts of seven months, which were illegally extended to two years, and were forced to tour Germany. Daeninckx concludes that the ex-serviceman was exchanged for alligators from a German circus. Thus, his hero and his companions are dehumanized, while the French and the Germans, former enemies, are united in a common act of racist spectacle. When they are placed as human exhibits in a zoo, the Kanaks are clearly seen as anomalous, inferior, a different species, "overexposed and pathologized before the disciplinary gaze of a ... voyeuristic public."[13] Significantly, by displaying the Kanaks as atavistic humans, France erased their participation in "modernity's greatest spectacle of carnage" and, thus, their contribution to the defence of the metropole.[14] Moreover, Wanakaeni's account shows that not only were the Kanaks forced to contribute to the world's first modern technological war by defending, and dying, for the colonial centre, but the war also caused the ruin of New Caledonia's agriculture and its depopulation. Daeninckx thus reconfigures global conflict at the margins of the empires, an approach relatively new in Great War studies.

WEST INDIAN MEMORIES OF PRIDE AND PROTEST

In their respective stories, "Uriah's War" and "Tamarindus Indica," two writers of Afro-Caribbean descent, Andrea Levy and Dionne Brand, commemorate the achievements of the British West Indies Regiment. This infantry unit, formed in November 1915, recruited over sixteen thousand men from the West Indies, the Bahamas, British Honduras, and Bermuda.[15] Their patriotism and imperial enthusiasm were soon severely challenged. As Richard Smith points out, "Under military law all black men, whether British subjects or otherwise, were regarded as aliens and could not rise above non-commissioned rank."[16] After they arrived in Seaford on the south coast of England,[17] the majority of soldiers in the British West Indian Regiment were used as labour battalions, performing duties that they found humiliating: "Despite the vital importance of such tasks, the West Indians regarded their status as deeply inferior, particularly as they were denied the opportunity to fire a shot in anger, although routinely serving within range of enemy shellfire and experiencing casualties as a result."[18] Only two battalions were deployed against the Turkish army in 1917 and 1918, in Jordan and Palestine, respectively.[19]

Andrea Levy's "Uriah's War" uses the form of direct address, a letter or testimony, to convey the war experiences of Uriah and Walker, two clerks from Jamaica who volunteer for the European front. The simplicity of form and language confers upon the text a documentary quality. The two young men are motivated to enlist by their loyalty to Britain and the Empire and by their faith in British supremacy: "You see, the Empire was our protector, that is how we thought. England was great, sort of thing. And she was under threat."[20] Walker, the more determined of the two, convinces Uriah to join the army, believing that their "sacrifice would see the black race uplifted,"[21] thus "underlining the implicit links between military service in the First World War and the expectation of citizenship," a perspective not uncommon among West Indian volunteers.[22] The initial incident at a pub in Seaford, where they disembark, shows the depth of the protagonists' attachment to the imperial idea. When a white customer objects that they should not fight for poor wages in a white man's war while politicians enrich themselves, paradoxically, the two Black men find his remarks embarrassing and incongruent. A shilling a day for a West Indian, if not for a white soldier, is after all a substantial sum of money. Moreover, their lack of sympathy for the white objector is caused by racial anxiety as they fear that they might be blamed for this incident. Most important, however, the two West Indian protagonists feel offended. Because of their loyalty and profound desire to prove themselves by fighting for King and Empire, they see the man's remarks as "belittling [their] patriotism."[23]

Levy's story shows the reaction of West Indian soldiers to the white commanders' conviction that they were not "of the calibre required of armed combat."[24] When Uriah and Walker learn that they will not be allowed to fight against the Germans, England's true foes, the latter is enraged. Nevertheless, they soon realize that to join the Egyptian Expeditionary Force is quite fortunate for their battalion because at least they will thus confront the Turks in front-line combat instead of performing labour duties at the Western Front:

"Now, who wanted to come all that way to be in a labour battalion? Running back and forth with shells and what-and-what for the front line. No rifle, no combat, but just as likely to die. That would have been a humiliation."[25] In this perspective, only the battlefield could provide a test of manhood. Despite the fear and carnage, Uriah and Walker are thus filled with pride after their gallant victory over the Turks. They are praised for their cheerfulness and devotion by "a big-big major,"[26] and are cited in dispatches, which challenges the stereotype of West Indian soldiers as unfit for war. Eventually, they feel included in the imperial family as they pat the backs of their comrades "from Britain, New Zealand, Australia, India, [and] Africa" at the Armistice.[27]

Nevertheless, this sense of glory and grandeur proves an illusion at the camp at Toranto, in Italy, where the historical soldiers of the British West Indies Regiment (BWIR) waited to embark on their journey back home. According to Smith, the discriminatory practices used against the BWIR at Toranto caused much unrest. West Indians were prohibited from using recreational facilities and were thus segregated from the white soldiers, which provoked their anger. They also rebelled against the commander of the camp, who ordered them to clean the latrines of Italian labourers.[28] "Uriah's War" refers to these historical incidents, which the protagonists interpret as insults to their masculinity and imperial pride. They are shocked to learn that all the West Indian Battalions, notwithstanding their achievements on the battlefields, are classified as "inferior 'native' units."[29] When a white sergeant orders them to clean the latrines, Walker tries to remonstrate, explaining that because of their front-line service in Palestine, the West Indians have been promised to be spared all degrading work. In the ensuing argument, the sergeant shouts at Walker "that [they] were better fed and treated than any nigger had a right to expect."[30] Walker then throws a pail at the man's head. As a result, "the most gallant and courageous soldier" is submitted to Field Punishment Number One.[31] This scene is deeply ambivalent: Levy depicts Walker's suffering using Christian undertones of crucifixion, yet Uriah also compares Walker to a slave shackled in fetters,[32] thus placing the punishment in a strictly racist context. In a surprising narrative twist, the epilogue of the story is appended by Walker, for Uriah is shot when he tries to intervene on behalf of his friend.

Most important, Uriah's tragic death and the racist treatment of the West Indians at the front result in Walker's disillusionment and shift of loyalties. His address is a testimony of injustice and discrimination against those who volunteered to defend Britain and the Empire. His tone becomes that of an open accusation for the "insult to the patriotism of the loyal people of the West Indies. Uriah and me did not fail you. We were British soldiers. But you have failed to recognize our contribution."[33] Significantly, however, Levy's portrait of Walker suggests that the Great War became a catalyst for the emergence of anglophone Caribbean nationalism.[34] In the novel's didactic ending, Walker rejects the ideal of Britain as the Mother Country and pledges to struggle so that democracy and freedom, the most important principles for which the Great War was fought, are implemented in Jamaica, his true home. "Uriah's War" thus illustrates "the emergence of a distinct racial consciousness brought about by the hardships of war and by discrimination at the hands of the military establishment."[35] Grieving for the loss of his friend is therefore displaced in favour of Walker's discourse of sacrifice, a vehicle to fight against racism and to demand improvements. Levy's story thus illustrates the complex processes of un-forgetting, commemorating the death of 1,130 Jamaicans in the war,[36] yet at the same time proposing an activist protest, linking Jamaican memory of the Great War with military service and strategic struggle for independence and democracy.

WEST INDIAN MEMORIES OF SHAME AND DISHONOUR

Dionne Brand's "Tamarindus Indica" furthers in another direction this reflection on Black masculinity in another direction during the Great War. The protagonist of the story, Samuel Sones, is a mixed-blood descendant of African slaves and Indian indentured workers in Trinidad. He is encouraged by his mother to rise above his position in life and thus erase the stain of illegitimacy and colour. He volunteers for the war when he sees his white childhood friend, Michael De Freitas, in uniform in 1916. He believes that his service in the army will both improve his existence and the lives of the Trinidadian people: "A man could strive. And he had been let into the Second West India Regiment, proving that men of colour were improving their situation and would be repaid for their duty."[37] His attitude reflects that of many West Indians who believed that Britain would "dispense equity and justice" after they had proven themselves in combat.[38] When the efforts of the governor of the island, entreating the British to accept the Second West India Regiment, are eventually met with "*condescending* assent,"[39] these expectations of "post-war redemption"[40] are a source of personal and collective pride for the protagonist of the story: "He had come all the way here to serve the mother country, Great Britain ... He was in his physical prime, boarding the ship first to Great Britain then to Palestine. Private Samuel Gordon Sones of the Second West India Regiment."[41]

The realities of front life prove a devastating disappointment for Sones as his regiment is assigned labour duties. More so than Levy, Brand places the demeaning treatment of the West Indian volunteers in the historical context of plantation work, showing that the assignment of labour services to Black soldiers raises in their minds the spectre of slavery: "One night he wrote home to his mother, 'We are treated neither as Christians nor as British citizens but as West Indian 'niggers' without anybody to be interested in us or look after us.'"[42] Instead of the glory he expected, Sones feels increasingly belittled by the orders he can barely understand. With profound insight Brand explores the deepening depression of her protagonist, who becomes anxious, alienated, and homesick. He also feels estranged from his comrades, who are united by their anger against the racist practices of the army that does not allow them "to do combat against European soldiers because of their colour."[43] His personality is erased in the machinery of the army, and, unlike Levy's Uriah and Walker, Sones no longer finds reassurance in a sense of common mission.

Consequently, when the Black troops are sent to Palestine to confront the Turks, Sones does not experience any feelings of elation or heroism. On the contrary, he is weakened by fear and exposure. His ordeal on the battlefield proves terrifying and annihilating to the self. The protagonist's ultimate act of rebellion has therefore to be interpreted taking into consideration the historical context, his personal itinerary, and physical and psychological exhaustion. When, after the battle, De Freitas orders him to fill the officers' water canteens, Sones knocks the white man down. This act of protest against white authority, embodied by his childhood friend, seems a desperate attempt to protect his masculine valour against the institutional racism of the British Army. As a result, he is court-martialled, sent back home for misconduct, and spends two years in a military jail. This experience fills him with profound shame, and, even after he is released, he obsessively blames himself for aspiring above his position in life.[44] Sones avoids women and mortifies himself, every day walking in the sun in his frayed army uniform:

> He was filled with so much self-loathing every time he remembered the Second West India Regiment, he tried to root out that small place inside him that led him to it. Root out that small pain that never grew any bigger but was like a tablet of poison ... [H]e understood that the insult would stay with him no matter if he knocked De Freitas down or killed him. And he understood that it was his fault. All of it. He deserved it for pushing himself up and thinking that he was more than he was.[45]

Brand thus demonstrates that, while for the white soldier death in battle is equated with tragedy and mourning, for the Black soldier "exclusion from combat becomes the focal point for the emotions of loss."[46] Importantly, "Tamarindus Indica" contests the official West Indian memory of the 1914–18 conflict, dominated by the veterans' conflation of pride with their struggle for self-determination. The exploration of dishonour and the disgracing effect of the protagonist's front experience and martial punishment supplements the official story of war heroism.

CONCLUSION

Negotiating between collective amnesia and the desire to commemorate the lost (his)stories of the Great War, "Un petit air mutin," "Uriah's War," and "Tamarindus Indica" exemplify the tension between "the racial Other as the embodiment of uncontrolled violence and the racial Other as ally."[47] In Daeninckx's story, the desire to commemorate an unknown facet of the Great War is most clearly visible. In a fascinating way, Daeninckx situates the Kanak experience during the 1914–18 conflict within a larger context of racial exploitation and global postwar transformations. The documentary aspect is also important in Andrea Levy's "Uriah's War," which further explores racist mistreatment of West Indian volunteers. By portraying protagonists telling (his)stories, verbalizing their memories of the past in the form a semi-ceremonial address, both Levy and Daeninckx highlight the importance of communicative memory, a unique source of information about Black soldiers in the First World War. Levy fuses the Jamaican memory of the Great War with veterans' pride and the struggle for independence from Britain. By contrast Dionne Brand's "Tamarindus Indica" explores the affective potential of shame and self-blame, illustrating the collapse of the imperial ideal of Mother Country in a West Indian perspective. This vision accord with Smith's statement that, "In contrast to contemporary mainstream British grieving processes which denote war as a tragedy, the commemoration of the West Indian war contribution tends to inculcate mourning as masculine (dis)honour, forgetting and shifting geographies of home."[48]

"Un petit air mutin," "Uriah's War," and "Tamarindus Indica" are based on rigorous historical research, yet, by giving voice to Black protagonists – in the form of explicit address or implicit introspection – they also move "from the disciplinary realm of history which is 'objective in the coldest, hardest sense of the word,' … toward the domain of memory which is 'subjective in the warmest, most inviting senses of that Word."[49] "Un petit air mutin," "Uriah's War," and "Tamarindus Indica" thus prove that what is significant in investigating the forgotten past is not only a question of historical accuracy but also the potential for an aesthetically powerful and ethically moving expression of meaning.[50] The legacy of colonial encounters fictionalized by Didier Daeninckx, Andrea Levy, and Dionne Brand oscillates moreover between the desire to commemorate a Black presence on the European front and the need to question the racist logic that used the Black soldiers' bodies for the war effort while simultaneously denying the full humanity of *troupes indigènes*.

NOTES

1 Santanu Das, *Race, Empire and First World War Writings*, ed. Santanu Das (New York: Cambridge University Press, 2013), 4.
2 Richard S.Fogarty, *Race and War in France: Colonial Subjects in the French Army, 1914–1918* (Baltimore: Johns Hopkins University Press, 2008), 7.
3 Joe H. Lunn, "France's Legacy to Demba Mboup? A Senegalese Grito and His Descendants Remember His Military Service during the First World War," in *Race, Empire and First World War Writings*, ed. Santanu Das (New York: Cambridge University Press, 2013), 108.
4 Jan Assman, *Pamięć kulturowa: Pismo, zapamiętywanie i polityczna tożsamość w cywilizacjach starożytnych*, trans. Anna Kryczyńska-Pham (Warsaw: Warsaw University Press, 2008), 71.
5 Das, *Race, Empire and First World War Writings*, 1.
6 The term "Black" refers in this chapter to people of various ancestry, mostly perceived as "black" by white soldiers during the Great War. "Black" is therefore a signifier of skin colour but also refers to various contexts of racialization briefly discussed in this chapter. On the

difficulties of defining race, see Das, "Introduction," in *Race, Empire and First World War Writings,* 1–11; David Theo Goldberg, "Introduction," in *Anatomy of Racism*, ed. David Theo Goldberg (Minneapolis: University of Minnesota Press, 1990), xi–xxiii.

7 Katharine Hodgkin and Susannah Radstone, "Introduction," in *Contested Pasts: The Politics of Memory*, ed. Katharine Hodgkin and Susannah Radstone (London: Routledge, 2003), 1.

8 Daeninckx's famous novel *Le Der des Ders* (1984), combining historical with detective fiction, also explores unknown facets of the First World War, revealing dishonest financial schemes in France during the war. His novel *Cannibale* (1998) fathoms the history of the infamous human zoo at the colonial exhibition of 1931. See the interview with the author, "Daeninckx 3 Cannibale," 24 February 2011, https://www.youtube.com/watch?v=onRoLUJjxCQ.

9 See Sylvette Boubin-Boyer, "La Nouvelle-Calédonie durant la Première Guerre mondiale," in *Révoltes, conflits et Guerres mondiales en Nouvelle Calédonie et dans sa région*, ed. Sylvette Boubin-Boyer (Paris: L'Harmattan, 2008), 9–41.

10 Marcel Claude, *La révolte Kanaks de 1917*, 31 March 2015, https://www.youtube.com/watch?v=q3Xauxv3xzY.

11 Didier Daeninckx, "Un petit air mutin," in *Le Chemin des Dames: De l'événement à la mémoire*, ed. Nicolas Offenstadt (Paris: Edition Stock, 2004), 443. "The truth is that the bush had become a desert, the fields had lain fallow, diseases such as the plague or leprosy invaded weakened bodies" (my translation).

12 Alexandre Rosada, *Exposition Coloniale Kanak 1931*, 28 June 2014, https://www.youtube.com/watch?v=4DJRcSEkftI.

13 Anne McClintock, *Imperial Leather: Race, Gender and Sexuality in the Colonial Contest* (New York: Routledge, 1995), 42.

14 Mark Whalan, "Not Only War: The First World War and African American Literature," in Das, *Race, Empire and First World War Writings*, 284.

15 Richard Smith, "Post-Colonial Melancholia and the Representation of West Indian Volunteers in the British Great War Televisual Memory," in *The Great War in Post-Memory Literature and Film*, ed. Martin Löschnigg and Marzena Sokołowska-Paryż (Berlin: De Gruyter, 2014), 388; Richard Smith, ""Heaven Grant You Strength to Fight the Battle for Your Race': Nationalism, Pan-Africanism and the First World War in Jamaican Memory," in Das, *Race, Empire and First World War Writing*, 265.

16 Smith, "Post-Colonial Melancholia," 390.

17 Smith, "Heaven Grant You Strength," 268.
18 Smith, "Post-Colonial Melancholia," 390.
19 Ibid.
20 Andrea Levy, "Uriah's War," in *Six Stories and An Essay* (London: Tinder Press, 2014), 115.
21 Ibid., 127.
22 Smith, "Heaven Grant You Strength," 274.
23 Levy, "Uriah's War," 114.
24 Smith, "Post-Colonial Melancholia," 391.
25 Levy, "Uriah's War," 117.
26 Ibid., 121.
27 Ibid.
28 Smith, "Post-Colonial Melancholia," 391.
29 Levy, "Uriah's War," 122; Smith, "Heaven Grant You Strength," 269.
30 Levy, "Uriah's War," 124.
31 Ibid., 125.
32 Ibid.
33 Ibid.
34 Smith, "Post-Colonial Melancholia," 391; "Heaven Grant You Strength," 270.
35 Smith, "Heaven Grant You Strength," 271.
36 Ibid., 265.
37 Dionne Brand, "Tamarindus Indica," in *At the Full and Change of the Moon* (New York: Grove Press, 1999), 87.
38 Smith, "Heaven Grant You Strength," 267.
39 Brand, "Tamarindus Indica," 76, emphasis added.
40 Smith, "Heaven Grant You Strength," 268.
41 Brand, "Tamarindus Indica," 76.
42 Ibid., 87. This is the statement of an anonymous Black sergeant cited also by Smith in "Post-Colonial Melancholia," 391.
43 Brand, "Tamarindus Indica," 87.
44 For a development of these ideas, see Anna Branach-Kallas, "Narratives of (Post-)Colonial Encounter: The Old World in Contemporary Canadian Great War Fiction," in *North America, Europe and the Cultural Memory of the First World War*, ed. Martin Löschnigg and Karin Kraus (Heidelberg: Universitätsverlag WINTER, Heidelberg, 2015), 100–2; Anna Branach-Kallas, *Corporeal Itineraries: Body, Nation, Diaspora in Selected Canadian Fiction* (Toruń: Nicolaus Copernicus University Press, 2010), 172–3.
45 Brand, "Tamarindus Indica," 95–6.
46 Smith, "Post-Colonial Melancholia," 392.

47 Heather Jones, "Imperial Captivities: Colonial Prisoners of War in Germany and the Ottoman Empire, 1914–1918," in Das, *Race, Empire and First World War Writings*, 176.
48 Smith, "Post-Colonial Melancholia," 386.
49 Iwona Irwin-Zarecka, quoted in Graham Carr, "War, History, and the Education of (Canadian) Memory," in *Contested Pasts: The Politics of Memory*, ed. Katharine Hodgkin and Susannah Radstone (London: Routledge, 2003), 61.
50 Elizabeth Butler Breese, "Claiming Trauma through Social Performance: The Case of *Waiting for Godot*," in *Narrating Trauma: On the Impact of Collective Suffering*, ed. Ron Eyerman, Jeffrey C. Alexander, and Elizabeth Butler Breese (Boulder-London: Paradigm Publishers, 2013), 234.

5

Selective Remembering and Motivated Forgetting: The Primacy of National Identity in Australia's Differential Memorialization of Its Wars

Sheila Collingwood-Whittick

> Institutional memorials, whether shrines and monuments, or moments, like public holidays, tell us which people and deeds merit commemoration.[1]

In a review of Henry Reynolds's *Forgotten War*,[2] historian Raymond Evans draws attention to the imposing statue of a soldier from the First World War at the centre of Bundaberg, a coastal town in the state of Queensland. Standing atop a lofty column, the life-sized, marble digger looks down on Bourbong Street,[3] a busy thoroughfare thought to owe its name to the Indigenous toponym "Bier-rabong," an Aboriginal word meaning "plenty dead." As Evans explains, the "present course" of Bourbong Street runs close to the site where a group of local Aboriginal people were massacred in the 1860s – a frontier atrocity that "no one now remembers and many would be quick to deny ever occurred." [4] Bundaberg's "revered" war memorial is thus read by Evans as epitomizing a paradigmatic feature of Australia's psycho-social landscape – namely, "ritual remembrance beside long-standing forgetfulness."[5] It is this paradox at the heart of Australia's memorial practices that is explored in the following discussion.

To highlight the cognitive dissonance underlying settler Australians' differential memorialization of the wars in which they have been involved is by no means to dismiss the psychological impact on the nation of the loss of thousands of its young soldiers at Gallipoli in 1915. As Ann Curthoys contends: "The importance of this story to the Australian imaginary can hardly be over-estimated, so bound is it with the trauma of the Australian experience of the first world war."[6] Underpinning and illuminating the ever-increasing importance accorded to this story is, this chapter seeks to show, a number of historical phenomena quite unrelated to Gallipoli itself.

"'LEST WE FORGET' RUBBING SHOULDERS WITH 'WHY DON'T THEY JUST GET OVER IT.'"[7]

There are no national, and "less than a handful" of local,[8] monuments in Australia that commemorate the tens of thousands of Indigenous victims of the Frontier Wars.[9] Indeed, the only gesture the Australian War Memorial (AWM) in Canberra makes towards recognizing the existence of Indigenous people is a sculpted frieze that depicts Aboriginal "natives" alongside other species of Indigenous flora and fauna as part of Australia's natural habitat. To quote historian Alan Stephens: "Indigenous warriors who fought invaders for their homeland, their families, and their way of life, have been officially defined out of [Australia's] war commemoration history."[10]

By contrast, no centre of population on the continent is too tiny to have its own cairn, obelisk, cenotaph, statue, or memorial avenue in remembrance of the *non-Indigenous* dead of the various *overseas* conflicts in which Australia has been involved since the end of the nineteenth century. Historian Ken Inglis estimates that there are probably more war memorials per capita in Australia than in any other country.[11] After the First World War, Australia "raised the largest memorials of any nation."[12] "The Great War Memorials built in Sydney, Melbourne, and Canberra, as well as the Anzac statue in seemingly every public park in Australia, were," Peter Hoffenberg observes, "the first truly national monuments on Australian soil."[13] "The war experience was," he underlines, "at the heart of [Canberra's] urban design ... 'Anzac park' [being] the 'central axis of the city.'"[14]

The deployment of troops in the Gallipoli campaign of 1915 represented, of course, Australia's first *major* military engagement overseas. Though it was a "military disaster,"[15] Australians did not choose to remember that the terrible loss of life incurred was almost certainly the result of inexperienced troops and "poor military preparation."[16] Rather, remembrance of Gallipoli centres on the Anzacs' "glorious bravery in the face of the odds, proving to others the worth of Australians and Australia's role in the world."[17]

The germination of this "official and collective" memory can be directly traced to the exalted rhetoric employed by the Australian prime minister Billy Hughes when addressing Anzacs in London on the first anniversary of the fatal Gallipoli landing. "Soldiers," he declared, "your deeds have won for you a place in the temple of the immortals of the world and hailed you as heroes ... The deathless story of Gallipoli will yet be sung in immortal verse, inspiring yet unborn Australians with the pride of their race, courage, tenacity of purpose and fearlessness."[18] Hughes's vision was diligently nurtured by Charles Bean, Australia's official war correspondent at Gallipoli, the author of the first *Official History of Australia in the War of 1914–18*, "the man who has published more words and reached more readers than any other Australian historian,"[19] and the founder of the massive and imposing AWM.[20]

Though Bean "did more than anyone to turn the digger into a national legend,"[21] his main objective seems to have been to persuade the world that the Anzac's "reckless valour in a good cause, [his] enterprise, resourcefulness, fidelity, comradeship, and endurance,"[22] were exemplary of Australia's settler "race" as a whole. Bean's emphasis on the importance of "race" was not unusual in Australia. It was commonly believed in the army, for instance, that "military comradeship ... rested on pride of race."[23] Moreover, as Ball reminds us, Bean was not just a fervent advocate of the idea that "Australian physical and social conditions were improving the British 'stock,'" his "whole epistemology was built on the principles of European and especially British ascendancy."[24]

To understand why Bean's attribution of Anzac heroism to the men's settler ancestry had the resonance it did in white Australia, we need to bear in mind that, in the early decades of the twentieth century, there existed both a residual embarrassment about the nation's convict past and a long-standing cultural cringe in relation to the "mother country."[25] Social Darwinists had even expressed fears of racial degeneration having resulted from "the fatal influence of a warm but enervating climate."[26]

Indeed, many settler Australians had expressed widespread anxiety at the outbreak of the Great War as to how the country's young, inexperienced soldiers would compare with battle-hardened troops from other countries. Yet, paradoxically, "colonists yearned for the sort of blooding on an international stage that would prove their racial vigour and exorcise their convict inheritance."[27] In the circumstances, Bean's confirmation that Anzacs had distinguished themselves on the field of battle came as an immense relief to settler society. And the debt owed to the dead and war-damaged soldiers was immediately honoured by the sacralization of their memory.[28]

Australian bush poet "Banjo" Paterson summed up the nation's boundless pride and gratitude in "We're All Australians Now." Significantly, the term "mettle," used in a key verse of the poem,[29] was the very concept that Bean, the "laureate of Anzac,"[30] chose to highlight in *his* panegyric of Anzac troops. "The achievement of the men lay not in the Allies' military prowess," proclaimed Bean, "rather it lay *in the mettle of the men themselves.*"[31] Intimately linked to this seminal notion was his ideological vision that "it was on the 25th of April, 1915, that the consciousness of Australian nationhood was born."[32] More than seventy years later, Australia's political leaders would make copious use of this persuasive origin myth to kick-start Anglo-Australia's faltering sense of national identity. By the beginning of the twenty-first century, when Scates carried out his study of young Anzac pilgrims, the dominant feeling among them was, he noted, that Gallipoli constituted "that defining moment in our history."[33]

THE EVOLUTION OF ANZAC DAY COMMEMORATION

The twenty-fifth of April was officially named "Anzac Day" in 1916,[34] the year when the first memorial marches took place and church services were held throughout Australia to honour the grieving nation's fallen soldiers. It was not until the 1920s, however, that some of today's most important commemorative rituals were established. These include Anzac Day beach services at Gallipoli (1925), the symbolic ceremony of the dawn service (1927) at which "the landing at Gallipoli became a visual metaphor for the beginning of Australian nationhood,"[35] and the first "organized" pilgrimages to Gallipoli (1929).[36]

But though all these ceremonies were widely attended for decades, attendance figures towards the end of the 1960s were inevitably shrinking, and it was generally assumed that, as the last of the veterans and their immediate families died, public demonstrations of remembrance would themselves wither away.[37] Then, in the mid-1980s, against all previous expectations, Anzac Day began to enjoy what has since become a spectacular renaissance. As Chris Sheedy remarks, "It is almost impossible to imagine, in this period in which Australians' attitudes towards anything Anzac has reached an almost religious level of zeal, that just a few decades ago the commemoration was expected to simply die off."[38]

In 1984, Anzac Day marches throughout Australia attracted an estimated fifty thousand participants. Twenty years later, there were more than that number taking part in the dawn service at Melbourne's Shrine of Remembrance alone.[39] Between 1995 and 2005 the number of Australians visiting the Gallipoli battlefield quadrupled (from five thousand to twenty thousand).[40] And on the one hundredth anniversary of Gallipoli, attendance records at dawn services were smashed as a total of 275,000 people throughout the country participated in this profoundly symbolic ritual.[41] As for government and corporate spending on the First World War centenary commemorations, Australia's outlay of AU$552 million is currently running at more than double the amount allocated by all five of the remaining allied nations combined.[42]

Since "the re-energising of Anzac has," according to one prominent journalist, "become the central organising principle of Australia's past and how the nation interprets its future,"[43] the phenomenon has, naturally, attracted a great deal of academic discussion in Australia, with a variety of factors adduced to explain it. One commonly held view is that Peter Weir's film *Gallipoli* (1981) played a determining role in "re-establishing Anzac as the pre-eminent myth of Australian nationhood."[44] Not only was it "the country's highest-ever grossing film at the time ... It dominated the Australian Film Industry awards, taking nine out of thirteen major honours."[45] In the year of the film's release, the number of visitors to the AWM increased by 50 per cent, and education packs were distributed to Australian schools to assist with the teaching of the film.[46]

Another widely shared opinion is that today's Anzac celebrations are "a sort of civil religion in an almost post-Christian society that no longer delivers ancient certainties to young people who are in search of nourishment for the spirit."[47] Sociologist Gary Bouma, for example, sees Anzac Day rituals as "an excellent example" of the new spirituality Australians have invented to replace more conventional and now outmoded religious practices.[48] Scates's young interviewees describe the Anzac Day ceremony on the Gallipoli

Peninsula as a "spiritual" experience, "as close to a sacred day as Australians ever get."[49] Like Scates, who considers this experience to be an "emotional catharsis,"[50] Holbrook defines it as "an *emotional* experience [that] bears many of the elements of religious worship. It allows people to feel part of a broader community of worship and functions as some kind of vicarious emotional clearing house."[51]

While conceding that "emotion and affect, through the medium of the individual story, have been central to renewal of Anzac," Christina Twomey sees the myth's resurgence as resulting from "international developments in the late twentieth century around the rise to cultural prominence of the traumatised individual." It is, she asserts, "trauma [that] attracted an audience back to Anzac."[52]

But though socio-cultural phenomena of the kind cited above certainly helped generate the "orgy of memorialisation" that Anzac Day has become,[53] they do not explain *why* the settler colonial psyche was so overwhelmingly receptive, so predisposed to respond to such stimuli in the latter half of the twentieth century.

ANZAC DAY AS THE RE-BIRTH OF AUSTRALIAN NATIONAL IDENTITY

The first of a series of major developments leading to the angst-ridden debate on national identity[54] that arose in Australia in the 1970s was "Britain's unmistakable downgrading of the racial ties of empire – most vividly symbolized in the turn towards Europe – that brought home to Australians the fact that the global affections of Britishness were distinctly one-sided."[55] The sense of emotional divestiture provoked by the motherland's abandonment of its traditional role in nurturing what Australians' felt to be their core identity was further exacerbated by the demise of the "White Australia" policy. Formally implemented at Federation, this policy was "unambiguously geared towards preserving the 'British' character of Australian society from the external threat of migration."[56]

Hard on the heels of these two radical reorientations in Anglo-Australia's self-understanding came discussions about the approaching bicentennial of British settlement on the continent, an event that encouraged the kind of melancholic introspection reflected in the rhetorical question posed by a popular national newspaper: "Could it be that the central theme we are supposed to be celebrating, 'a sense of nationhood' or 'national identity,' simply doesn't exist?"[57] But, though Australians were undeniably destabilized by the sudden evaporation of their "Britishness," the psychic tremours emanating from the recrudescence of a past they did not want to remember were, it could be argued, the main cause of the serious fissures that subsequently appeared in the nation's collective psyche.

In 1968, anthropologist W.E.H. Stanner compared his compatriots' relationship with their colonial history to a "cult of forgetfulness practiced on a national scale."[58] Devotees of such ritual disremembering consequently found themselves on the defensive once the impressive corpus of frontier historiography that Stanner's analysis had engendered began to be publicly debated.[59] For this new vision of Australia's past (dubbed "black armband history" by its many detractors) sent a succession of shock waves through settler colonial society.

For their part, Indigenous Australians, who, as early as 1938, had marched in protest against the 150th anniversary of the First Fleet's landing in Australia, had begun referring to "Australia Day" as a "Day of Mourning." In 1970, many of them, wearing red headbands to commemorate the spilled blood of their peoples, had again demonstrated against the official celebrations of the "Captain Cook bicentenary" – an event *they* saw as marking the beginning of a genocidal war. Once preparations for the bicentennial celebrations of "settlement" got under way, therefore, Aboriginal activists responded with the trenchant slogan "White Australia has a Black History."[60]

Adding to this disturbing turn of events, there occurred from the 1980s onwards a series of national inquiries and legal decisions "officially" recognizing the grievous injustices to which Australia's largely forgotten Aboriginal population had been subjected since the end of the Frontier Wars. Suddenly, the benign identity non-Indigenous Australians had imagined for themselves no longer seemed so solidly grounded. As previously suppressed moral qualms about the nation's colonial past began to surface, Australia's "socially organised amnesia" developed symptoms of fatigue stress.[61] "Memory," as Hamilton puts it, "[had] successfully unsettled the past."[62]

Briefly, the causes of this collective destabilization included a decade-long land-rights case,[63] which concluded with the High Court of Australia's profoundly unsettling decision to uphold the Meriam people's claim to native title. Following that seismic event came two highly publicized government inquiries, both of which exposed Australia's appalling human rights record with regard to its Indigenous peoples. The first, which took place in 1987, was a

Royal Commission charged with investigating the shockingly high numbers of Aboriginal deaths in custody. Eight years later, a second major national inquiry was launched, this time to examine the inhumane policy and practices behind the forced removal of children of Aboriginal descent (subsequently known as the "Stolen Generations") from their families.

What both investigations brought to light was that, in the eyes of many Aboriginal people, the federal government's repeated assaults on their culture appear as mere variants on the devastating war that has been waged against them since the frontier era. It is an impression strongly supported by the lexicon of warfare employed by many non-Indigenous historians, anthropologists, environmental philosophers, and Indigenous writers when describing the kind of treatment to which Indigenous Australians have been subjected since the end of frontier killings. After all, as Waanji novelist Alexis Wright reminds us, it was in the twentieth century that "the Aboriginal man was put into … prison camps, like prisoners in the two world wars."[64] Even today, as the narrator of *Plains of Promise* spells out, many Aboriginal people are still the victims of "an undeclared war. A war with no name."[65]

Hence anthropologist Deborah Bird Rose's observation that, when seeking to describe their experience of colonization, her Indigenous interlocutors often speak of Europeans as "settl[ing] on the blood and bones of the Aborigines they killed." And, she stresses, "their words are literal rather than metaphorical."[66] It is an arresting image, one not easily forgotten. It is striking, therefore, to see it re-occur in David Williams's analysis of colonial guilt in the United States of America. Explaining why white Americans refuse to acknowledge the brutality of that country's colonization, Williams suggests that to do so would be tantamount to recognizing that the nation "was born in blood, and it feeds on land choked with the bodies of its victims."[67] It would seem, then, that for many people of Anglo-Celtic descent, facing up to the truth of their forebears' so-called "peaceful settlement" of Australia represents the same psychological threat that, argues Williams, so unnerves non-Indian Americans. For, apart from the profound, trans-generational trauma it inflicted on the continent's Indigenous populations, the violence of the Frontier Wars was on such a scale and of such extreme cruelty that it can be said to have traumatized, albeit in a different sense, settler society itself.[68]

One need only examine the post-frontier war behaviour of settler Australians to recognize the symptoms commonly displayed by people who have been party to traumatic events. The frequent obliteration of historical records in the late nineteenth century; the exclusion of Aboriginal peoples from the Constitution of 1901; the carefully doctored, peaceful, Aboriginal-free history Australia invented for itself after Federation:[69] in short, the impenetrable pall of silence that hung over more than a hundred years of atrocities until W.E.H. Stanner shattered it in 1968[70] – all these practices testify to the powerful urge to erase the abominations of the frontier era from settler consciousness. To quote Lyndon Ryan, the Australian history professor leading a research team that is currently investigating and mapping frontier massacres: "The whole thing about massacres is the code of silence that is imposed in the immediate aftermath. Many people are too frightened to speak. Aboriginal people won't speak. They might get killed themselves."[71]

Thus, from Federation onwards, observes David Hansen: "Aborigines in general and the war in particular began to be excluded from public consciousness, and the history of the frontier retold as a struggle with the land itself and with the natural elements ... Indeed, for most of the 20th century, academic, general and school classroom histories barely mentioned Indigenous Australians, let alone the well-established mechanisms of their displacement and destruction."[72]

Of acute relevance here are several of Paul Connerton's observations about forgetting. Discussing the kind of forgetting that, he suggests, is "*constitutive in the formation of a new identity*," he emphasizes "the gain that accrues to those who know how to discard memories that serve no practicable purpose in the management of one's current identity and ongoing purposes."[73] It was precisely in this situation that settler Australians found themselves at Federation. Like the South East Asian societies to which Connerton refers, Australia's Anglo-Celtic population was in the process of "creating a new and shared identity in a new setting."[74] In the circumstances, "[not] to forget might have "provoke[d] too much cognitive dissonance."[75] On the other hand, Connerton remarks, burying/erasing/muzzling the past does not guarantee that we no longer remember it – "we cannot ... infer the fact of *forgetting* from the fact of silence.[76]

For Australia's settler descendants, the mass killings that continued to take place in parts of Western Australia, Queensland, and the Northern Territory well into the 1920s were hardly conducive to the "fact of forgetting."[77] On the contrary, it could be argued that the generations of Anglo-Australians alive during and after the First World War could not *but* be reminded of earlier atrocities as the last remaining areas of the continent were violently wrested from Aboriginal possession.

As Ryan reports, "massacres became more violent, systematic and calculated over time" and, consequently, "some of the most violent episodes in [Australia's] colonial past took place well into the 1920s."[78] Accounts left "by perpetrators as well as survivors" of the ten massacres carried out after the First World War are described as "utterly horrifying," leading Ryan to conclude that "the hatred of Aboriginal people after World War I seems to have intensified." As she understands it: "The Boer war and World War I ... have a big impact on the way massacres are carried out. And that's to do with better gun technology, like long-range rifles."[79]

For the post-First World War settler population, then, Bean's exculpatory suggestion that the admirable "mettle" Anzacs had shown at Gallipoli was an *inherited racial characteristic* came at just the right time. It offered them a welcome alternative to the identity irremediably linked to the shameful memory of the "fitful war of extermination waged upon the blacks"[80] until 1928.[81] Celebrating the memory of the Anzacs' "sacrifice" provided "the soft light of remembrance that abets the loss of memory and therefore *the evasion of responsibility*."[82] As Michael Kammen observes, "We arouse and arrange our memories according to our psychic needs."[83]

WILLED FORGETTING AND STRATEGIC REMEMBRANCE

Explaining the "significance of geography" in the construction of "the sense of the nation," David Storey emphasizes people's need for a "national geography built around particular places and utilizing explicit territorial allusions."[84] But, for settler descendants, there are, *au fond*, no places in Australia that can be unequivocally identified as depositories of national pride. All the sites associated with their ancestors' arrival in and occupation of the continent are inextricably linked to the dispossession and massacre of the land's Aboriginal inhabitants. If the Anzac myth has acquired such phenomenal importance today, it is precisely because it roots national identity in a place and an epoch totally disconnected from those killing times of the frontier. To quote Martin Ball, Anzac mythology is "a means of repressing the secrets of Australia's genesis. By wrenching the origin of the nation into the twentieth century, the invasion and colonization are hidden from view. The Aboriginal population is conveniently forgotten. The convict stain is wiped clean."[85]

By the end of the 1960s, just as Anzac Day celebrations were dying out, the state of ***willed*** forgetfulness to which the nation had so eagerly surrendered in the first half of the century began to be rudely disrupted. For, while Australians in the 1970s no longer had the first-hand knowledge of frontier atrocities that their predecessors had harboured, the consciousness-raising events referred to earlier in this chapter had nevertheless eroded "their sense of innocent national selfhood."[86] Being several decades removed from the Frontier Wars, and having been fed a sanitized version of Australia's colonization, today's non-Indigenous Australians cannot be considered (as earlier generations conceivably might) as victims of either perpetrator or bystander trauma.

Yet, like the non-Indian Americans that Williams describes, many of them continue nonetheless to be haunted "by the disturbing sense that their present was bought with a bloody past."[87] And it is this feeling that has generated today's spectacular resurgence of enthusiasm for the Anzac myth. The increasing amounts of energy and money devoted to commemorating Gallipoli are a direct reflection of the ardour of Anglo-Australia's desire to avoid having to acknowledge the nation's real origins. Current Anzac Day celebrations can thus probably best be understood as a potent screen memory whose essential function is to blot out reminders of the "brutal, bloody and sustained confrontation that took place on every significant piece of land across the continent."[88] For Bruce Scates et al.: "[The] Anzac centenary serves as much as a blockage as it does a mnemonic marker. Its prominence reduces other, earlier, aspects of Australian history to the status of a faded prelude."[89]

Worth remembering here are Freud's remarks on motivated forgetting. The repression of memories was not, he observed, a response solely confined to trauma victims: it was also a means of "turning away from unpleasurable perceptions," "a flight-reflex in the presence of painful stimuli."[90] In 2004, Freud's analysis was validated by US researchers who succeeded, thanks to brain-imaging scans, in identifying what *they* defined as the "biological mechanism that exists in the human brain to block ***unwanted*** memories."[91] Similar findings have resulted from research conducted by social psychologists. One recent survey showed, for example, that whenever an ingroup has been found to have committed "morally unacceptable" actions, one of the most common "group-protective strategies" its members adopt to maintain a positive image of that ingroup is collective "forgetting."[92] "Socially organised amnesia" of this kind is, to quote Hamilton, "one of the most powerful forces that shape national remembering."[93]

CONCLUSION

Given the historical circumstances outlined earlier, Prime Minister Bob Hawke's decision to take a group of First World War veterans to attend the dawn service with him at Anzac Cove in 1990 can be regarded as a group-protective strategy par excellence.[94] But it was John Howard who put the crowning touches to this evasion reflex. Angered by the damaging impact he felt "black armband history" was having on Australia's self-image, he exhorted his compatriots throughout his term of office to ignore the "professional purveyors of guilt"[95] and to feel "comfortable and relaxed"[96] about the "very *generous and benign*" history of Australia.[97] On the one hand, he defiantly asserted that there was "no tradition of conquest or imperial ambition in Australia,"[98] on the other, he resurrected Bean's idea of Gallipoli as the birthplace of Australia's national identity.

In the fractious context of the History Wars, identifying Australian nationhood with the sacrifice of young Australian lives at Gallipoli was a stroke of genius for, as Ben Wellings judiciously observes, "the sacrality of the dead and the mute force of their sacrifice, made it very difficult for a political critique of Anzac nationalism to emerge."[99] Gallipoli also provided settler descendants with the kind of narrative of which, Ann Curthoys informs us, they are such avid consumers. In fact, she claims, it is "the principal victimological story that Australians tell themselves over and over."[100] If Curthoys's analysis is correct, it seems reasonable to surmise that, in addition to the profound grief it provoked, the massive death toll at Gallipoli had quite another consequence for those of settler descent. For what it offered was the opportunity to shed the ignominious reputation they had inherited as *perpetrators* of violence and identify henceforth with the more honourable status of *victims* of war.

Post-Howard, a brief hope flickered among Indigenous Australians and their non-Indigenous supporters that Kevin Rudd's electoral promise of a national apology would help heal the trans-generational war wounds from which Aboriginal people still suffer today. But when it came, Rudd's long-awaited "Sorry" speech focused exclusively on the harm caused by the forcible removal of Indigenous children from their biological families. At the national level, the relentless war of dispossession that British settlers and their descendants waged against Aborigines for 140 years was and remains unacknowledged.

Today, as the nation indulges in an orgy of spending to commemorate its involvement in the First World War, the AWM persists in rejecting as "inappropriate" all calls to commemorate the first war to occur on Australian soil.[101] For the "white blindfold" myth that settler Australians "have no songs of strife, / Of bloodshed reddening the land,"[102] is still the only view of the past that the majority of settler descendants are willing to countenance. In twenty-first-century Australia, celebrations of Anzac valour at Gallipoli remain a "substitute, surrogate, or consolation for something that is missing"[103] – notably, suppressed memories of the Frontier Wars and the atrocities that characterized them.

NOTES

1 Clare Wright, "Forgetting to Remember," *Griffith Review* 48 (April 2015), https://griffithreview.com/articles/forgetting-remember/.
2 Henry Reynolds, *Forgotten War* (Sydney: University of New South Wales Press, 2013).
3 "The title [digger] – and it is one – describes the volunteer civilian members of the First *AIF* [Australian Imperial Force]." See Graham Seal, "Tradition, Myth and Legend," in *Inventing Anzac: The Digger and National Mythology* (St Lucia: University of Queensland Press, 2004), 1–9.
4 Raymond Evans, "The Fight to Remember," *Sydney Morning Herald*, 10 August 2013, http://www.smh.com.au/entertainment/books/the-fight-to-remember-20130808-2rj2m.html.
5 Ibid.
6 "Expulsion, Exodus and Exile in White Australian Historical Mythology," *Journal of Australian Studies* 23, no. 61 (1999): 11.
7 Evans, "Fight to Remember."

8 Reynolds, *Forgotten War*, 238.

9 "The absence of such monuments testifie[s] to the 'silence' about the past essential for the mythology of Australia as 'The Quiet Continent.'" See Peter H. Hoffenberg, "Landscape, Memory and the Australian War Experience, 1915–18," *Journal of Contemporary History* 36, no. 1 (2001): 122, https://doi.org/10.1177/002200940103600105. Significantly, the rare monuments Indigenous communities have themselves erected to mark atrocities committed in the colonial era are regularly vandalized. See Reynolds, *Forgotten War*, 46.

10 Alan Stephens, "Reconciliation Means Recognising the Frontier Wars," *The Drum*, ABC, 7 July 2014, http://www.abc.net.au/news/2014-07-07/stephens-reconciliation-means-recognising-the-frontier-wars/5577436.

11 Marilyn Lake, "Beyond the Legend of Anzac," lecture given on ABC Radio National, 26 April 2009, http://www.abc.net.au/radionational/programs/hindsight/beyond-the-legend-of-Anzac/3145132. For an analysis of what distinguishes these local memorials from the "great national sites of remembrance," see Jay Winter "War Memorials: A Social Agency Interpretation," in *Remembering War: The Great War between Memory and History in the Twentieth Century* (New Haven: Yale University Press, 2006).

12 Martin Ball, "Lost in the Monumental Landscape," *Australian Humanities Review* 12 (December 1998–March 1999), emphasis added, http://australianhumanitiesreview.org/1998/12/01/lost-in-the-monumental-landscape/.

13 Hoffenberg, "Landscape, Memory," 125.

14 Ibid., 126.

15 Lake, "Beyond the Legend."

16 Note James Brown's observations on this point in "Anzac instincts" in *Griffith Review* 48 (2015), https://griffithreview.com/articles/Anzac-instincts/.

17 Matt McDonald, "Lest We Forget: The Politics of Memory and Australian Military Intervention," *International Political Sociology* 4 (2010): 289, https://doi.org/10.1111/j.1749-5687.2010.00106.x

18 See "Honour for Anzacs," *New Zealand Herald*, 4 May 1916, http://paperspast.natlib.govt.nz/cgi-bin/paperspast?a=d&d=NZH19160504.2.63.

19 Ken Inglis, "The Anzac Tradition," *Meanjin* 100 (March 1965), https://meanjin.com.au/blog/the-Anzac-tradition/.

20 Seal, "Tradition, Myth," 3.

21 Richard White, *Inventing Australia: Images and Identity, 1688–1980* (Sydney: Allen and Unwin, 1981), 126.

22 C.E.W. Bean, *Anzac to Amiens* (Canberra: AWM, 1946), 181, quoted in Ann Curthoys, "Expulsion, Exodus and Exile in White Australian

Historical Mythology," *Journal of Australian Studies* 23, no. 61, (1999): 1–19, 12.

23 See Inglis, "Anzac Tradition."

24 Martin Ball, "The Story of the Story of Anzac" (PhD diss., University of Tasmania, 2001), 143.

25 Interestingly, Anzac specialist Bruce Scates suggests: "The same urge that leads distant descendants to scour shipping lists in search of convict ancestors leads hundreds every year to the killing fields of Gallipoli and Flanders." See Bruce Scates, "In Gallipoli's Shadow: Pilgrimage, Memory, Mourning and the Great War," *Australian Historical Studies* 119 (2002): 5, June 2021. https://doi.org/10.1080/10314610208596198.

26 Carolyn Holbrook, "Remembering with Advantages: The Memory of the Great War in Australia," *Comillas Journal of International Relations* 2 (2015): 19, https://doi.org/10.14422/cir.i02.y2015.002.

27 Tom Griffith, "Counting the Fallen of a War without Uniforms," *Sydney Morning Herald*, 20 December 2001. Reynolds, by contrast, underlines Australia's yearning for "acknowledgement and even acceptance in the eyes of the world." See Reynolds, *Forgotten War*, 252.

28 See Graham Seal, "Anzac: The Sacred in the Secular," in *Sacred Australia: Post-Secular Considerations*, ed. Makarand Paranjape (Melbourne: Clouds of Magellan, 2009), 213. In 1916, the federal government took steps to prevent the name "Anzac" being profaned by use "in connection with any business, trade, calling or profession" (ibid.).

29 "The mettle that a race can show / Is proved with shot and steel / And now we know what nations know / And feel what nations feel" (from: Andrew Barton ("Banjo") Paterson, "We're All Australians Now," published as "An Open Letter to the Troops," in *The Bulletin*, 1915, https://englishverse.com/poems/were_all_australians_now.

30 Lake, "Beyond the Legend."

31 Quoted in Lake, "Beyond the Legend," emphasis added. A number of "inconvenient truths," missing from the legend that subsequently developed, run directly counter to the mythical Anzac character first crafted by Bean. The most significant of these is the massacre of Bedouin tribespeople by members of the Anzac Mounted Division at the Palestinian village of Surafend in 1917. See Paul Daley, *Beersheba: A Journey through Australia's Forgotten War* (Carlton: Melbourne University Press, 2009).

32 Quoted in Ken S. Inglis, "Bean, Charles Edwin, Biography," *Australian Dictionary of Biography*, http://adb.anu.edu.au/biography/bean-charles-edwin-5166.

33 Scates, "In Gallipoli's Shadow," 11.

34 Vecihi (John) Basarin, "Battlefield Tourism – Anzac Day Commemorations at Gallipoli: An Empirical Analysis" (PhD diss., Deakin University, 2011), 28.
35 Inglis, *Sacred Places*, 330.
36 Basarin, "Battlefield Tourism," 22. Scates's extensive research on the resurgence of interest in Gallipoli has resulted in numerous publications, which are listed here: https://history.cass.anu.edu.au/people/bruce-scates#acton-tabs-link--tabs-0-row_2-3.
37 Holbrook, "Remembering with Advantages," 23.
38 Chris Sheedy, "Peter Weir's Film *Gallipoli* Launched Anzac Day into Event Australia Celebrates," *Sydney Morning Herald*, 24 April 2015, http://www.smh.com.au/national/WW1/peter-weirs-film-gallipoli-launched-Anzac-day-into-event-australia-celebrates-20150421-1mpmcm.html.
39 Anna Clark, "An Epic Forgetting," *Sydney Review of Books*, 28 August 2014, http://www.sydneyreviewofbooks.com/forgotten-war-black-war/.
40 Caroline Winter, "Tourism, Social Memory and the Great War," *Annals of Tourism Research* 36, no. 4 (2009): 612, https://doi.org/10.1016/j.annals.2009.05.002.
41 See Deborah Gough, "Anzac Day 2015: Dawn Services around Australia," *Sydney Morning Herald*, 26 April 2015, http://www.smh.com.au/national/WW1/Anzac-day-2015-dawn-services-around-australia-20150425-1mt2am.html.
42 Ian McPhedran, "Government Spending More Than $8800 for Every Digger Killed during World War 1," *News Corp Australia Network*, 4 September 2015, http://www.news.com.au/national/Anzac-day/government-spending-more-than-8800-for-every-digger-killed-during-WW1/news-story/34808367386af87773c8e4326d2a46e8.
43 Paul Kelly, "The Next Anzac Century," *Australian*, 23 April 2011, http://www.theaustralian.com.au/national-affairs/the-next-anzac-century/story-fn59niix-1226043226240?nk=a3d9cc50b4e0153f9d0aa822b855488f-1465291672.
44 Carolyn Holbrook, "Historically Speaking: Reflecting on the Anzac Centenary and Memorialisation," speech given to the Professional Historians Association, Emerald Hill Library and Heritage Centre, South Melbourne, 5 April 2016, http://honesthistory.net.au/wp/holbrook-carolyn-Anzac-centenary-and-memorialisation-speech-to-pha-vic/. Like several other historians, Holbrook links the Anzac Day revival to a renewed interest in family history in the 1970s, one example being Bill Gamage's influential *Broken Years* (1974), in which the Gallipoli tragedy was recounted "in a highly emotive style" through the use of Anzacs' letters and diaries (ibid.). See also Curthoys, "Expulsion, Exodus and Exile," 11–13.

45 Holbrook, "Remembering with Advantages," 25.
46 Stuart Ward, "'A War Memorial in Celluloid': The Gallipoli Legend in Australian Cinema, 1940s to 1980s," in *Gallipoli: Making History*, ed. Jenny Macleod (London: Frank Cass, 2004), 59–72.
47 Ken S. Inglis, "They Shall Not Grow Old," *Age*, 30 April 2005, https://www.theage.com.au/national/they-shall-not-grow-old-20050430-ge02j8.html.
48 Gary Bouma, *Australian Soul: Religion and Spirituality in the Twenty-First Century* (Melbourne: Cambridge University Press, 2006), 23.
49 Scates, "In Gallipoli's Shadow," 8.
50 Ibid., 20.
51 Holbrook, "Historically Speaking."
52 Christina Twomey, "Trauma and the Reinvigoration of Anzac: An Argument," *History Australia* 10, no. 3 (2013): 106.
53 The image is borrowed from Paul Connerton, who describes the Great War as having given rise to "an orgy of monumentalization." See Paul Connerton, "Seven Types of Forgetting," *Memory Studies* 1, no.1 (2008): 69.
54 The "national obsession," as Richard White calls it. See White, *Inventing Australia*, viii.
55 James Curran and Stuart Ward, "A Commendable Emptiness," in *The Unknown Nation: Australia after Empire* (Carlton Melbourne University Press, 2010), 15.
56 Ibid., 11–12.
57 Tony McLellan, "What Exactly Are We Celebrating in 1988?," *Sydney Morning Herald*, 16 March 1986, quoted in *Mistaken Identity: Multiculturalism and the Demise of Nationalism in Australia*, ed. Stephen Castles, Bill Cope, Mary Kalantzis, and Michael Morrissey (Sydney: Pluto Press, 1992), 101–2.
58 W.E.H. Stanner, *After the Dreaming: Black and White Australians – An Anthropologist's View*, the Boyer Lectures, 1968 (Sydney: ABC, 1969), 67.
59 Military-social historian Peter Stanley refers to "a string of historians" as having "from about the mid-1970s documented the fact, spread and impact of frontier conflicts." "On Anzac Day, We Remember the Great War But Forget Our First War," *Conversation*, 24 April 2014, https://theconversation.com/on-Anzac-day-we-remember-the-great-war-but-forget-our-first-war-23246.
60 In 1988, the protest march attracted forty thousand people – the largest number to attend such an event since the Vietnam moratorium. At the end of the march, prominent Indigenous activists Gary Foley, Burnum Burnum, and the Reverend Charles Harris all addressed the assembled crowd.

61 Alastair Thompson, Michael Frisch, and Paula Hamilton, "The Memory and History Debates: Some International Perspectives," *Oral History* 2, no. 22 (1994): 41.

62 Ibid. No mere clash of dissonant scholarly opinion confined within the cloistered spheres of academe, the virulent debates engaged in by journalists, politicians, and other public figures still flare up periodically in the media today.

63 In 1982, a group of Murray Islanders led by Eddie Mabo began legal proceedings to demand recognition of their traditional land rights by the State of Queensland and the Commonwealth government.

64 Alexis Wright, *Plains of Promise* (Brisbane: University of Queensland Press, 1997), 74.

65 Ibid.

66 Deborah Bird Rose, *Hidden Histories: Black Stories from Victoria River Downs, Humbert River and Wave Hill Stations* (Canberra: Aboriginal Studies Press, 1991), 35.

67 David Williams, "In Praise of Guilt: How the Yearning for Moral Purity Blocks Reparations for Native Americans," in *Reparations for Indigenous Peoples: International and Comparative Perspectives,* ed. Federico Lenzerini (Oxford: Oxford University Press, 2008), 246.

68 Note Dominick La Capra's observation, "not everyone traumatized by events is a victim. There is the possibility of *perpetrator trauma.*" See Dominick La Capra, *Writing History, Writing Trauma* (Baltimore: Johns Hopkins University Press, 2001), 79, emphasis added.

69 Violence "had been written out because it didn't fit in with the sort of favourable picture that historians wanted to create of the new nation," claims Henry Reynolds. See Henry Reynolds, "Race Wars Written out of Australian History," ABC, *7.30 Report*, 7 June 1999.

70 The "Great Australian Silence" was the much quoted phrase coined by Stanner in *After the Dreaming*, the series of Boyer Lectures the distinguished anthropologist gave in 1968.

71 Lorena Allam and Nick Evershed, "Forced to Build Their Own Pyres: Dozens More Aboriginal Massacres Revealed in Killing Times Research," *Guardian,* 17 November 2019, https://www.theguardian.com/australia-news/2019/nov/18/forced-to-build-their-own-pyres-dozens-more-aboriginal-massacres-revealed-in-killing-times-research.

72 David Hansen, "Not mentioned in Despatches," *Australian Review of Public Affairs*, November 2013, http://www.australianreview.net/digest/2013/11/hansen.html.

73 Connerton, "Seven Types of Forgetting," 63, emphasis in original.

74 Ibid.

75 Ibid.
76 Ibid. 68, emphasis added.
77 So far, Ryan has discovered more than twenty. See Allam and Evershed, "Forced to Build Their Own Pyres."
78 Ibid.
79 Ibid.
80 Carl Feilberg, *The Way We Civilise* (Brisbane: G & J Black, 1880), https://en.wikisource.org/wiki/The_Way_We_Civilise.
81 This was the year when the last known massacre in Australia took place near Coniston cattle station in the Northern Territory.
82 Nathalie Z. Davis and Randolph Starn, "Introduction," *Representations* 26, (Spring 1989): 1–6, emphasis added, https://doi.org/10.2307/2928519.
83 Michael Kammen, *Mystic Chords of Memory: The Transformation of Tradition in American Culture* (New York: Vintage, 1993), 9.
84 David Storey, *Territories: The Claiming of Space* (London: Routledge, [2001] 2012), 107.
85 Martin Ball, "The Pleating of History: Weaving the Threads of Nationhood," *Cultural Studies Review* 11, no. 1 (2005): 171.
86 Haydie Gooder and Jane M. Jacobs, "Belonging and Non-Belonging: The Apology in a Reconciling Nation," in *Postcolonial Geographies*, ed. Alison Blunt and Cheryl McKewan (New York: Continuum, 2002), 212.
87 Williams, "In Praise of Guilt," 247.
88 Military historian John Coates, quoted by Matt Peacock in "War Memorial Battle over Frontier Conflict Recognition." ABC, *7.30 Report*, 26 February 2009, http://www.abc.net.au/7.30/content/2009/s2502535.htm.
89 Damien Williams, Bruce Scates, Laura James, and Rebecca Wheatley, "The Anxious Anzac: Suggestions for a Metric Moment in Late Australia," *Matériaux pour l'histoire de notre temps* 1, nos 113–14 (2014): 146, https://doi.org/10.3917/mate.113.0142.
90 Quoted in Simon Boag, "Repression, Suppression, and Conscious Awareness," *Psychoanalytic Psychology* 27, no. 2 (2010): 166, https://https://doi.org/10.1037/a0019416.
91 Lisa Trei, "Psychologists Offer Proof of Brain's Ability to Suppress Memories," *Stanford Report*, 8 January 2004, emphasis added, https://news.stanford.edu/news/2004/january14/memory-114.html.
92 Michael J.A. Wohl, Nyla R. Branscombe, and Yechiel Klar, "Collective Guilt: Emotional Reactions When One's Group Has Done Wrong or Been Wronged," *European Review of Social Psychology* 17 (2006): 2. Particularly relevant here is the authors' observation that: "Group

members may sometimes be 'socially protected' from experiencing collective guilt by erecting a wall of silence around the negative past, but when the group's harmful past does come to the fore, additional social mechanisms can be activated" (ibid., 8).

93 Hamilton, "Memory and History Debates," 41.

94 Basarin recounts: "The event, attended by more than 5000 Australians, was televised live, to large audiences in Australia and hence began the growing interest in attending the ceremonies at Gallipoli on Anzac Day." See Basarin, "Battlefield Tourism," 25.

95 *Future Directions, Liberal Party of Australia* (December 1988), quoted in Mark McKenna, "Different Perspectives on Black Armband History," Research Paper 5, Parliament of Australia, 10 November 1997, http://www.aph.gov.au/About_Parliament/Parliamentary_Departments/Parliamentary_Library/pubs/rp/RP9798/98RP05.

96 "An Average Australian Bloke," interview with Liz Jackson on *Four Corners*, ABC, 19 February 1996, http://www.abc.net.au/4corners/stories/2011/08/08/3288524.htm

97 Quoted in McKenna, "Different Perspectives," emphasis added.

98 Mark Riley, "We're Not Warlike, That's Why We Fight: PM," *Sydney Morning Herald*, 11 November 2003, http://www.smh.com.au/articles/2003/11/10/1068329489034.html

99 Ben Wellings, "Lest You Forget: Memory and Australian Nationalism in a Global Era," in *Nation, Memory and Great War Commemoration: Mobilizing the Past in Europe, Australia and New Zealand*, ed. Shanti Sumartojo and Ben Wellings (Bern: Peter Lang, 2014), 53.

100 Curthoys, "Expulsion, Exodus and Exile," 11. Commenting on his interviews with young Anzac pilgrims, Scates notes: "The complaint that 'we were led like lambs to the slaughter' was common." See Scates, "In Gallipoli's Shadow," 12.

101 Note: "The Australian War Memorial was first advised internally to acknowledge the frontier wars way back in 1979. Our military historians accept that colonial conflict is part of our military history, but the Memorial still holds out." See Michael Green, "Lest We Remember." 24 February 2014, *The Wheeler Centre: Books Writing Ideas*, https://www.wheelercentre.com/notes/f261bb085eb4.

102 Banjo Paterson, *Song of the Future* (1889), *Bulletin*, 21 December 1889, http://www.middlemiss.org/lit/authors/patersonab/poetry/songfuture.html.

103 Davis and Starn, "Introduction."

2

Memories of Colonial Involvement and Civil Wars

6

The Gurkha with the Khukuri between His Teeth: First World War Postcards and Combat Representations of Nepalese and Indian Colonial Troops

Gilles Teulié

If postcards were a significant medium in the communication of information between the end of the nineteenth century and the 1920s, they have only recently caught the eye of researchers and become an object of scholarly study. Yet, in 1914, 800 million postcards were published in France and more than a billion in Germany,[1] which today represent primary sources of considerable interest. One reason for academia's past negligence, or even disdain, towards postcards was the difficulty of gathering a worthy and attractive corpus of postcards on a given theme. However, today, with the infinite potential offered by the internet to locate such documents, postcards scattered around the planet can be purchased with the click of a button. Nearly 54 million postcards are for sale on the world's largest website for collectors, and yet this represents only an infinitesimal proportion of what dealers, collectors, institutions, and museums have in stock.[2]

This chapter is framed by deltiology, the study and collecting of postcards, and considers postcard depictions of Nepalese and Indian troops (who fought on the Western Front in 1914 and 1915) when they were off duty or while they were travelling to the front. It also reflects on the representations of combat situations, not through photographs (which were unlikely to be taken in the midst of battles), but through drawings and sketches. Interestingly, contemporaries favoured the more vivid engravings taken from sketches rather than the static photographs. Maurice Halbwachs – a French sociologist who was thirty-seven years old in 1914 – exemplified this preference. For him, these drawings were even more "real" than photographs as they encapsulated in one picture all sorts of visions that told a complete story.[3]

Of the 250 postcards of British Imperial First World War troops produced in the French city of Marseilles, a massive 182 portray Indian troops, with only ten postcards showing Anzac soldiers (two New Zealanders and eight Australians), twenty-nine representing South Africans, and thirty featuring Anglo-Indian soldiers. The reason for such a discrepancy between Indian soldiers and all other Commonwealth troops can be explained by commercial and editorial considerations. Indian fighters (sepoys) caused a sensation in Marseilles and elicited the wonder of its inhabitants. The utter exoticism of the British Raj troops may have led postcard printers in the city to give pride of place to the arrival of these troops in their hometown.

How far can these postcards deepen our understanding of European representations of Imperial troops? In this chapter, I argue that European artists as well as their publishers, who transposed images from the Raj onto the Western Front to glorify war (e.g., by portraying gallant and dashing charges), reproduced colonial stereotypes. They likewise lured postcard buyers by offering awe-inspiring and morbid aspects of the conflict (such as night raids) for lucrative purposes. By doing so, they also created biased representations of the war. This chapter places emphasis on Gurkhas, Nepalese soldiers who have fought alongside British troops for more than two hundred years since the Anglo-Nepal(ese) War of 1814–16. I use the Gurkhas to examine what Lionel Caplan qualifies as the process of "militarizing the mind" during the First World War.[4] A European construct of the Nepalese and Indian trooper did exist, and, beyond an objectivity that was at best spurious, an imperial frame of mind prevail and was transposed from the colonies onto the Western Front.

Troops from the British Raj were sent to Europe to bolster "civilization," and they were endowed with all the virtues granted to exceptional allies coming to the rescue of the mother countries. Yet European propagandists, who presumably supported the war effort, contrived to find ways for the public to be thankful that these colonial troops were on their side and to almost feel sorry for the Germans who were to confront them. In doing so, they offered visions of a stereotypical ethnic identity and endowed Indigenous soldiers with characteristics that were so remote from reality that they become all but fictional characters. Imbued with imperial literature, European ideologists focused on metonymical elements inherited from colonial wars, and these became symbols of the victory they thought would soon be theirs. The Gurkha knife is a central element of the rhetoric of martiality that both preceded and followed two infantry divisions that had been urgently dispatched to Europe: the Meerut and Lahore divisions of Nepalese and Indian troops that reached Marseilles on 13 September 1914.[5]

THE KHUKURI, *THE WEAPON OF THEIR RACE*

The title "L'arme de leur race" is taken from the edition of the French newspaper *Le Miroir* that was published on 29 November 1914. It deals with the "famous and deadly hand-weapon of the Gurkha soldiers,"[6] "the Ghurkas' legendary machete-like knives."[7] The khukuri,[8] a curved forward-angle knife, is the traditional weapon of the Gurkhas, and it has retained its popularity to this day. It is so emblematic of the Nepalese people that most Gurkha regiment members bear it on their coat of arms, along with insignias defined by the heraldic language: "two upturned kukris meeting at the top, their naked blades crossing." The soldier and the weapon are intimately linked – to the extent that Gurkha soldiers allegedly refuse to fight without their cherished khukuri. Whether true or not, such stories are part of the legacy of the Gurkha myths that have accompanied the representations of Gurkhas in Western imagination to the point at which they have become part of British popular culture.

Indeed, the weapon became even more celebrated when Jonathan Harker sheared his "great kukri knife" through the throat of the evil eponymous character in Bram Stoker's famous 1897 novel.[9] Lionel Caplan, in his study of the representations of the Gurkhas in Western books written by British officers, points out that "when others use rifles and more sophisticated weapons, the Gurkhas seem to draw only their *khukuri* – the short curved knife – which is a general utility instrument in the Nepalese countryside, but is represented in the discourse as the national weapon."[10] I have been unable to trace any references to the Gurkhas in the French press or literature of the First World War that did not also mention the *khukuri*, as exemplified in children's literature: "Their favourite weapon is the Chûri [sic], a long bent knife which can plunge into the breast or stomach of their adversaries with an admirable dexterity."[11] Furthermore, the link between man and knife is supposed to be so strong that sometimes the object turns into a living creature: "Maybe the Prussian officer expected to set it on his wall as a trophy, but the koukhri returned to the Hindu's belt as a faithful dog returns to his master."[12]

Famous individual bladed weapons have survived the passage of time to reach us as elements of an idealized past: Excalibur, Durandal (the mythical sword of Roland, Charlemagne's nephew), and Hrunting (Beowulf's sword). Yet seldom has a weapon, shared by entire regiments and the population from which they originated, become so famous. The origin of the khukuri is lost in time, though some experts argue that the shape of the khukuri resembles the favourite sword of Alexander the Great's horsemen, the Greek and Macedonian kopis.

From 1903 onwards, the British Army provided Indian-made khukuris for its Gurkha regiments. The standard pattern of the khukuri evolved throughout the twentieth century and was later manufactured in Britain. As models approved by the British authorities, all five patterns are labelled "Mark," the first one, produced between 1903 and 1915, being labelled "Mark 1," abbreviated as "MK1." The military khukuri was the usual implement found in the hands of Gurkha soldiers during the First World War, but some used their own knives, which they had either brought from Nepal or crafted themselves. From an early age, Nepalese boys were trained to use khukuris as a tool, which helped them in their everyday life. Once they were adults, they learned to use the khukuri as both a tool and a weapon. Weighing nearly a kilo, the khukuri was heavy to handle but was efficient in the hands of a trained Gurkha.

One may wonder why such an association has preserved its vibrancy throughout time, particularly when, with the introduction of gunpowder, bladed weapons became obsolete on the battlefields. One explanation may be that European men's ideal martial attitudes were defined by the Greek warriors of the Iliad, who were singled out for personal fights against Trojan enemies, or by the Spartans, who defied the might of the Persian army at the battle of Thermopylae. Hand-to-hand fighting against fearful odds was more appealing than was defeating an overwhelming number of enemies using modern Maxim machine guns, as was the case in Sudan at the battle of Omdurman in 1898. Some conservative Victorians, such as Walter Pater, praised the Spartan model (*Lacedaemon*, 1892), and some of Britain's formidable enemies were so compared: the South African Zulus, for example, were called "Black Spartans." It is therefore not surprising that the Gurkhas would be given high martial status whereas the British Army's association with the Persian Army was detrimental to its prestige. "200 Spartan-Gurkhas who defended another Thermopylae were attacked by two thousand from [General Martindale's] army corps. They held firm and defeated them severely."[13] Notwithstanding his modern fighting capability (shooting well with a rifle, for example), it was his pugnacity in close combat that was put to the fore in European representations of the Gurkha soldier. Indeed, the British East India army in the Anglo-Nepalese War of 1814–16 did not easily overcome the Gurkhas.

Their reputation as fierce and sturdy fighters, not easily impressed or prone to despair, was shaped during that campaign. By 1914, this reputation was well established: "Gurkhas are the bravest and the most enthusiastic warriors among the Indian peoples."[14] Keen on having good fighters on their side to defend the famous North-West frontier, the British East India Company authorities recruited Nepalese soldiers as soon as the Anglo-Nepalese war ended. Specific regiments were named after the population; even though the proper name was originally the Gorkhali army (from the town of Gorkha, a Nepalese village northwest of Kathmandu), the new regiment became famous under the name of the 1st Ghurka Rifles. Except for the 9th Gurkha Rifles, Gurkha soldiers were not from the warrior caste (*Kshatriya*) but from that of the traders, cultivators, and herdsmen (*Vaisiya*).[15] The Gurkha regiments were engaged in the major British campaigns on the Indian sub-continent, including the Sikh wars of 1845–46 and 1848–49, during which they were considered a terror for the Sikhs. When the Indian Mutiny of 1857–59 took place, they gained fame among Europeans as they remained loyal to the British and defended white women and children. In 1914, the Gurkhas represented 10 per cent of the Indian Army that prepared for the Great War. Interestingly, this Indian Army was composed only of troops recruited in the North and Northwest of India (Punjabi Muslims, Sikhs, Rajasthani, Hindu Jats, Dogras, Garhwalis, Pathans, Afridis, and Gurkhas). This selection was based on the "martial race" theory deriving from Social Darwinism, according to which some peoples were considered to be born fighters (particularly the Scots, the Sikhs, and the Gurkhas).[16] This led recruiting officers to think that Southern Indians were subject to racial "degeneracy": "the martial race idea had at its core the belief that some Indians were inherently more warlike than others."[17] The Sikhs and the Gurkhas were singled out not only for their martial ability (the Sikhs distinguished themselves in the Crimean War) but also for their loyalty.

During the First World War, French journalists were either aware of this distinction or were influenced by British accounts that distinguished Sikhs and Gurkhas from other troops when they set foot in Marseilles in September and October 1914: "How efficient are these Indigenous troops? All those who have seen them fight have the highest opinion of them. They are recruited from the most warlike populations in India, the Punjab Sikhs for the cavalry, and the Nepal Gurkhas for the infantry. They are to be compared to our own Senegalese troops."[18] The same sentiment appears in other French newspapers: "The Sikhs and the Gourkhas are the best troops from India,"[19] or "As a soldier, the Hindu, particularly the Gurka and the Sikh, is far superior to any African or American Indian."[20] The recurrence of such information lacking prior verification accounts for the profusion of falsehoods, misconceptions, and fabrications during the Great War. The emphasis put on the importance of the khukuri in headlines such as "The 'Kukris' at Work: How the Gurkhas Terrorised the Enemy" may therefore be questionable.[21] This title is an example, among many, that insists on the close relation between soldier and weapon, but the question is: Did the Gurkhas became famous "because of their 'inherent' fighting characteristics, or did the kukri as a symbol of status and skill obscure and glorify the reality of combat?"[22] To answer this question we must examine how and why Gurkhas were mythicized during the First World War.

THE MYTH OF THE GURKHAS: A RIFLE IN ONE HAND AND A KHUKURY IN THE OTHER

A number of stories circulated during the First World War about friends and foes. Whether produced by information/propaganda services on both sides, or initiated by individuals, they spread like wildfire through the population concerned by the war and were at the heart of enduring stereotypes and myths.[23] One of them gives pre-eminence to "the deadly Kukri,"[24] the knife that was both a weapon and a tool and that, as such, also had (and still has) a spiritual function: it is used every year during the Dashain festival when family weapons are blessed with the blood of a slaughtered bull. This ritual may be the origin of the myth according to which a khukuri could not be unsheathed from its scabbard unless it was to shed blood. An Australian newspaper recalled this myth in 1915: "The Gurka tradition is that once drawn, the kukri must not be sheathed without drawing blood."[25] It is the lethal image of the knife that seems to attract local populations who meet the Nepalese for the first time:

> The Gurkha`s knife. The presence of the Indian troops in France is exciting the greatest interest in the towns through which they pass to their base. French soldiers, civilians, and girls all want to see the famous Gurkha knife and are somewhat appalled when the little natives of Nepal solemnly declare that their religion forbids them to draw the weapon without drawing blood. What seemed an insurmountable difficulty was overcome, however, when some British soldiers who had served in India were able to tell the curious that the knife would be shown if the spectators would allow a slight cut with it to be made in the tip of their finger in order to fulfil the letter of the law. A bandaged finger is now becoming quite popular with the French girls.[26]

Yet this myth was debunked during the war: "Contrary to popular belief, the Gurkha has no scruples about displaying this terrible weapon to the eyes of the curious."[27] With this example, we understand that misconceptions and false assertions were more common than the will to "tell the truth." To go a little further in the politico-military setting of the war, another myth foregrounded the idea that Gurkhas preferred to fight with their khukuris rather than with their rifles: "They advanced and began to fire, but in their excitement, they soon gave themselves no time to load. Most of them threw their rifles away and grasped the kukri (knife). The first lines of the Bavarians and Wurtembergers, who thought they were facing tigers, were already mown down."[28] Or again, the Gurkhas, "in a formidable hand- to- hand fight, let go of their rifle and seize their knife."[29] This stereotype of the Gurkhas attains mythic status when journalists explain that, as one hand is free, it can grasp the German soldier's bayonet: "Bayonet vs Kukri – Indians get wounded mostly in the left hand through grasping

German bayonets while using their terrible weapon."[30] Sometimes a misconception of what a real combat looks like leads to a nonsensical representation of battle, as in the case of a French account of a Gurkha charge: "20 metres from enemy lines, they stop, throw down their knives and kill for sure the first rank of the enemy."[31] They then charge with rifle and bayonet and later come back to fetch their khukuri as they cherish it above all else. Such erroneous representations of a Gurkha charge spread during the First World War because a good story that helps the war effort is always better than a true account. Rare were those who tried to debunk those stories; one exception is Colonel G.S. Ommanney of the 3rd Battalion of Gurkhas (Queen Alexandra's Own) who, during his thirty years' service in India, witnessed how the Gurkhas entered battle: "When he charges the Gurkha does not discard his rifle, as is generally supposed. He balances it in his left hand with the kukri in his right, firing with the left elbow pressed tight into the side and the left hand right up to the trigger guard."[32] The propaganda campaigns thus encountered some resistance in the Allied press but, as the extract below explains, not often enough:

> We pointed out at the time that the stories were improbable, and came from entirely irresponsible sources, such as anonymous corporals of field artillery and from correspondents of papers like the *Daily Mail* and the *Daily Express*. To the ignorant at home, and also in India, the words "Indian troops" call into memory such other words as "Sikhs," "Gurkhas," "Bengal Lancers." With the Sikhs are associated terrific bayonet charges, with the "Bengal Lancers" the whirlwind swoop of cavalry, and the Gurkha, of course, is armed only with a "kukri," some kind of sharp cutting instrument with which at one blow he sweeps off the head of his enemy. Naturally, therefore, when Indian troops are engaged in war, ignorant persons who wish to create a sensation make up stories of which Sikhs and bayonets, and "Bengal" Lancers and charges, and Gurkhas and kukris are the key-words.[33]

At the end of his comments, the journalist berates the Allies' propaganda (which he calls "special lie manufactories"), resenting that they should stoop to "German methods": "The facts we have mentioned may not be known to everybody, but the military censors must be perfectly aware of the absurdity of the stories that are clustering round the Indian troops. They know that Gurkhas do not cut off heads with kukris, and 'Bengal' Lancers do not charge dismounted. Why, then, have they let the stories through?"[34]

Some French journalists also reacted to the stories told about the Gurkhas' exploits with a more realistic and down-to-earth understanding of combat: "We were told 'You will see them with their *Kukris*! They can chop off the head of a bull with a single strike.' It is true. But just try to charge a German trench only armed with a Kukri! Unfortunately! Scientific war, 220 shells and *Minenwerfer* rarely allow the Gurkhas to display their skill, which is prodigious and their fighting qualities which are effective."[35] Combat reality, therefore, does not match an ideological martial framework that purports to extol the fighting qualities of the Allied troops. However, how far could the creator of such a framework go to praise the soldiers?

THE "CREEPING" GURKHA, A WILD ANIMAL?

One of the devices used to glamourize the Nepalese involved endowing the Gurkha trooper with positive animal qualities. For centuries, Europeans have compared Indigenous people to animals. This dehumanizing process enabled the former to confirm their dominant position and reaffirm the inferior status of the latter. This is obvious when the comparison is to an animal with negative connotations, such as a monkey or a snake, but it is also apparent, albeit in a less obvious way, when the comparison is to an animal associated with more positive values, such as strength (the lion), cunning (the tiger), or velocity (the leopard). Nevertheless, even if someone is said to run like a leopard or a cheetah, this seemingly positive comparison still states that the person is like "a beast." Rudyard Kipling was clearly racist when he used analogies that presented Gurkhas as puppies of their British masters, the latter "transforming inferiors into cherished 'pets.'"[36] "The little men hitched their *kukris* well to hand, and gaped expectantly at their officer as terriers grin ere the stone is cast for them to fetch. The Gurkhas ground sloped downward to the valley, and they enjoyed a fair view of the proceedings."[37] Kipling's comparison to puppies is not present in First World One descriptions of the Gurkhas, but other animal associations are rampant. In an article entitled "More Praise for Gurkhas," an Australian journalist portrays the Nepalese soldiers as cats: "The Gurkhas are born fighters. They are very small men, well knit, with a Japanese face. They are as nimble as cats."[38] There are obvious connections between English speakers and French journalists, as in the following examples: the Gurkhas have "a quick, gliding and smooth step which evokes that of a feline,"[39] or "men from the jungle, with feline ruses."[40] A French First World War children's literature booklet uses the same image: "The Gurkhas crawled towards the ammunition park like cats hunting for prey. The rifle in one hand, the kukri held between their teeth, they were slowly going forward, stopping to listen and to see as their eyes, used to the shadows of the night, could penetrate the darkness."[41] This stereotypical image may have inspired French artist E. Soukanech, who made a nuanced patina bronze statue of *Le terrible Gourkha*, height 9.5 centimetres, width 25.5 centimetres. It depicts a crawling First World War Indian soldier with a turban, holding a rifle in his right hand and a knife between his teeth. A shared imperial culture may explain the common images between French and British, yet there is a shift to another even more positive animal comparison.

This comparison was all the easier as Nepal is a natural environment for the Bengal tiger. Gurkhas were seen as good fighters because they were excellent hunters, as shown in the following example, in which a Gurkha kills a German soldier: "his cry stays in his throat, as he has just met Akka's koukhri. Never had the tiger hunter struck faster nor more accurately."[42] From this, , it becomes easy to transform the Gurkha into his former prey: "yes, they have the rifle and the bayonet, but all possess another fearful weapon, the kukri, a small sword with a slender, large and heavy blade ... When they leap like tigers, with this knife between their teeth, nothing can resist their terrible impulse."[43] The French booklet *Tommies et Gourkas* compares the two outstanding forces in the Indian Army, both praised as being the crème de la crème: "the Sikhs, first rate horsemen, look much more like Europeans. They are handsome tall bearded men, who wear a high turban. Without being ferocious like the Gurkhas who must never let themselves be taken alive, they are formidable fighters. Their name means 'lion' and they deserve it; the Gurkhas are rather like tigers."[44] In this excerpt, the positive analogy with the tiger is associated with the adjective "ferocious," which might be ambiguous when applied to humans as the display of uncontrolled feelings is linked to madness or bestiality.

6.1 A Gurkha in Marseilles in 1914 (French postcard).

In Scandinavian mythology, the Berserkers, Odin's mad warriors who used to bite their shields before battle (see the Lewis Chessmen at the British and Edinburgh Museums), were described as ferocious and nearly demented. Their representation is similar to that of the Gurkhas in the First World War press: "there are very few who care

6.2 Postcard of Gurkhas in Marseilles (published in Paris).

to face a Gurkha maddened with the lust of war,"[45] or as shown by the unequivocal title of an article widely published in 1915, "Blood-Lust of the Gurkhas." The descriptions are blood-curdling and really emphasize the folly of soldiers who are close to being animalistic: "[They] threw themselves on the enemy with a bizarre cry that resembled the bellowing of wild animals," there was "a savage-clamour," and, "like wild beasts that the smell of blood had maddened, the Gurkhas continued their terrible task into the night. The heavy blades of the 'kukri' plunged into the breasts, sliced the necks, carving, amputating; it was a battle of madness of which it is impossible to give any idea." Here, the writer gives the point of view of the German victims so that the reader can grasp the full horror of what is happening: "Terrified by the apparition of these sombre-visaged demons, who threw themselves on them, strangling them and cutting their throats, the Germans turned in furious flight."[46] In European war literature, it is not only the Gurkhas who become hellish creatures but also their weapon: in a nineteenth-century French

6.3 A Non-combatant posing as a fighter.

6.4 Exoticism and fear of the "dreadful war knife."

translation of the passage by Kipling quoted previously, a comment is added to enlighten French readers about the khukuri, along with a fantasy war cry that is not mentioned by Kipling: "Ulu lu-lu-lu! Shouted the Gurkhas running down the slope with a joyful clatter of the *Kukris*, the infernal weapon of the Gurkhas."[47]

These distorted images of the Gurkhas evoke other negative images from the Victorian era – images that were downplayed by First World War propagandists who preferred to praise the Gurkhas for their war effort. In the nineteenth century, the reader was exposed to such comments as: the "Gorkha" population is "tough and savage";[48] the Doum people of the Central African Republic became Christians due to their hatred of "their ancient tyrants, the Gourkhas";[49] "the Gourkhas, inhabitants of Naipaul [sic], … devoted themselves to Shiva's cult, the Goddess of destruction."[50] A French author using the pen name North Peat stated that British authorities should be careful not to entrust the defence of the British Raj to the Sikhs and the Gurkhas, even though the troops of these ethnic groups had remained loyal to the East India Company and had defended European populations during the Indian Mutiny. He claimed that the Sikhs were "cruel and vindictive." He thought that they fought the mutineers not because they loved England but because they hated Muslims. As for the Gurkhas, he added, Sir C. Napier considered them as "horrid little savages." To make his point, he then quoted Malcolm Ludlow, who argued: "the English may use these people, but should not dare imagine that their wife and children would be more secure with the Sikhs and the Gurkhas then they were amongst the Rajputs." North Peat concludes by warning his readers that England should not indulge itself in an excess of confidence by trusting "inferior casts."[51] This negative image of the uncivilized person was rekindled during the First World War as the Gurkhas' efficacy against the Germans prompted former colonial fears of the Indigenous cutthroat. In *Tommies et Gourkas,* the Gurkhas crawl with their bloody knives between their teeth after having slit the throats of six German sentries. It fills a white soldier, John Bisquitt, with awe to see how the Gurkhas dealt with these sentries.[52] Das Santanu ably explains this phenomenon: "The non-white soldiers then, whether brown or black, mobilized racial fantasy: the roots go back to nineteenth-century racial science, but how does the war affect these impulses? A popular story applied to the non-white soldiers, applied to the Indians, Spahis and Moroccans, was that

6.5 Red Cross soldiers from India in Marseilles.

they chopped off the ears of the German soldiers and made them into a garland around their neck. Heather Jones has argued that the colonial atrocity stories provided 'a safe space for articulating fear, wartime violence [and] killing.'"[53]

It is through knowing how such images framed European minds that we may understand what attracted French civilians, who met Indigenous troops eager to check the stories they had heard in imperial literature. The following example concerns a most unusual (and therefore precious) comment that was written on the back of a photograph postcard representing English artillery non-commissioned officers being congratulated by French civilians in Toulouse. The unknown Frenchman wrote to his correspondent about those he called "Hindus": "They are charming. I don't know if I told you that I went to meet one of them. It was on Sunday. He was giving us strong handshakes. He knew but two words in English, but he smelt like a savage, Pouah![54] And they say they like water.[55] He showed me how he would deal with the Boches. [He] took out his knife from his pocket and pretended to slit his throat, 'Germans' and he smiled. He seemed intelligent and happy with his demonstration."[56]

Although just reporting his meeting with a sepoy, the French writer of the postcard revealed a racist attitude towards the "other" as, for him, an unclean man was a "savage." Many colonials, who thought that they were "civilized" and who expressed the idea that "cleanliness is next to Godliness," shared this contempt for the sepoy. Furthermore, this writer seems to emphasize that the sepoy was a cutthroat because he was happy with what he could do to a German soldier. Yet the lack of intimate knowledge of Indian and Nepalese peoples led to confusion over a slouch hat and a turban, metonymical objects that contributed to the construction of an imagined Indigenous "other."

GURKHAS IN POSTCARDS, MISTAKEN IDENTITIES?

The reassuring message that the Germans feared Indigenous troops, the Gurkhas in particular, was meant to lift the morale of the French population. For the benefit of wounded soldiers, *Le Petit Marseillais*, on 3 November 1914, advertised postcards representing the Indian troops who had just landed in the French port of Marseilles: "Everybody will want to preserve an interesting souvenir of the passage in Marseilles of that gallant army from India, which, as stated by dispatches, provokes such awe in German ranks." The people of Marseilles, who were sensitive to such messages, repeated them in their correspondence, as demonstrated by a postcard written by a Mme Petit from Marseilles on 18 December 1914, showing Gurkhas parading in her hometown. In her message, she underlines the importance of Nepalese martial qualities: "A look at the troops of the Indian armies in transit in Marseilles. The Gurkas, short men who are much appreciated and much dreaded by the enemy armies on the battlefield."[57]

Thus, postcards were social constructions that linked many interested parties (photographer, publisher, buyer/sender, addressee, and visitor). If the postcards are factual, they depict Indian soldiers and offer prosaic captions such as "Indian troops" or "Our friends the Indians." Some participate more explicitly in martial rhetoric, with such captions as "War of civilisation against barbarity,"[58] and "The Hindu troops are going to teach a 'Kultur' lesson to the Germans."[59] The civilized/barbarian dichotomy is a widespread topos in First

World War propaganda that often engages in role reversal. For instance, a French sketch postcard depicts a Senegalese soldier guarding German prisoners and saying to visiting civilians, "Have you come to see savages?,"[60] thus making the Indigenous troops part of the civilized populations while transforming the Germans into savages.

If we take a broad view of the French postcard production devoted to Nepalese and Indian troops in the First World War compared to other British troops, we discover that the sepoys have pre-eminence over any other ethnic soldier. Just as the postcards in Marseilles massively preferred and represented sepoys, so the rest of the French production preferred sepoys to Anzacs, Springboks, Canadians, or Anglo-Indians, to name but the main imperial contributors to the war effort. The graphic power of the sepoys is naturally linked to the exoticism that surrounds them as they are so unlike white soldiers, who are almost identical in appearance to British metropolitan troops. This French "Indian" postcard corpus, composed of several hundred photograph and sketch postcards, may be accounted for by the fact that local French printers chose to profit from the passage of Indian troops through their towns. Indeed, the vast majority of the cards represent sepoys posing in front of or inside the trains in which they are travelling, the alternative being cards showing them posing in the cities in which they are stationed.

The cards therefore often depict soldiers (smiling or not), all very dignified, standing to attention in front of the camera or going about their business, parading, drilling, and being visited by King George V. We can trace the whereabouts of Indian troops in France by looking at these postcards and determining whether they originate from Marseilles, Toulouse, Orléans, Dijon, Nantes, Paris, Le Mans, Nancy, Montauban, or Limoges. From this corpus of several hundred cards I single out a dozen that are interesting for my argument concerning the khukuri. They all have in common the presence of a khukuri, which a sepoy is exposing to the camera. We have seen that the khukuri was featured in European First World War representations as a fearful weapon in the hands of the Gurkhas. Yet the sepoys posing with a khukuri in the postcards are roughly equally divided between Gurkhas and Indian soldiers who are wearing a turban (see figures 6.1 to 6.4). In fact, the Gurkhas were not the only Indian Amy regiments to be supplied with a khukuri. Soldiers of the Garhwal Brigade (Meerut Division) also received one,[61] as stated by George Morton-Jack: "only Garhwalis and Gurkhas carried the khukuri, a bone-handled knife with a single-edged and curved eleven-inch blade."[62] George Morton-Jack might have simply referred to "combat troops" as, in fact, prior to the First World War another Indian regiment, the Army Bearer Corps, also had the khukuri as part of its gear.

Members of that corps came from the *Kahar* caste of carriers (of doolies and palkies). According to A.E. Milner, the way they carried a *dooly*, half-walking, half running, made them perfect stretcher-bearers as the wounded person was carried smoothly and not bumped around.[63] He also states that they wore the same uniform as the "native army" except for their badge "and a 'Kukree' in place of a bayonet."[64] With this remark, the emphasis is put on the fact the bearers are non-combat troops: they have no rifle and thus do not need a bayonet. Here, the khukuri is therefore not a weapon but a tool used for their work as stretcher-bearers. To come back to the sepoys-showing-their-khukuri French postcards, the common denominator is that the khukuri is well exposed to the gaze of the spectator as the soldiers are holding it before them if they are standing to attention or resting it on their chests if they are relaxing. In one instance, the Gurkhas are holding the khukuri above their heads in a threatening attitude, as if about to attack the photographer.[65] The French photographers clearly asked the sepoys to strike these martial poses so that the viewers of the postcards could imagine, with delight and maybe with awe, who the Germans were going to face. Yet it is interesting to note that soldiers with turbans, other than the Gurkhas easily identified by their slouch hats, are portrayed in such attitudes. I have been able to trace four photograph cards with Gurkhas holding a khukuri (figures 6.1 and 6.2) and seven cards representing turbaned sepoys (figures 6.3 and 6.4). Interestingly, they were all taken in Marseilles, where the troops landed and rested some days before leaving for the front – except for one picture, which was taken and printed in Toulouse and another that was published in Paris.

As the people of Marseilles were the first French citizens to see the Indian troops, we may infer that they wanted to match what they saw with the stories they had heard prior to the arrival of the troops. Hence the captions of the postcards focus on the same stereotypical representations of the sepoys with their supposedly favourite weapon: "The Hindus in France: Gurkhas and his [*sic*] dreadful war knife" (figure 6.1);[66] "Our Allies in Marseilles. Gurka soldiers

6.6 Postcard of Indians surprising the Prussians.

and theirs [sic] knives" (figure 6.2);[67] "The Hindus in France, Soldier and the Kukri (dreadful war knife)" (figure 6.3).[68] The association of man and weapon in these postcards conveys a lethal combination. The underlying statement seems to be that, without one or the other, they are of no efficiency, but together they become outstanding German killers. What the sender of document 3 had in his mind is to be understood from the text he wrote to a Mlle Claire in Toulon (Var), in which he states that he hopes the Indian Soldier "will not frighten you with his big knife." Some postcards go even further as the men are no longer the focal point, even though they are still holding the khukuri: "The Hindus in France. The dreadful war knife" (figure 6.4).[69] From the same perspective, the Toulouse postcard provides didactic details about the knife, which implies that people were keen on the object and the knowledge and myths surrounding it: "Hindu Troops: A multi-function weapon: a bayonet and a kitchen knife for camp duties, also used to break lianas when travelling."[70]

6.7 "Gourkhas" [*sic*] slaughtering sleeping German soldiers.

This accurate description of the khukuri as both a tool and a weapon shows two different approaches to the representations of the Gurkhas – one sensational, the other more pragmatic. Notwithstanding, one cannot but think that both representations are part of the same ideological pattern that consists in exposing to the French public the horrible danger that the Germans are going to face once these ill-named "Hindu" troops have reached the theatre of war. Yet veracity was not what the postcard editors were seeking. Indeed, if the Gurkha postcards (figures 6.1 and 6.2) do express coherence between text and image (a khukuri in the hands of a Gurkha can become a "dreadful war knife"), the postcards that depict turbaned sepoys with a khukri that is described as a lethal weapon are wrong (figures 6.3 and 6.4). A close-up on the belt of these soldiers shows that they are members of the Army Bearer Corps, who are, as stated earlier, non-combat troops. Their khukuri is therefore a tool and not a weapon: it should not be labelled a "dreadful war knife." The

6.8 Bilingual French postcard, "A surprise from the Gurghas [*sic*]."

metonymical turban, the best element with which to identify an Indian soldier for the 1914 French population, along with the fact these men had their own khukuris, explains why the publishers convinced the soldiers to pose in front of the camera.

We have further proof of this with another Marseilles postcard that portrays some twenty-five posing Indians, those in the front row sitting and those in the second row standing (figure 6.5). Some are smiling, one has adopted a martial, threatening attitude with clenched fists. Most of them wear a red-cross armband,[71] and one of them, who is sitting in the centre of the postcard, proudly shows his khukuri to the camera while smiling. For the French public, unaware of the distinctions that characterized the different Indian and Nepalese ethnic groups, a turbaned sepoy could be one of the well-known Sikh soldiers whose martial skills and fierceness in battle were acknowledged in France to be second to none. Members of the public would not know that Sikhs were not armed with the famous khukuri, and thus they were not sensitive to the manipulative process of the postcard editors who distorted reality for their own benefit by portraying a sepoy with a "khukury between his teeth" (figure 6.3). The French public's confusion between Gurkhas and other Indian soldiers is furthered by another postcard from this corpus: "The 1914 European war. 403. In the Indian camp. These brave soldiers have shown their value during bayonet fights and night surprises."[72] It is true that, on many occasions, Indian soldiers were engaged in hand-to-hand fighting in the trenches, particularly at the battle of Festubert (15–27 May 1915). However, like hunting animals, they are also praised for their night attacks, at which, according to the literature, they were gifted. This card is transitional as it leads us to reflect on the combat representations of Indian troops in French as well as in German postcards.

POSTCARD SKETCHES OF FIGHTING GURKHAS

If we look at the postcards that display known regiments from the Raj, such as Bengal Lancers or Sikhs, we find what the Australian journalist quoted above was complaining about. Indeed, the Bengal Lancers are represented as an irresistible cavalry force that systematically sweeps the Germans out of the way, while the Sikhs do the same through infantry charges.[73] In one instance (figure 6.6), Sikh soldiers are charging with rifle and khukuri, while one of them (in the bottom left hand corner) is running with his knife between his teeth just as depicted in the press description analyzed above.[74] A fair number of drawings representing fighting Indian troops are in fact devoted to night attacks, and the vast majority mention Gurkhas, even though the viewers could seldom recognize one. In a first example, it is worth noting that the soldiers depicted look more like Middle Eastern soldiers than Asians: "At night, the 'Gourkhas' took over an enemy trench slaughtering all the 'Boches' with their dreadful 'Yatagan'" (figure 6.7). What is more, a yatagan is a weapon produced in the Ottoman Empire and does not look like a khukuri, yet the confusion seems to be a frequent one: "Gurkhas, sturdy men, who behead a *bullock* (a great Indian ox) with a single stroke of a kukri (yatagan)."[75] Another example is entitled "1914 … IN BELGIUM – A surprise from the Gurghas [sic]" (figure 6.8).[76] It makes the same mistake of calling turbaned sepoys Gurkhas. The drawing seems to have been first published in a French newspaper, which, once again, puts the khukuri at the heart of the representations: "Our engraving shows one aspect of the combat methods of these Eastern soldiers. Like English soldiers, they wear a khaki uniform, but they keep their high national turban. They are armed with a rifle and bayonet-knife. Furthermore, they

6.9 French postcard, "An Indian Raid" with explanations below.

have the national weapon, the 'Kukry,' which is a short and broad sword, so well sharpened that with a well-placed single blow, it can chop off the head of the enemy."[77]

The stereotypical ethnic identity of the Gurkhas is once again confused with that of other Indian troops. This confusing interplay between the two groups can be explained by the proximity of their birthplaces (North Eastern India and Nepal, respectively) and by the fact that the Bearer Corps, the Garhwals, and the Gurkhas were officially allowed to have a khukuri and also had a common culture:

> One of our colleagues evoked a passage of the Ramayana, the sacred book of the Indians, which sums up in a few lines the warrior precepts of these soldiers: "slip into the shadow like a tiger in the jungle, hold your breath, stop the beating of your heart, put out the fire of your gaze. In the thick of the night, you must be the pick of the rock or the dead stump. You must remain still, one night, one day, one night and one day again. Then when you are confident that the enemy has fallen asleep, hit him in the throat and bleed him to death." Our engraving is the perfect illustration of these precepts.[78]

6.10 German postcard, "Charging Indian Troops at Ypres (Ypern)."

Nepalese share with Indians the *Ramayana*, the ancient Hindu epic poem written in Sanskrit. Intriguingly, I have not been able to find the source of the above quotation, originally in French. The question remains as to whether or not it was a convenient "invention" on the part of a French journalist. At this stage, what may be understood from the use of such a reference is that the fighting qualities of the First World War Indian troops are part of a cultural heritage that dates back to antiquity. According to the journalist, the French artist's representation of a night attack perfectly matches the *Ramayana*'s description and confirms the comparison of the Gurkha (and Indian) to the tiger: "One of the officers of these untameable men declared a few days ago: 'They are like tigers in the night.' It is the way our artist has represented them." Similarly, "An Indian Raid" (figure 6.9) by artist R. Bataille depicts what must have been seen as a gruesome portrayal of the killing capacities of the Indian soldiers referred to as "Gurkhas" in the caption of this postcard. Indeed, what is represented is the instant six Gurkhas spring as one out of the darkness and slit the throats of six German sentries guarding an ammunition park. The caption uses the same martial rhetoric discussed above: the soldiers are "lurking in the shadows," they observe the sentries, and, when the time comes, they "crawl, their knife between the teeth," and eliminate the sentries. The gory killing represented in the postcard is probably a legacy of the French pre-war blood-curdling sensation magazines, which were devoted to crimes, executions, and bloody accidents (the equivalent of the British "Penny dreadful magazines"). If turbaned soldiers are often referred to as Gurkhas, in at least one instance we have a reversal of this pattern, with Gurkhas (recognizable by their short pants, slouch hats, and Asian features) being named "Hindus" as they storm a German trench armed with the khukuri: "No 68 France: during a night attack, Hindu soldiers, holding their knives, storm German trenches."

The Germans themselves seemed to be confused as to what Gurkhas really looked like. Sikh soldiers are renamed "Gurkhas" in the German postcard: "Der europäische Krieg 1914/1915. nr. 13: Kampf mit indischen Truppen (Gurkhas) bei Ypern" (The War in Europe. Battle with Indian Troops at Ypres) (figure 6.10). Once again, this postcard is a good example of the absence of verisimilitude as far as combat is concerned. We can see several Indian soldiers holding a knife in one hand and a rifle in the other, just as in the press descriptions analyzed above. Yet the fact that these soldiers are seen to be crawling discredits the representation. What seems to have been important for the German artist was to ensure that the postcard viewer would remember this: the Indians do not play fair, they creep up from behind the enemy and surprise him, and they do not act bravely as in a gallant charge. Another German example, "Heranschleichende Indische Gurkhas" (Creeping Indian Gurkhas) was published in an illustrated newspaper in Leipzig. It looks as though the common mistakes that we have noted in Allied representations can also be observed in German representations. The "Gurkhas" are once again Sikh soldiers holding a fairly recognizable khukuri. It is night in the countryside somewhere on the Western Front, and two of the soldiers are kneeling on lookout, while a third is creeping behind them with his khukuri (nearly one kilo) held between his teeth.

This stock phrase, "l'homme au couteau entre les dents" (the man with a knife between his teeth), is a long-familiar and enduring French expression, meaning someone who wants to get revenge. It is no positive image, as the knife clenched between the teeth evoked for French people bloodthirsty pirates who needed their hands to be free to hold the rope that would enable them to swing onto an enemy ship. In the French Empire, the crawling killer with his knife held between his teeth became associated with the dangerous Indigenous attacker who, in the nineteenth century, would pounce on sentries at night. During the First World War, some French cartoons (including postcards) portrayed Senegalese soldiers with a bloody knife between their teeth holding the head of a German soldier with his helmet still on ("La moisson de Boudou Badabou"), which accounted for the widespread idea that Africans were "savages" and killers. In between the two world wars, the Bolshevik with a knife between his teeth became a stock character of anti-communist propaganda as he was the man who wanted to take other people's property (see Andrien Barrere's 1919 poster *Comment voter contre le Bolchevisme*). During the Second World War, French communists caricatured Hitler by representing him with a knife between his teeth. Nevertheless, circumstances change opinions. What was seen as a terrifying and negative image when the viewer felt threatened by the man with a knife between his teeth became acceptable when a similar man "was on our side" and preyed on the German enemy: "The Kaiser will undoubtedly despise [him] as he despises the English army … But we shall see how the Boches behave in front of [his] 'Kukry.'"[79]

CONCLUSION

What traces and recollections remain of the Gurkhas' individual or collective participation in the First World War? The postcards discussed in this chapter are memories of war involving Indigenous peoples, what I have elsewhere called the "first circle of memory."[80] They tell the story of Western representations of the colonized populations who came to Europe to defend the metropolis in the same way as do other forms of expression, such as propaganda posters. Even though the Gurkhas, like other forces from the Empire, are highly praised in Western martial discourses, they are nonetheless considered as lesser people: "What is striking, however, is a certain discursive tension between the depiction of these sepoys as loyal, gallant, fearless soldiers of the Raj, and yet somewhat subhuman, supposedly lacking the 'natural' qualities of the white man: qualities of leadership, decision, maturity."[81] I would like to argue in conclusion that the European reading public readily accepted a biased vision of Indigenous soldiers because for so long they had been assailed by stereotypes. They were therefore on familiar ground when they looked at these postcards and probably had an impression of déjà-vu. The Indigenous soldiers were associated with animals and were thus close to savagery. The Gurkha, like the Tamil fighters of Sri Lanka, were associated with the tiger. However, this apparently positive comparison can quickly degenerate into a negative one as the tiger can be bloodthirsty as well as fierce and thus imply a dangerous association between the soldier and a cutthroat, signifying the divide between the "civilized" soldier and the "savage."

As agents of propaganda, postcard publishers during the First World War, whether consciously or unconsciously, engaged in a global strategy that firmly reinforced cultural stereotypes in the public's mind. The Gurkhas were praised and were meant to be larger than life.[82] They served a clear purpose in Allied martial rhetoric, but, in the process, they were exoticized, stigmatized, and caricaturized. The European public embraced this idealized but false vision of the Gurkhas without seeing that, beneath the apparent praise, lay an imperial attitude that dissolved the positive imagery of ferocity so as to expose the negative imagery of savagery. Because Europeans were afraid of the cruelty of the "savage," it was comforting for them to think that it would be the Germans who would have to face it (in the form of the Gurkhas) rather than the Allied troops. This accounts for such comments as "I'd sooner have them as friends then foes," made by an Australian soldier who fought alongside the Gurkhas at Gallipoli.[83] Many years later, the reaction is the same as a journalist from the *Daily Mail* quotes a British soldier in Afghanistan in the title of his article: "As a Gurkha Is Disciplined for Beheading a Taliban: Thank God They Are on Our Side!"[84] The Falklands War rekindled the same propaganda devices, which, according to some, had a martial effect: "'By the time they arrived on Mount Tumbledown, the Argentinians had seen pictures of Gurkhas sharpening their kukris and read all these stories about them eating their prisoners,' says Major Corrigan. 'So when the Gurkhas actually appeared, all they found were empty trenches.'"[85] A martial myth once launched is difficult to suppress as there is no way of stopping its spread. The Gurkhas and their khukuris are a perfect illustration of this perennial myth.

NOTES

1 Pierre Brouland and Guillaume Doizy, *La Grande Guerre des Cartes Postales* (Paris: Hugo Images, 2013), 17.
2 https://www.delcampe.net/fr/collections/.
3 Quoted by Anne Becker, *Voir la Grande Guerre, un autre récit 1914–2014* (Paris: Armand Colin, 2014), 18.
4 Lionel Caplan, *Warrior Gentlemen: "Gurkhas" in the Western Imagination* (Providence, RI: Berghahn Books, 1995), 4.
5 John Parker, *The Gurkhas: The Inside Story of the World's most Feared Soldiers* (London: Bounty Books, 2005), 80.
6 David Bolt, *Gurkhas* (London: White Lion Edition, 1975), 20.
7 Alex Schlacher, *Arc of the Gurkha. From Nepal to the British Army* (London: Eliott and Thompson, 2014), 101.
8 The word is spelled in a number of other ways: "khukri," "khukury," "kukri," "kukery," "kookrie," "kookree," "kukoori," "koukhri," "cookri," "cookery," and so on.
9 Bram Stoker, *Dracula* (Free eBooks at Planet eBook.com), 539.
10 Caplan, *Warrior Gentlemen*, 134.
11 A. Norec, *Tommies et Gourkas* (Paris: E. Rouff éditeur, collection "Patrie," 1917), 12.
12 Henri Pellier, *Le serment des trois hindous* (Paris: Libraire Larousse, Les livres roses pour la jeunesse no. 219, 1918).
13 Adolphe Dubois de Jancigny et Xavier Raymond, *Inde* (Paris: Firmin Didot Editeurs, 1845), 555.
14 *Le Monde Illustré*, 8 July 1916.
15 Byron Farwell, *The Gurkhas* (New York: Norton and Company, 1990), 21.
16 See Heather Streets-Salter, *Martial Races: The Military, Race and Masculinity in British Imperial Culture, 1857–1914* (Manchester: Manchester University Press, 2004).
17 David E. Omissi, *Indian Voices of the Great War: Soldiers' Letters, 1914–1918* (Houndmills, Basingstoke: Palgrave Macmillan, 1999), 2.
18 "L'armée de l'Inde," *Le Petit Provençal*, 22 September 1914.
19 *Bulletin de la Société de Géographie Commerciale de Paris*, no. 1102, 1896.
20 *La Croix*, no. 9749, 23 December 1914.
21 *Zeehan and Dundas Herald*, 29 January 1915 (Tasmania).
22 Martina Sprague, *Kukris and Gurkhas: Nepalese Kukri Combat Knives and the Men Who Wield Them* (Copyright 2013 Martina Sprague), 66.

23 Troy R.E. Paddock, *World War I and Propaganda* (Brill Academic Pub., 2014); Stephen Badsey, "*'The German Corpse Factory': A Study in First World War Propaganda* (Helion and Company, 2016); Aviel Roshwald, *European Culture in the Great War: The Arts, Entertainment and Propaganda, 1914–1918* (Cambridge: Cambridge University Press 2002); Brian Lund, *Propaganda in the First World War on Old Picture Postcards* (Reflections of a Bygone Age, 2014); Harold D. Lasswell, *Propaganda Technique in the World War* (Martino Fine Books, 2013 [1938]).
24 *Western Mail (Perth)*, 30 April 1915, 4.
25 *North Western Courier* (Narrabri, NSW), 14 July 1915.
26 *Avon Gazette and Kellerberrin News (WA)*, 19 June 1915, 4.
27 *Register* (Adelaide), 6 November 1915.
28 *"The Indians in battle. The Kukri preferred to the Rifle," Avon Gazette and Kellerberrin News (WA)*, 30 January 1915, 4.
29 *La Guerre Mondiale* (Geneva), 18 September 1914, 16.
30 *Northam Adviser* (WA), 10 February 1915, 3.
31 *La Lanterne* (Bordeaux), 6 November 1914.
32 "Gallant Gurkhas," *South Eastern Times (Millicent, SA)*, 16 April 1915, 6.
33 *"Indians' exploits. Gurkhas and the Kukri. An indignant Protest," Argus (Melbourne)*, 22 December 1914, 5. T*his article was reproduced in at least four other newspapers before the end of January 1915.*
34 Ibid.
35 Maurice Dekobra, "L'armée des Rajahs," *La Vie Parisienne*, 23 January 1915, 61.
36 Caplan, *Warrior Gentlemen*, 3.
37 Rudyard Kipling, *The Drums of the Fore and Aft* (New York: Brentano's, 1898), 55.
38 *Daily Telegraph* (Launceston, TS), 16 January 1915, 3.
39 *Le Petit Parisien*, 12 October 1914.
40 Maurice Thierry, "Une ruse d'Indiens," in Mme J. Durand, *Nos diables bleus*: Récit de la guerre dans les Vosges (Paris: Librairie Larousse, Les livres roses pour la jeunesse, no. 1, n.d.), 44.
41 Charles Guyon, *Nos Amis les Anglais pendant la guerre* (Paris: Librairie Larousse, Les livres roses pour la jeunesse, no. 155, n.d.), 28.
42 Pellier, *Le serment des trois hindous*, 30.
43 Guyon, *Nos Amis les Anglais*, 24; Pellier, *Le serment des trois hindous*, 30.
44 Norec, *Tommies et Gourkas*, 12.
45 *Leader* (Orange, NSW), 12 December 2016.
46 The same article was published in several newspapers, including the

Sunday Times (Sydney), 14 February 1915, 1; and the *Globe and Sunday Times War Pictorial* (Sydney), 20 February 1915, 16.

47 Rudyard Kipling, "Les deux tambours," in *Revue des Deux Mondes* (Paris: Bureau de la revue des Deux Mondes, 1891), 611.

48 Jules Radu, *Nouvelles Méthodes* (Paris: Privately printed by the author, 1868), 335.

49 Pierre Maël, "Terre de Fauves," *Le Journal de la Jeunesse* (Paris, Hachette et Cie, 1894), 35.

50 Auguste Lagravère, "L'Eléphant," *Les deux Mondes Illustrés: Le Journal des Grands Voyages*, 10 October 1880, 92.

51 North Peat, "La Révolte de l'Inde d'après des témoignages d'anglais," *Revue Contemporaine* (Paris 1858), 33.

52 Norec, *Tommies et Gourkas*, 19–20.

53 Das Santanu, "Writing Empire, Fighting War: India, Great Britain and the First World War," in *India in Britain South Asian Network and Collections, 1858–1950*, ed. Susheila Nasta (London: Palgrave Macmillan, 2012), 28–45, 36.

54 French interjection to express disgust.

55 The writer of the postcard failed or refused to acknowledge that transit soldiers in 1914 could seldom take a shower.

56 "1ere Série de la Guerre 1914. 4 Toulouse. *On félicite les sous-officiers d'artillerie anglaise au passage dans une gare*. Phototypie Labouche Frères, Toulouse." All postcards mentioned or reproduced belong to private archives.

57 "CAMPAGNE 1914. Nos Allies de passage à Marseille, Editeur Barabino, 13 rue d'Aubagne Marseille."

58 Photograph postcard: "*Guerre de la Civilisation contre la Barbarie (1914). Femmes indiennes tissant des colliers de jasmin pour offrir aux soldats partant pour la guerre*, Edition de Luxe. P.G. Evrard. Paris."

59 Photograph postcard: "20 Guerre 1914, Phot. Blanch. Collection Jove."

60 Drawn illustration postcard: "I.M.L. no. 2058. D'après Jonas et *L'Illustration*."

61 The Garhwali population comes from the northeastern state of Uttrakhand in India, close to Nepal.

62 George Morton-Jack, *The Indian Army on the Western Front: India's Expeditionary Force to France and Belgium in the First World War* (Cambridge: Cambridge University Press, 2014), 30.

63 A.E. Milner, "The Army Bearer Corps," JR *Army Med Corps*, London, 1906, 685–9, 687. http://jramc.bmj.com/content/6/6/685.full.pdf+html.

64 Ibid., 689.

65 Photograph postcard: "*La guerre de 1914. 24 Soldats Indiens Gurkhas avec leur arme terrible, prêts à l'attaque. 24 Gurkhas with their formidable weapons ready to make an attack*. B. & C. Paris."
66 "Coët Edt., phototypie Roure, 31 rue du Petit St Jean, Marseille."
67 "ELD éditeur, Paris."
68 Ibid., note 65.
69 Ibid.
70 "Hindu Troops. 2e Série no 10. Edition historique de l'ancienne Photographie Provost, 15 rue Lafayette, Toulouse."
71 "*A Group of Indus* [*sic*] *Camping in Marseilles*. Edition des P.T.T. des Bouches du Rhône."
72 "Artaud Nozais, Nantes."
73 Lancers: "Leon Hingre 1914. *Soir de bataille près d'Ypres après la charge des Indiens*, Paris, A. Royer, 'Galerie Patriotique'"; "*Les Uhlans aux prises avec les Lanciers Indiens*, Paris, A.H." Infantry: "*Les Indiens à l'attaque*, 49, Editée par l'Ecole Technique Supérieure de Représentation' R. Turbigo, 57, Paris"; "*Indiens chargeant à la baïonnette*, 'Geo' 1914."
74 "Jean-Robert, *Guerre de 1914 – Les Indiens surprennent les Prussiens.*"
75 *Le Collectionneur de timbre-poste*, 1 January 1915.
76 French and English caption ELD editor (Paris).
77 "Our Eastern friends. Indian soldiers in active combat," *Le Petit Journal illustré*, 25 October 1914.
78 Ibid.
79 Ibid.
80 Postcards, when immediately published after the picture had been taken, were, I argue, the "First circle of Memory" – that is, they were among the first memorial devices that preserved what people had witnessed, beyond their life expectancy. See Gilles Teulié, "The First Circle of Memory: First World War Postcards of British Imperial Troops in Marseilles," in *Commemorating Race and Empire in the First World War Centenary*, ed. Ben Wellings and Shanti Sumartojo (Aix-en-Provence, Liverpool: Presses Universitaires de Provence, Liverpool University Press, 2018) 107–29.
81 Santanu, "Writing Empire," 35.
82 Caplan, *Warrior Gentlemen*, 135.
83 *Register* (Adelaide), 6 November 1915.
84 Robert Hardman, MailOnline, 20 July 2010, http://www.dailymail.co.uk/debate/article-1296136/As-Gurkha-disciplined-beheading-Taliban-Thank-God-side.html.
85 See Parker, *Gurkhas*, 218; and Caplan, *Warrior Gentlemen*, 134.

7

The Humour of an Indian Soldier's Memories of the First World War in M.R. Anand's *Across the Black Waters* (1939)

Florence Cabaret

Across the Black Waters (1939), written by Indian author Mulk Raj Anand,[1] who lived in London and in several other European places from 1925 up to 1945, foregrounds the Indian soldiers' participation in the war. This period was punctuated by his involvement not only with the 1926 coal miners' strike, the Ghandian campaign for Indian independence, the All-India Progressive Writers' Association, and the Bloomsbury group but also with the Republicans during the Spanish Civil war, which testifies to his "conviction of the inextricability of politics and literature."[2] Today, *Across the Black Waters* remains famous for being one of the rare fictional accounts of a sepoy's experience of the First World War[3] – all the more so as "it was the first time [Indian soldiers] had fought in a white men's war,"[4] as most Indian regiments had fought in Mesopotamia, East Africa, and the Middle East rather than in France and Belgium, as is the case in this novel. Along with *The Village* (1939) and *The Sword and the Sickle* (1942), *Across the Black Waters* is the second part of a trilogy telling the story of Lalu Singh, a young Punjabi villager, from the beginning of the twentieth century to the 1930s. In *The Village*, Lalu Singh flees India's rural life by joining the Indian Army, hoping to become emancipated from the fate of Indian peasants plagued by religion, landlords, and money lenders. *Across the Black Waters* focuses on Lalu Singh's memories of the French and Belgian campaigns from the moment his regiment arrives in Marseilles in 1914 until its members are arrested by the Germans in 1915 as they wage war in the Flanders trenches of Festubert. Eventually, in *The Sword and the Sickle*, Lalu Singh returns to India after two years in a prisoners' camp in Germany,[5] bringing back memories that modify his representation of Europe as a place of equality and justice, and help him understand that, in India, the war effort contributed to exacerbating the problems of exportations and land concentration.

Two-thirds of *Across the Black Waters* deals mainly with the foot soldier Lalu and his comrades' discovery of France, Belgium, and their inhabitants, so that the narrative is less concerned with war action proper than are several later chapters, which are dedicated to trench warfare. As a novel, *Across the Black Waters* is not a first-hand autobiographical account but,[6] rather, a fictional autobiographical narrative written twenty years after the war by a writer who had heard testimonies of soldiers back from Europe when he was a child living in Punjab,[7] and who was also familiar with British war poets and the historiography of the First World War.[8] A severe indictment of war, the novel has often been praised not only for its humanism and documentary description of enduring Indian soldiers but also for its pre-postcolonial account of the British Empire's demise under Indian eyes.[9] Thus, as early as 1939,[10] Anand was writing against the grain of the Eurocentric perception of the First World War,[11] prefiguring the "global turn" in First World War studies,[12] which is ironically embodied in the late shift of illustrations for two different front covers of *Across the Black Waters*.[13]

Following in the footsteps of the "cultural turn" in First Word War studies, as illustrated by such recent works as *Laughter in the Trenches: Humour and Front Experience in German First World War* (2012),[14] and *Humor, Entertainment, and Popular Culture during World War I* (2015),[15] I would rather focus on a less solemn and little known dimension of *Across the Black Waters* and study the impact of the recurring presence of humour in an overall tragic narrative, demonstrating that the war memories brought back to India by Lalu Singh are not just traumatic memories of sacrifice and powerlessness.[16] More pointedly, Anand uses humour to write back to three types of collective memories of the Indian troops involved in the French and Belgian front during the First World War. Humour allows him to expose the British colonial theory of "the martial races," to shatter the representation of Indian troops as a homogeneous group of anonymous and impenetrable faces in favour of the more realistic view of a network of complex and changing individuals, and to reverse the Orientalist notion of exoticism so as to integrate new Indian characters into the European mindset and imagination.

DEBUNKING THE THEORY OF THE "MARTIAL RACES"

When Anand chose to fill in the gap of what British fiction had overlooked in its portrayal of the European part of the First World War, it did not mean that he wanted to particularly extol the bravery of Indian troops. Indeed, his novel is rather a behind-the-scene narrative, where Anand unveils the other face of the colonial theory of the "martial races,"[17] and where humour is a means of reconsidering the stereotype of Sikhs as highly principled, fierce, and manly fighters, as described by General MacMunn in 1933.[18]

Humour first participates in Anand's integration of Indian soldiers into the bulk of all the troops that waged this war since he stages them not only as sharing what is known as the tradition of virile "military humour" – but also, in this specific context, what has been documented as "trench humour." Bawdy humour is very quickly introduced in the novel when Lalu's regiment is cantoned near Orleans and the group of comrades go on a discovery expedition to the town, during which they encounter "a few shops displaying strings of dirty brown sausages" (42). A giggling discussion is immediately triggered between Lalu and the other Indian soldiers about shape, size, being a North Indian and being a Sikh. Towards the end of the novel, on their way to Festubert, Ribald sings about village girls (212–13), provoking contagious laughter among the soldiers, illustrating the role of humour as providing relief from the tension of the war and in renewing morale, which is part of male bonding and survival strategies in times of war. Ludicrous episodes, such as the scene at the Orleans brothel (44), or comic interludes, such as the snow-ball mock fight near Festubert (227), are portrayed along with scatological humour downplaying the fear of death in the mud (97, 99, 101) and emotional jokes breaking the solemnity of farewell embraces (230). All of this offers a humanizing description of soldiers that appears to be closer to the nuanced image we form of their British and French counterparts than to the photographs of stern and war-like sepoys marching through the streets of Marseilles.

By resorting exclusively to oral and visual forms of humour, Anand puts into perspective the stereotypical image of superior fighters devoted to the cause of the British Empire – a ready-made image that some sepoys mock in the novel. Such is the case when, near Ypres, Captain Owen explains to Lalu's regiment the positions of the other Indian troops so that his sepoys have a better sense of a war in which advances and retreats are constant on both sides. The older lance corporal Kirpu cannot help commenting: "So, Sahib, we brave lions have come just in time to stop the enemy ... Owen Sahib laughed. The reputation of Uncle Kirpu as a humorist made the very sound of his voice infectious with light heartedness" (96). Interestingly, the humorous reference to the invincible lion, a translation of the middle name "Singh" of male Sikhs and a metaphorical representation of the valour each Sikh believer should embody, also points to the distance Kirpu takes not only from the colonial concept of the "martial races" but also from the Sikh belief itself. The intricacy of the British and Sikh representations of bravery is simultaneously targeted by Kirpu's humorous remark, which he makes publicly, in front of other sepoys and in front of their British officer. The fact that Captain Owen responds positively to the joke feeds the humane description of this British officer in the novel and contributes to the balanced depiction of the relationships between some sepoys and some of their British officers. Here, intragroup humour, used as a humour of esteem, contributes to the solidification of the group through approval and reduces ethnic and social distance.[19] The character of Captain Owen is described as sharing several humouring moments with his men (73, 90) and showing awareness that humour is no less a sign of civilization than war. We may think of a conversation during which Owen resorts to humour to oppose the "martial race" rhetoric used by another sepoy, who tries to flatter the British officer by praising the civilizing mission of the war: "'Oh well, buck up, Kushi Ram,' said Captain Owen, coming over and patting the Babu with a deliberately happy air. 'Civilization also means a sense of humour, you know. Don't let us fall victims to the mere solemnity of civilization'" (176). Again, this metalinguistic comment underlines the novel's recourse to humour as a way of qualifying the "martial race" theory from different viewpoints.

Still, it is the character of Kirpu who, through humour, most discredits the stereotype of the martial races as he stands at the centre of many of the regiment's joke relationships. While he regularly taunts his comrades about their proud perceptions of themselves, he frequently debunks the image of Indian soldiers as dedicated warriors by pointing to their weaknesses. Such is the case when he describes with magniloquence and antiphrasis Dhanoo's snoring in the trenches as a ferocious noise that will surely deter the enemy,[20] or when he deflates the show-off speech of the son of another NCO by saying that he remembers him as a naked, puny little child who could hardly defend himself from crows pecking at his buttocks.[21] These various humorous moments obviously flesh out a variety of characters who are not conceived as homogeneous back-up props or impervious cannon fodder for the benefit of the British colonizers but, rather, who are reminisced and staged in all their complexities and specificities.

HUMOUR AS A SERIOUS AND HUMANIST NARRATIVE DEVICE

Humour thus provides the novel with a tragic underlying narrative arc about submission and subordination as the character of Kirpu eventually commits suicide after being threatened with court martial by the NCO Harvindar Lok Nath, who can no longer stand Kirpu's provocative humour. Kirpu evolves from being perceived by other Indian NCOs as an agent of integration and cohesion to being suspected as a potential agent of discord and sedition, pointing to the double function of humour as a mode both of inclusion and of exclusion.

The ambivalence of Kirpu's role grows increasingly perceptible as he oscillates between a comforting and reassuring fatherly figure and a more unsettling Falstaffian character.[22] Whereas, when they first arrive in France, his sometimes dry humour is aimed at individual comrades as a way of deriding them and minimizing their excitement or anguish about their new living conditions,[23] or as a way of showing the superstitious nature of their religious practises,[24] his humour later becomes more and more targeted at other Indian NCOs. The latter, contrary to the British captain, feel much more threatened by Kirpu's trespassing on military propriety as Kirpu had just recently been promoted to a Lance Nak – so that now, like them, he is supposed to be the buttress maintaining order and discipline among the rank and file. Such is the case when Kirpu says, in a proverbial and falsely bland way, that the death of the Indian officer Lachman would not be a great loss – quite the contrary, actually, as it would enable more men to have access to food rations. And this leads usually shy Dhanoo to join in the contagious ring of black humour.[25] Simultaneously, Kirpu's experience helps him understand that it is for lack of efficient back-up artillery that the sepoys fail to regain an enemy trench, or that they cannot get rid of water in the trenches because the ground is water-logged (111). Thus, if humour tones down the truth of his statements, its aggressive and polemic nature gradually takes over. This occurs, for instance, when Havildar Lok Nath "sens[es] the ambiguity of Kirpu's description of fleas and lice" (189) harassing soldiers and feeding on them just as they fed on their officers. Or, more eloquently, it occurs when Kirpu mimics an NCOs hyperbolic speech to obviously criticize the military code of honour (188) and launches into excessive praise of the polishing of boots for the coming parade "because the boots are the most important thing even if you forget obedience, courage, loyalty and the rest" (190). At this stage, the image of the martial races has been ferociously marred by Kirpu, whose moral influence over the men has always been greater than that of his superior Havildar Lok Nath. Taking advantage of the death of the NCO who favoured and protected Kirpu, Lok Nath eventually punishes him for his defying humour. While Lalu tries to defend Kirpu by pleading that "it is just his sense of humour" (200), Lok Nath retorts that he has had enough "of his humour, his snivelling, dirty, demoralizing talk" (200), indicating that he knows "who will have the last laugh" (200). Thus, we can say that if humour "reduces the scathing bitterness in the novel,"[26] it also "eloquently protest[s] against the army code which provides a lot of comfort to the officers at the expense of the ordinary soldiers ... without any official or moral sanction."[27]

Here again, Anand refuses to gloss over the taboos and deficiencies that characterize, but are obviously not specific to, Indian regiments. His use of humour in this war narrative is also a means of emphasizing the internal tensions and potential changes at work within these uprooted small societies known as regiments.[28] Kirpu the humorist may thus appear as a dissenting figure in the making, threatening military obedience while, more symbolically, recalling and prefiguring other forms of dissent on the part of Indian people. One may indeed remember that the sepoy members of the martial races had been identified as such due to their faithfulness to the British during the traumatic 1857 Sepoy Mutiny, which is now regarded as a first step in the march towards India's independence. Such an interpretation of Kirpu's metaphorical role as a humourist echoes throughout contemporary works that state that, for Indian soldiers, "the war was a shaping agent for a new political and social consciousness, based on their experience of fighting side by side with British and other soldiers from the Empire, as well as encountering the local populations in France and Britain."[29] By undermining respect towards established power relations, Kirpu's humour moves away from consensus and points to alternative relations between Indians themselves as well as between Indian and British people. Yet Kirpu's death foregrounds the strength of power structures in India and Indian regiments in 1914. The novel thus never discards the idea that Kirpu decided to take his own life so as to protect his relatives from the shame of a court martial and the subsequent financial catastrophe this would entail for parents who would no longer receive their son's military pension. Nonetheless, by presenting the protagonist Lalu as an intelligent follower of Kirpu's humorous and challenging take on life, *Across the Black Waters* reminded the 1939 reader that, as early as 1914, Indian civil disobedience (if not military disobedience) was on the way, that it partly originated in the First World War, and that it had amplified over the past two decades.[30] What's more, as the narrative ends with the regiment's capture by the Germans, it prefigured the year 1940, "when Britain was engaged in a renewed conflict with Germany [just] after the evacuation from Dunkirk."[31] The novel thus brought to mind nagging memories of the Indian sacrifices that had been made to support the First World War effort and how this remained a bone of contention between Indian and British leaders as the promises of self-government that had been made in return for India's help during the war had been postponed from one roundtable to another – and had not yet been honoured in 1939.

HUMOUR AS A MEANS OF QUESTIONING THE ORIENTALIST NOTION OF "EXOTICISM"

Humour not only reorients our viewpoint on martial races and possible sedition on the part of colonized people but also reorients conventional representations of French and Belgian civilians not only to enhance their exoticism in the eyes of the Indian soldiers but also to underline the many common points between Indians and Europeans.[32] In this sense, humour may again be described as an inclusive trope showcasing a common sense of humanity between people who thought they belonged to separate worlds.

Linguistic oral humour is very much present in those scenes of encounter in which French and Belgian civilians are viewed as those who speak a "soft but unintelligible lingo" (33), to quote an ironic use of a typical colonial phrase describing Indigenous people abroad. Humour plays on various levels as Anand also appeals to British audiences by exoticizing French as a foreign funny language, reproducing an experience shared by both British soldiers and Indian soldiers when confronted by a common sense of loss in this alien land. Though Indian characters struggle humorously with their misunderstanding of the French language (13–14), they are also shown to be smart enough to pick up words that may hurt a certain British national pride. Such is the case when, at Orleans, the sepoys come face to face with "the figure of a young girl with a sword in her hand, her head thrust heroically forward and her whole body speaking of some brave deed which she had performed" (34). When compared to the inscription (an oddly spelled form of "Jeanne d'Arc") at the bottom of the statue, the obscure word "Jindarc," used by French people, suddenly reminds Lalu of a history lesson about Joan of Arc that "he had read at the Church Mission School at Sherkot" (34). Lalu explains the meaning of the statue to his comrades, who are astounded to discover that "the Angrez Sahib and the *Francisis* were enemies at one time" (34) and that "a girl *Jarnel* … drove out *Angrezi* Army" (33): "The maid seemed to become a heroine like the Rani of Jhansi. Lalu felt the blood coursing in his veins with the ambition to follow her on the path of glory" (33). The conflation of an Indian revolutionary heroine who jeopardized the Raj in 1857 and of a French female military leader who drove the British out of France during the Hundred Years' War is humorously used by the narrator not only to conjure up the dreaded image of Indians revolting against the British but also to question gender roles in times of war, whether in Europe or in India, not to mention existing tensions between European allies in 1939.

Still, humour circulates from one group to another and the narrator organizes a network in which no particular group is the regular butt of the joke. Thus, Indian and European characters exchange roles and are often shown as entertaining each other. When Indian soldiers are being sketched by a French woman in Marseilles, the woman becomes "a strange animal" and a true object of entertaining spectacle – no less than the sepoys who discover "the caricatures of each other … as they came to life on paper, happy as children to see the sketches" (19). Another scene describes soldiers delighting themselves by consciously impersonating the Orientalist trope of the snake charmer to amuse a group of French children and French soldiers. If the narrator plays the falsely naïve observer when he concludes "soon there was complete understanding between East and West" (21), he nonetheless regularly comes back to such successful moments of shared humanity triggered by humour. While a waiter working in a French café becomes a "juggler" and a "clown" (35) for the greater pleasure of Lalu and his comrades, the Indian sepoys themselves provoke affectionate laughter on the part of French customers who watch them for the first time experiencing the pleasure of several glasses of Cognac (38).

Obviously, humour between Indian soldiers and French and Belgian civilians is not always a way of laughing *at* the other but, more often than not, a way of laughing *with* the other in a game in which positions are not fixed but can be shared and exchanged. Humour is therefore conceived as a sign of recognition between people who begin to share the memory of the same moments – both those that were sad and those that were happy. Several comedic scenes underscore the performative power of humour, which often succeeds in gathering soldiers and civilians. Consider the episode when the young Belgian boy André asks Lalu about the need for Sikhs to "have to roll that beard and the long hair on your heads in funny knots" (177), a question that reminds Lalu of his own scepticism towards what he regards as superstitious rituals. The intellectual proximity Lalu unexpectedly feels with such a young boy both startles and delights him: "'You are my twin brother,' Lalu said to him in the ensuing laughter" (177). But this closeness is itself the sign of a non-conformist mind as we know that, before leaving India, Lalu was banished by his family for having cut his beard and hair, and having shared a meal in a Muslim shop. Like Kirpu, Lalu remains a singular character, ready to find humour in situations in which others would see sacrilege, blasphemy, and sedition.[33] Indeed, the Christmas scene describing a moment of fraternization around shared laughter and cakes between the British, the Germans, and some daring Indian soldiers is abruptly cut short by "three English Sergeants and Lance-Corporals jumping across the parapets and calling to the *Goras* to come back, with vague, angry gestures" (250). Between soldiers from antagonistic armies, proximity through humour and laughter must not be tolerated as it may too explicitly hint at the absurdity of the war – which the novel aptly recalled to its 1939 readers.

CONCLUSION

A running thread throughout *Across the Black Waters*, humour is part and parcel of the writing against the grain strategy employed by a novel willing to address the fact that the existence and memories of the sepoys in the First World War have been ignored. First, humour contributes to the alternative narrative of an Indian novel that appropriates, through comedic passages, what European literature in 1939 had mainly memoralized as "essentially" a Western tragedy. Second, humour contributes to the novel's progressive and humanist dimension in that it is a means of questioning and undermining stereotypical and fabricated memories of the sepoys. In this sense, however, it inscribes itself within the European vein of the anti-heroic novel about the First World War, which focuses on the way this war destroyed the ideals as well as the lives of courageous soldiers. Third, by staging defiant characters whose rebelliousness is located in humour, the novel provides 1939 readers with a metaphorical understanding of a non-violent but effective mode of emancipation on the part of Indian protagonists. Finally, it comes as no surprise that Anand, with his provocative use of humour so notoriously popularized by Salman Rushdie, is often described as having paved the way for the 1980s boom of the Indian novel in English.

NOTES

1 Mulk Raj Anand, *Across the Black Waters* (New Delhi: Orient Paperback, 2000). All quotations from the novel are taken from this edition.
2 "Making Britain: Discover How British Asians Shaped the Nation, 1870–1950," Open University, http://www.open.ac.uk/researchprojects/makingbritain/content/mulk-raj-anand.
3 "Much has been written by Western authors about their respective soldiers' contributions and sacrifices. However, the Indian soldier's contribution has remained by and large ignored. This novel, described by British literary critic Alistair Niven as Mulk Raj Anand's best since 'Untouchable,' fills in the gap to a great extent." See Randeep Wadehra. "The War Novel by Mulk Raj Anand," *Tribune*, 6 August 2000, http://www.tribuneindia.com/2000/20000806/spectrum/books.htm#5.
4 See Rozina Visram, "Soldiers of the Empire in Two World Wars," in *Ayahs, Lascars and Princes: The Story of Indians in Britain 1700–1947* (London: Routledge, [1986] 2015), 120.

5 This part of Lalu Singh's is therefore overlooked in the trilogy.

6 As Santanu Das reminds us, many Indian soldiers were "recruited from non-literate or semi-literate backgrounds." See Santanu Das, *Race, Empire and First World War Writing* (Cambridge: Cambridge University Press, 2011), 7.

7 In the dedication of the novel, Mulk Raj Anand also mentions that he was the son of a Punjabi subedar in the Indian Army (the equivalent of a captain in the British army) in the Dogra Regiment.

8 For more information on the various books written just after the war about Indian soldiers' involvement in the First Word War, see Florian Stadtler, "Historiography 1918-Today (India)," in *International Encyclopedia of the First World War*, http://encyclopedia.1914-1918-online.net/article/historiography_1918-today_india.

9 Santanu Das, "The First World War and the Colour of Memory," *Guardian*, 22 July 2014.

10 On the occasion of the First World War Centenary other books have also underlined the centrality of Anand's novel in a transnational perception of the First World War:
– Das, *Race, Empire and First World War Writing,* which Das conceives as "embed[ding] the experience and memory of the First World War in a more multiracial and international framework" (1).
– Pete Ayrton, *No Man's Land: Writings from a World at War* (London: Serpent's Tail, 2014). Pete Ayrton's anthology, which introduces an excerpt from *Across the Black Waters* (i.e., when the sepoys arrive at Marseilles), clearly states his desire to decentre First World War fiction when he writes that the book "is located not just in France and Belgium, but also on the Italian Front, the Balkan Front, the Eastern Front and the war at sea. We hear new voices from Armenia, India, Africa and New Zealand and Ayrton paints a global mosaic of destruction, suffering and lost innocence."
– John Dennehy, "Anthology Presents New Voices from the First World War," *National*, 13 March 2014, viewed 20 October 2020, https://www.thenationalnews.com/arts-culture/books/anthology-presents-new-voices-from-the-first-world-war-1.303966.

11 "In many ways, the war is both known and totally unstudied. Indeed, the Western Front, as Patrick French wrote in a recent article for *Granta*, is a misnomer: it was a global front – 'Everyone was there.' The British Empire went to war against Germany, not merely 'Great Britain.' The white Dominions – New Zealand, Canada, Australia – made a blood sacrifice to the British Empire. The First World War has a powerful presence in Australian lore – Gallipoli endures as a powerful symbol of Australian

nationhood – as it does in Canada. Lesser known is the role that Indians played in the war – it's rarely noted that one third of the British sector of the Western Front in the winter of 1914/15 was held by Indian troops. Indian soldiers suffered and died by the thousands in the Mesopotamian theatre. By war's end, 74,000 Indian soldiers had perished. Academics have explored such neglected angles, but a non-fiction epic in the popular style about the war and the empire remains to be written." See Dennehy, "Anthology Presents New Voices from the First World War."

12 "The recent global turn in first world war studies and commemorative events is partly propelled by Europe's changing image of itself: we live in multicultural societies. Baroness Warsi, leading the colonial war commemoration in Britain, recently noted: 'Our boys were not just Tommies – they were Tariqs and Tajinders too. They came from many nations and held many different faiths.' Warsi's is an important caveat, as much to far-right parties keen to whiten the war as to ethnic and religious groups who may want to hijack its pluralities." See Santanu Das, "The First World War and the Colour of Memory," *Guardian*, 22 July 2014.

13 In 2000, the New Delhi publishing house Orient Paperbacks still chose a photographic collage showcasing barbed-wire, trenches, and white Tommies engrossed in the reading of a map. Conversely, the First World War centenary edition of *Across the Black Waters*, published by Shalimar Books in October 2014, decided to use photographs of First World War sepoys, foregrounding a triptych of Indian Sikh soldiers staring at the reader or mounting guard, and staging the dignity and military obedience of these men rather than the war-like and potentially heroic context of the first illustration, which completely ignored the protagonists of the novel. This latest publication lies at the heart of the Salt of the Sarkar Project, aiming at reaching a new generation with this marginalized part of the First World War by promoting the reading of Anand's novel and the publication of Indian soldiers' letters (http://saltsarkar.co.uk/). The project also envisages the creation of a new theatrical adaptation of the novel in 2018, which had already been brought to the stage in 1998 by the Mán Melá Theatre Company at the Hackney Empire for the eightieth anniversary of the First World War.

14 Jakub Kazecki, *Laughter in the Trenches: Humour and Front Experience in German First World War* (Newcastle upon Tyne: Cambridge Scholar Publishing, 2012), 2.

15 "The way that we discuss this war in the twenty-first century is solemn, respectful, and most certainly humourless." See Karen Randall, "Preface," *Humor, Entertainment, and Popular Culture during World War I*, ed.

C. Tholas-Disset and K. Ritzenhoff (London: Palgrave Macmillan, 2015), xii. See also: "It has been conceived as a way to recreate an aspect of the war that remains little known, because the conflict is usually synonymous with devastation and trauma" (Tholas-Disset and Ritzenhoff, *Humor, Entertainment, and Popular Culture during World War I*, 5).

16 As already pointed out by C.J. George: "The much required tragic relief in the novel is provided by Uncle Kirpu's sullen humor and citation of pain-petalled proverbs." See C.J. George, *Mulk Rak Anand, His Art and Concerns: A Study of His Non-Autobiographical Novels* (New Delhi: Atlantic Publishers, 1994), 96.

17 Such a theory is described as the "belief that some groups of men are culturally or biologically predisposed to the arts of war." See Heather Streets, *Martial Races: The Military, Race and Masculinity in British Imperial Culture, 1857–1914* (Manchester: Manchester University Press, 2004), 1.

18 George Fletcher MacMunn, *The Martial Races of India* (London: Low, Marston and Co., 1933).

19 William H. Martineau (1972), "A Model of the Social Functions of Humor," quoted in Kazecki, *Laughter in the Trenches*, 23–4.

20 "And they waited again in suspense, relieved only by the uncanny sight of Daddy Dhanoo dozing where he had sat mumbling the name of God to the rosary of his heart.
'Woe to the enemy if they see such warriors as Dhanoo,' said Lalu. …
'*Ooh, hain…*' Daddy Dhanoo said, opening his heavy-lidded eyes and smiling apologetically.
'Nothing, nothing, there is no talk,' Kirpu assured him. 'You go to sleep; only don't snore, or you will frighten the enemy'" (114).

21 "'*Yus, yus*, interrupted Subah falling into English as the ability to speak the language of the Sahibs was reputed to increase his sense of importance among the Indian ranks. My father, the Subedar Major Sahib, was a special orderly officer to Lord Kitchener Sahib…'
'*Ohe*, what can you remember of that?' said Uncle Kirpu crudely demolishing the bluff that Subah was seeking to impose on them. 'You were only a kid, as big as my little finger, running about the regiment naked with your little *looli*, and the crows used to peck at your bottom…'
Everyone laughed at this. And, as Uncle Kirpu was well-known for his caustic wit and, as one of the first sepoys to join the regiment, was allowed to say anything to anyone, whether British officer, Indian officer, N.C.O. Or sepoy. Subah joined in the laughter, though a pale blush of embarrassment covered his red face, and his eyes glanced furtively from side to side" (25–6).

22 I owe this comparison with the character from Shakespeare's *Henry IV* to C.J. George: "Uncle Kirpu's 'Falsataffism' towards honour and war appears many times in the novel." See George, *Mulk Rak Anand*, 98.

23 "'Slowly, slowly, gentleman, Franceville, is not running away,' Kirpu said, blinking his mischievous eyes, and shaking his sly, weather-beaten face in a mockery of Lalu's haste."
'Being a man of many campaigns, you feel there is nothing new,' Lalu teased.
'I don't feel peevish and shy as a virgin, as you do, son,' said Uncle Kirpu and patted Lalu on the back affectionately" (10).

24 About female virgins in the Muslim paradise (100); about the Sikh tuft knot as an apt device to know where one's head is located (111).

25 "'He is an officer and is allowed to drink eight bowls of milk and to eat sixteen loaves of bread,' mocked Kirpu. 'There would be no grief at his death, for hunger would flee.' 'Don't cut such inauspicious jokes,' said Dhayan Singh from farther up" (101).

26 "The laughter-provoking repartee between Lalu and Uncle Kirpu reduces the scathing bitterness in the novel." George, *Mulk Raj Anand*, 96.

27 "Behind his sardonic humor, Mulk Raj Anand is eloquently protesting against the army code which provides a lot of comfort to the officers at the expense of the ordinary soldiers … [T]hrough the words of his mouthpiece Uncle Kirpu [Anand] criticizes the slavery to which the sepoys are submitted by the officers, without any official or moral sanction." See George, *Mulk Rak Anand*, 97.

28 Taking up Mary Douglas's definition of a joke as "an opportunity for realizing that an accepted pattern has no necessity" (*Implicit Meanings: Essays in Anthropology* [London: Routledge, 1975], 96), I here point to the challenging nature of humour whose incongruity she describes as an "anti-rite" function showing "the sheer contingency or arbitrariness of the social rites in which we engage." See Simon Crichley, "Humour as Practically Enacted Theory, or, Why Should Critics Tell More Jokes," in *Humour, Work and Organization*, ed. Robert Westwood and Carl Rhodes (London: Routledge, 2007), 24.

29 Florian Stadtler accounts for the viewpoint offered by Shyam Narain Saxena in *Role of the Indian Army in the First World War* (New Delhi: Bhavna Prakashan, 1987). See Stadtler, "Historiography 1918–Today (India)."

30 The third novel of the trilogy illustrates how Lalu uses his various experiences of the war in France and Belgium to try and bring changes to his own village.

31 "Indeed, *Across the Black Waters* was a timely reminder to the British public of the wider Indian contributions made to the war effort, when India's support was called on once again in Britain's hour of need." See Stadtler, "Historiography 1918–Today (India)."

32 This reorienting of the novel's viewpoint on the First World War and the way European literature frequently tended to represent it starts with the arrival of the sepoys in the south of France before ending in the northern trenches, which succeeds in reorienting the British and French focus on the Somme and its surroundings. This redirection is itself mirrored by the inverted trope of the colonial journey "into darkness," exoticism, and conquest with the crossing of the "black waters" from India to Europe.

33 See the description of the Tommie defecating like the Indians (89). See also the humour that is shared by a Christian priest, Tommies, and Lalu but that excludes other sepoys who cannot understand English (252–3).

8

Picturing Control: The Visual Representation of the Kenya Emergency

Keith Bell

In 1998, I was working through the photograph collection of the RCMP Historical Collections Unit, "Depot" Division, in Regina, Saskatchewan, in search of images for an exhibition I was co-curating on representations of western Canadian colonization called "Plain Truth."[1] Among a pile of old police accident photographs, intended to introduce new recruits to the gruesome side of policing, I noticed an album of the kind once used for family photos – black cover and interior pages, images mounted on photo-corners, and captions written in white ink. My family has a stack of similar volumes dating back to the early years of the last century. Out of curiosity, I opened the album, only to find that there certainly were families and family homes here, but most of the people in the images were dead. For these were, for the most part, scene-of-crime documents.

The RCMP Heritage Centre, or Mountie Museum, in Regina is in the middle of the Canadian Prairies. But I knew at once that the photos in the album had come from elsewhere. In fact, they had come out of the earliest days of my own childhood, in Nairobi, Kenya. I could immediately place the mutilated corpses and ransacked farmhouses pictured in the album as white settler victims of the Mau Mau uprising in Kenya, some of the total of thirty-two people of European ancestry who were killed there between 1952 and 1955. I was born in 1950, so I was only two when the State of Emergency was declared. We lived on the edge of Nairobi, and my memories of the time recall a normal, peaceful colonial childhood – though there was the time in the living room of our house on what was then called Caledonian Road when my father allowed me, aged about four, to hold an unloaded automatic pistol. It had been loaned to him by my uncle, who had brought it back with him as a souvenir from the war against the Japanese in Malaya, where he had served as an officer in the King's African Rifles. My father carried the gun on patrols with the urban Home Guard, who enforced the nighttime curfew that confined African domestic workers to the servants' quarters in our whites-only residential neighborhood.

During the daytime, I could also watch the Royal Air Force Lincoln bombers flying overhead as they returned from their missions against the Mau Mau insurgents in the forests of the Aberdare Hills and the slopes of Mount Kenya. Later, when we built a new house in a different part of Nairobi, it was equipped with a system of internal doors that could be locked to block off access to the bedrooms upstairs, giving an opportunity to raise the alarm in the event of attack. It was not until I was well into my teens and no longer living in Kenya that I was able to overcome my fear of going outside in the dark, the time when the insurgents had usually chosen to carry out their raids.

None of this tension was evident in my family's photo albums. These were kept by my mother, a trained photographer and a careful recorder of our family life in Kenya, whom one might have expected to include some visual reference to the events of the Emergency. There was, after all, at least one skirmish between the security forces and insurgents in the bush-choked valley overlooked by our house on Caledonian Road. Not surprisingly, there had also been a raid on my Aunt Mary Grafton's farm in Limuru, the centre of insurgent activity in the early 1950s. But there is no overt trace of discomfort in the albums, where the 1950s appear to have been uniformly tranquil, as reflected in domestic scenes of colonial settler life. We see my sister and me playing in the driveway of our house, or (in my sister's album, this time) my mother and me sitting on the front lawn, taking in a grandstand view of the very valley where gunfire had been exchanged not long before and where, according to my father, bullets had struck a neighbour's house.

Although the Emergency was not formally depicted in my family's photos, it remains an invisible presence, an ominous undercurrent of memory. At the time these photos were taken, military activities and skirmishes were discussed daily in the papers, on the radio news, and in cinema newsreels, and rumours constantly circulated around the small, tightly knit settler community. For instance, in a letter dated

21 April 1954, my grandmother passed on to my mother information received from my Aunt Mary about "another sweep" of the countryside. "Sweeps" were police and army operations in which anyone suspected of Mau Mau activity could be detained and taken to a holding camp for interrogation and possible trial.

Given this traumatic context, it is not surprising that, when I happened upon the album in the Mountie Museum, so far from my birthplace – unexpected and out of context – it came as a shock. Who had assembled these photographs? How had they ended up in the archives of the Royal Canadian Mounted Police, especially as they were clearly official Kenya Police images of a vicious colonial war in which Canada had never played any part? What purpose had they been intended to serve? A number of years later, with these new questions in mind, I decided to look at the official photographic representations of the Kenya Emergency of 1952–60, a subject that became my sabbatical research topic during 2014–15.

POLITICAL AND MILITARY BACKGROUND

In 1952, the governor of Kenya, Sir Evelyn Baring, declared an official State of Emergency in the Crown Colony in response to the rapid spread of the insurgency known as Mau Mau, staged primarily by members of the majority Kikuyu, Meru, and Embu ethnic groups who mainly occupied Central Province. The conflict was inspired by resistance to the British colonial presence, including the appropriation of land, and the authorities responded with all the means at their command. The Emergency was finally declared over in 1960, by which time the insurgents had been crushed in a brutal campaign that resulted in many thousands of deaths, including an estimated twelve thousand Mau Mau fighters killed in skirmishes and ambushes, and the execution of the extraordinary number of 1,090 insurgents for various "crimes," ranging from murder to the possession of weapons.[2]

At the beginning of the Emergency, the Colonial Government had at its disposal several battalions of the King's African Rifles, which consisted of African soldiers commanded by white officers, and the Kenya Regiment, consisting of European regulars and reservists. As alarm grew among the settler community, these modest forces were quickly reinforced by at least one regular British Army battalion and units of the Royal Air Force. In addition, the government frequently called upon the Kenya Police, the Kenya Police Reserve, and the less well trained and equipped tribal police forces, the latter having been first raised in the 1920s to support the authority of the chiefs in the tribal areas. Subsequently, a Home Guard was set up, based on a system developed during the Malayan Emergency (1947–60). Home Guard posts, staffed by villagers – initially ill-trained but increasingly better organized and armed – were established in the areas affected by the insurgency, namely, Central Province. And a program of "villagization," modelled on the Briggs Plan in Malaya, forced the Kikuyu population, over a million in all, into 804 rigidly controlled, fortified settlements, thereby separating Mau Mau sympathizers from the increasingly isolated insurgents in the forests of the Aberdare Range and Mount Kenya. However, success in overcoming the insurgency came at a price for, as Caroline Elkins and others have demonstrated, these villages were effectively concentration camps where conditions were dire. Not only were they overcrowded and unhealthy, they were also largely cut off from the scattered smallholdings that normally provided subsistence for the villagers.[3]

From the beginning of the Emergency, the security forces reacted with extraordinary violence, especially the Kenya Police Reserve and elements of the Kenya Regiment and King's African Rifles. In addition, settlers engaged in their own "informal" anti-terrorist activities and application of summary justice. Furthermore, the internal battle between the "loyalist" (i.e., pro-government) Kikuyu and the Mau Mau insurgents led to torture and killings on a large scale. For example, the Lari Massacre, in which a hundred loyalists (including Chief Luke) were killed, also resulted in the deaths of some five hundred or more suspected Mau Mau and Mau Mau sympathizers, who were killed more or less indiscriminately by security forces and Home Guard units in the retaliatory massacre that took place immediately afterwards.[4]

In an attempt to properly organize the security forces and rein in the irregular activities of the settlers, the British government appointed General Sir George Erskine as general officer commanding, East Africa Command, to take operational control of all anti-terrorist operations, including the police, the African loyalist forces, and the Home Guard. Erskine immediately issued an order to the security forces to stop "'beating up' on the inhabitants of this country just because they are the inhabitants." In addition, he issued an order requiring each individual under his command to "stamp at once on any conduct which he would be ashamed to see used against his own people."[5] While Erskine achieved some improvements, he never completely eradicated the unofficial brutality.

Erskine's task was made all the more difficult because Governor Baring, caught between the powerful settlers' organizations and the Colonial Office, was unwilling to act forcefully. "Erskine tends to be too decisive and HE [Governor Baring] not decisive enough," reported the Hon. H.C.P.J. Fraser, MP, who had been sent to Kenya to report to the Colonial Office in October, 1953.[6] Similar difficulties were soon encountered by the new commissioner of police, Arthur Young (formerly commissioner of the City of London Police, and with experience in Malaya), who was appointed on a two-year secondment from Britain in March 1954 to knock some discipline into the police. Unfortunately, Commissioner Young found himself up against a colonial mentality rooted in a huge sense of settler entitlement and a firmly entrenched belief in white racial superiority. This unpalatable status quo was reinforced by the district administrators and rural magistrates – white settlers with deep roots in the farming communities – who consistently frustrated Young's attempts to convict government officials who committed atrocities against Africans. Frustrated by what David Anderson calls "a culture of impunity," in which the police forces acted without any fear of proper control, Young resigned in December 1954, just ten months into his term.[7]

MESSAGE MANAGEMENT

From the beginning of the Emergency, both the authorities in Kenya and the Colonial Office in London recognized the need to exert as much control as possible over the news media. It was not difficult to paint the insurgents as primitive and "subhuman," as an MP in the British House of Commons described them as late as 1959. Oathing ceremonies and the murder of settlers in their homes, as well as hundreds of killings in the Kikuyu community, could be spun to suggest that the colonial side held the moral high ground. This strategy could be employed even when the authorities were unambiguously in the wrong. Thus, following the Hola Camp massacre of 1959, in which eleven detainees were beaten to death by camp guards for refusing to engage in illegal forced labour (murders for which no one was ever prosecuted), Sir Charles Markham, a leading member of the Kenya Legislative Council, declared in the *Times* that the detainees in the camps were "hard core Mau Mau ... beyond the pale and beyond redemption" and "the very depths of humanity." As political scientist Harold Lasswell once observed, in modern nations, "every war must appear to be a war of defense against a menacing, murderous aggressor. There must be no ambiguity about who the public is to hate."[8]

All the while, the authorities sought to present life in Kenya as carrying on normally, with the security forces steadily gaining command of the situation. A department of information had existed in Nairobi since at least 1948, and this office was now expanded, with a branch in London to generate information and images favourable to the administration. The photography section was enlarged by an additional European photographer and two Asian darkroom technicians to answer the increased demand for images for the press office and for use overseas. The press office worked hard to place photographs and stories internationally. Two examples from Ministry of Defence files give an indication of the work of the Kenya and London offices:

> *Overseas comment on Kenya*
> A useful counter to the day-to-day reports of Mau Mau depravity in Kenya was a well-illustrated article, supplied by the press office on a new building development in Nairobi. April 1955, *Journal of Industrial Architecture,* Johannesburg.
>
> An article by the Press Office, Department of Information about the new European Settlement Board farms at Mau Narok appeared in the *Sphere* Jan 21st 1956. The article and photographs covered three full pages.[9]

The British had made extensive use of official government photography during the Second World War, controlling the representation of the conflict through a combination of censorship and images produced by specially organized units attached to the military. These photographs were then fed to the British press and distributed abroad with the intention of maintaining morale at home and presenting the British and Empire war effort and moral position as in every way superior to those of the Axis powers. One of the most accomplished of these units was No. 5 Army Film and Photographic Unit (AFPTU) commanded by Captain Edward Malindine, which photographed the newly liberated Bergen-Belsen concentration camp in April 1945. Their photographs of the camp and, in particular, the images of Joseph Kramer, camp commandant; Irma Gresse, a notorious guard; and Elizabeth Volkenrath, the head wardress, were particularly effective. These images were widely circulated in the British press, including a number that appeared on the front page of the *Daily Express* on 21 April 1945, with the Kramer image under the headline "The Shackled Monster of Belsen."[10] Not surprisingly, the Kenya and Colonial Office press campaign employed strikingly similar strategies. Like the AFPTU's shots of Joseph Kramer, images of captured Mau Mau leaders such as General China and Dedan Kimathi show these charismatic insurgents stripped of all their power, placed in humiliating positions, and surrounded by their captors and crowds of curious onlookers.

The Kenya Information Office in London was located in East Africa House in Trafalgar Square and was mainly intended to manage publicity within the United Kingdom. The office was run by Granville Roberts, who, as Kenya public relations officer, played an important part in organizing the political spin for Governor Baring and the Kenya administration. Roberts was a former *East African Standard* reporter with an intimate knowledge of the situation in Kenya – perhaps too intimate, as C.Y. Carstairs, reviewing the London office in 1952, commented, "he suffers from a strong 'settler' bias, of which he is conscious and he does his best to counteract."[11] Roberts was the author of a small booklet entitled *The Mau Mau in Kenya*, published in 1954 and priced at two shillings.[12] This publication looks remarkably like the series of small, illustrated books brought out by British information services during the Second World War with titles such as *The Desert Air Force* and *Russian Convoys.*

Roberts's pamphlet combines an informative text – the origins of the Mau Mau, their oathing ceremonies – with numerous illustrations, either police photographs or images produced by the information service photographers. His narrative follows the government line, seeking to identify the Mau Mau as brutal and retrogressive, lauding the loyalist Kikuyu as brave, and identifying murdered Christian Kenyans as martyrs. He approvingly quotes the conclusion of a parliamentary delegation to Kenya that had declared that "Mau Mau intentionally and deliberately seeks to lead the Africans of Kenya back to the bush and savagery and not forward into progress." Progress was only to be achieved under British administration, where the economic possibilities of the country might be fully exploited. Roberts declares, "It is the intention of all right thinking men in Kenya that this menace to the future of the country shall be overcome as quickly as possible."[13]

The numerous photographs that accompanied the text reinforced this interpretation of events. He was careful to illustrate the cooperation of loyalist Kikuyu with the administration and the efforts of the security services to protect loyalists from attack. "Successes," such as trials of suspects and no fewer than three images of the capture of General China, emphasized the progress of the counterterrorism campaign. In addition, Roberts was unsparing in exposing the brutal effects of the Mau Mau attacks, in particular the assault on Lari village, which he presents as an example of the "ferocious savagery ... typical of the Mau Mau."[14] A photograph of two dead children takes up half of the booklet's front cover, and scenes of the mutilated bodies of African people and farm animals fill six of the sixteen pages of illustrations.

It is hard to imagine who would want to buy such a publication, and it is not clear where the booklets were displayed for sale. The publisher, Hutchinson, was a major company with branches in cities including Melbourne, Sydney, Auckland, Bombay, Cape Town, New York, and Toronto, so it is possible that distribution was spread over a wide area. Concern for the potential readership may, however, account for the absence of images of murdered white settlers, who, in addition, did not fit the narrative of increasing military successes. The war is largely characterized as an internal dispute between Kikuyu factions, with the security forces working to defeat the insurrection and restore peace.

Despite Granville Roberts's best efforts, government attempts to control the photographic representation of the insurgency were not always entirely successful. For example, in early November 1954, an *East African Standard* photographer secured a photograph of the recently captured Mau Mau leader General Kaleba at the airport. The paper published the photo on 9 November. Kaleba and a small "gang" had attacked the Leakey farm in North Nyeri in October, killing members of this prominent family in unusually grotesque ways. This was a particularly sensitive event, especially among the settler community, and H. Dent, secretary of defence, wrote to the commissioner of police in Nairobi, remarking on the "unfavourable comment ... attracted" by the publication of the photograph and insisting "that in the future you will take steps to prevent such photographs being taken."[15] The commissioner replied that the photographer had arrived "unnoticed" and that police action to "destroy the camera [did he mean film?] would have been undesirable and illegal in the circumstances."[16] Illegal or not, one month later, on 9 December, we find the commissioner issuing a directive to his regional commissioners of police stating that "publicity was given to this Mau Mau leader which was undesirable" and instructing them to make sure "everything possible is done to prevent such photographs of terrorists or any other photographs of persons in custody or police/security Force activity being taken. They [the press] should make proper arrangements to obtain the photographs through the Government Information Services, and every step will be taken to prevent undesirable publicity being given to terrorists except through the official channels referred to."[17]

JOHN TIMMERMAN'S ALBUM

This brings us back to the remarkable album in the RCMP museum in Regina. Accession records show that the album, accompanied by some home-made Mau Mau weapons, was donated to the museum by a man named John Timmerman. A member of the RCMP from 1932 to 1940, he had gone on to serve with the Intelligence Corps, Canadian Army, from 1941 to 1945. He then joined the Special Branch of the British War Office and later became a member of the Allied Control Commission in the British zone in Germany. There, he was in charge of reorganizing the Criminal Investigation Department of the West German Police and was West German representative on the International Police Commission. In 1951, he was brought into Kenya to use his Special Branch experience in reorganizing the Criminal Investigation Department of the Kenya Police and was subsequently made senior assistant commissioner of the Kenya Police in Nairobi.[18] After leaving the colony in 1955, he returned to Canada, where he led security and intelligence liaison at the Department of External Affairs. In the 1970s, he ended his impressive career as the Canadian consul general in Chicago.

In 1954, however, Timmerman was stationed in Nairobi. That year, he returned home to Canada on a three-month leave and gave extensive interviews to the *Ottawa Journal* ("Foremost Battler against Mau Maus Visiting Ottawa")[19] and to Bernard Dufresne of the *Ottawa Citizen* ("Fights Mau Mau: Terror Shadows His Kenya Beat").[20] During the *Citizen* interview, he showed Dufresne his collection of Mau Mau firearms as well as his "scrapbook" containing, in the reporter's words, "photographs of Europeans and Africans killed by Mau Mau." Using "local boys" like Timmerman to spread the anti-terrorist line was a strategy of the Kenya Information Service and the Colonial Office,[21] and it is probable that Timmerman was taking time out from his leave to make the case for the suppression of the Mau Mau at a period when objections to the brutality of the campaign were beginning to surface in public.

Timmerman's strategy for presenting the war followed the usual approach favoured by the authorities, depicting the Mau Mau as vicious, regressive, and primitive. "You have no idea how evil and ruthless the Mau Mau society is," he informed Dufresne. "It takes a lot of training to keep from losing your head when you see the body of a child mutilated by these madmen," he continued: "it could be one of my own boys." Dufresne's article was subsequently circulated to other newspapers throughout Canada. Later on during the visit, after, remarkably, showing the unpleasant contents of the album to members of his family in Winnipeg, Timmerman left the guns and album with the RCMP assistant commissioner, S. Bullard, OC "D" Division, to forward to the Depot Division Museum in Regina for inclusion in its collection.[22] There, they were exhibited in a showcase until October 1963, when the album and weapons were removed at the request of A S. Band, the museum curator, who felt that the photographs in the album were "of such a beastly form that they cannot be exposed to public view." Band was also clearly anxious that the album might be associated by visitors with the RCMP, stating that it "is not of interest or in any manner concerned with our force."[23] As Kenya achieved independence in December of that year, it is also possible that it was considered inappropriate at the time to continue to show the country in such an unfortunate light.

Timmerman's apparent charm and his attractive family, together with his disturbing photo album, clearly made him a convincing spokesperson for the British campaign against the Kenya insurgency. However, his popularity did not extend to many non-Europeans in Kenya. Among his detractors was Dudley Thompson, a Jamaican-born, Pan-Africanist lawyer who, in an early attempt to forestall the Mau Mau uprising, had gone to Timmerman with a proposal for a meeting between the colonial government and representative African politicians. According to Thompson, Timmerman turned the idea down cold, saying that the administration was "not going to sit down and talk with murderers, cattle thieves and barbarians." Instead, Timmerman said, "We are going to teach them a lesson that they will never forget in this country." Profoundly disgusted, Thompson warned Timmerman that he had missed a vital, last minute opportunity to avoid a disaster.[24] (When disaster did strike, Thompson would go on to assemble the international legal team that defended Jomo Kenyatta against trumped up charges of being a member of the Mau Mau.)

Later comments place Timmerman in an even more unfavourable light. The prominent Kenyan historian Bethwell Allan Ogot is reported to have been scathing in his assessment of the man. Talking about police beatings of suspects, Ogot is said to have commented that this was "rampant especially in Nairobi where Mr John Timmerman, the notorious CID Chief (the Himmler of Kenya as he was called) … presided over the torture chambers," along with his henchman, G. Heine.[25] For his part, Thompson described Heine as "particularly evil" and noted his dismissal "in disgrace from the [police] service."[26]

There is no indication who was responsible for compiling the album Timmerman brought with him to Canada: it is not signed and the majority of the images are police crime-scene photos. However, many of these photographs were acquired, at some stage, by the Kenya Information Service, so perhaps that office was involved. In any case, the images were clearly selected to reinforce what Timmerman had to say to his Canadian press interviewers about the "bestial" Mau Mau and their brutal assaults on the "peaceful" white settlers. In addition, the events depicted probably show some of the cases that he would have overseen in his role as assistant commissioner.

Timmerman's album provides an overview of the two years following the declaration of a state of emergency in 1952. The first page consists of a kind of "rogues gallery," made up of head-and-shoulders photographs of five of the purported "leaders of the Mau Mau," arranged with the alleged kingpin, Jomo Kenyatta, in the middle, cut out in three-quarter view (see figure 8.1). His portrait is surrounded by mug shots of four other men – Bildad Kaggia, Kung'u Karumba, Fred Kubai, and Achieng Oneko – as smaller images in the corners of the page. (For some unknown reason, the sixth man accused, Paul Ngei, is not pictured.) Bizarrely, this stylized format mimics

8.1 Opening page of John Timmerman's album, 1954.

the layout of many early photograph albums, in which senior family members are arranged in a similar pattern on the opening page.

The next section of Timmerman's album features typed accounts of Mau Mau oathing ceremonies, based on interrogations of suspects and informers, and accompanied by photographs of the items used in the rituals. This presentation is clearly intended to establish the debased nature of Mau Mau practices, and it is clear from Dufresne's article that he was suitably alarmed and disgusted by what he heard and read. But the real shocker came next, in layouts of photographs from murder sites, featuring the bodies of dead white settlers, sometimes accompanied by those of their servants, together with interior and exterior views of their houses. The photographs of the deceased are particularly disturbing because the victims had mostly been hacked to death with "pangas," bush knives ordinarily used for clearing land and cutting firewood. These were the photographs that Dufresne mentioned in his article, calling them "pictures to prove his [Timmerman's] point – pictures not fit for general publication."

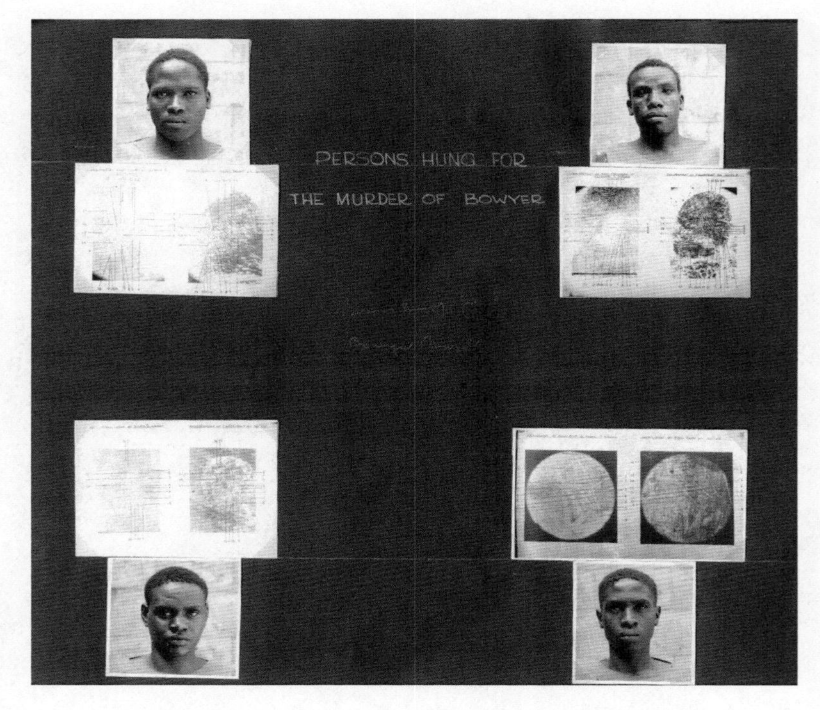

8.2 "Persons Hung for the Murder of Bowyer," from John Timmerman's album, 1954.

As if to eliminate any possibility of misinterpretation, the same point was driven home by extensive typed accounts of the attacks, also found in the album, which provide information about the subsequent arrest, trial, and sentencing of the attackers. For example, in the case of the particularly notorious killing of the Ruck family – Roger Ruck, Dr Esme Ruck, their six-year-old son, and an unnamed Kikuyu gardener – the judge's sentencing comments are also included. "The bestiality and savagery of the murders," the text reads, "beggars the power of words." Together with Timmerman's comments, this statement suggests the degree of anger and hatred that existed among many settlers, including the white-run police and judiciary, against large parts of the Kikuyu population at the height of the Emergency in 1953.

When I first came across the album in Regina, the images that most disturbed me were the ones that showed four-year-old Andrew

8.3 "Captured Photographs of Gangs in Forests," from John Timmerman's album, 1954.

John Stevens, lying dead, virtually beheaded, beside his tricycle in the driveway of his home, where he was killed on 3 April 1954. I, too, rode my tricycle outside our house; Andrew Stevens was the same age as me. And yet, in the same way that our family album showed no overt evidence of the Emergency, I recall no sense of imminent threat. My parents must have shielded me, as best they could, from panic as they carried on carrying on.

Accompanying the scenes of the European murders, Timmerman's album also contains images of some of the men convicted of the crimes. One page, headed "Fingerprints of Persons Hung for Ruck Murders," simply displays two photographs of fingerprints, implying the guilt of the invisible and, by now, dead individuals sentenced for the killings. Another page, "Persons Hung for the Murder of Bowyer," shows four head-and-shoulders shots of the convicted men, each accompanied by photographs of their police fingerprint sheets (see figure 8.2). Their faces are impassive, yet we know that, here again, we are gazing upon the dead. Their executions, anticipated by the photographs, are confirmed by the captions as having subsequently taken place.

The final and, in some ways, most surprising section of the album presents photographs of groups of Mau Mau fighters, clearly taken at their own request, in which they look relatively relaxed and confident (see figure 8.3).

Throughout the Emergency, the colonial authorities had been able to maintain a commanding control of the visual narrative. In their photos, the military and police are shown performing the roles of captors, protectors, and fighters. Meanwhile, the white and loyalist African populations are presented as victims of a criminal outrage and, finally, as the beneficiaries of British rule, albeit for a brief period before Kenyan independence. The insurgents, on the other hand, were severely constrained in their ability to represent their cause. There were few opportunities for them to show themselves to the world since photographic equipment, like modern weapons, was in short supply. This situation became worse as the insurgents were increasingly driven back into the forests of Mount Kenya and the Aberdare Range, but from time to time someone brought in a camera, and the insurgents posed for a group photograph. Most of these images probably disappeared, but some survived, often seized by the security forces when insurgents were captured and their camps raided. Some of these images made their way into Timmerman's album, with the subjects' role now reversed from "freedom fighters" to objects of identification and pursuit. Several of the men and women pictured are marked with letters in white ink, suggesting that the photographs had provided the police with a useful means of identifying the insurgents.

Thus, much as my family's private albums convey a sense of normalcy in a time of violence and distress, so the authorities used photography to confirm and extend their control of the upheaval. Whether in the work of the Kenya Information Service or in John Timmerman's album, the message is always clear. The bad guys are "savage and bestial," and the good guys are in charge. In this carefully crafted telling, the insurgents are undeserving of compassion or justice, while the loyalist Africans are always brave, and white settlers are seen continuing the necessary work of civilization, despite the constant threat to their safety. It would take half a century before this white supremacist interpretation of the conflict was shaken through the unflinching scholarship of Caroline Elkins, David Anderson, and others in overtaking the narratives and cover-ups by which the British Colonial Office and the colonial administration in Kenya presented the situation during the period of the Emergency and beyond. Meanwhile, my own encounter with Timmerman's album, together with subsequent interviews and archival research, has stirred specifically personal memories. Much more significantly, for the first time, Kikuyu survivors of British torture are now having their own memories accepted as truth in UK courts, thanks to a successful suit for compensation from the British government in 2013 and ongoing legal action.

ACKNOWLEDGMENTS

The author is deeply grateful to the following people: Jodi Ann Eskritt, RCMP Historical Collections Unit, Regina, Saskatchewan; the staffs of the Kenya National Archives, Nairobi, Kenya, and the National Archives, Kew, London, United Kingdom; Graham Ward, Bath, United Kingdom; Sheena and Henry Rolph, Norwich, United Kingdom; and Candace Savage, Saskatoon, Saskatchewan.

NOTES

1 Dan Ring, Keith Bell, and Sheila Petty, *Plain Truth* (Saskatoon: Mendel Art Gallery, 1998).
2 See two excellent accounts of the Kenya Emergency: Caroline Elkins, *Imperial Reckoning: The Untold Story of Britain's Gulag in Kenya* (New York: Henry Holt, 2005); and David Anderson, *Histories of the Hanged: Britain's Dirty War in Kenya and the End of Empire* (London: Weidenfeld and Nicolson, 2005).
3 Elkins, *Imperial Reckoning*, 233–74.
4 Anderson, *Histories of the Hanged*, 54–61.
5 General George Erskine, 23 June 1953, as quoted in Anderson, *Histories of the Hanged*, 259.
6 Huw Bennett, *Fighting the Mau Mau: The British Army and Counter-Insurgency in the Kenya Emergency* (Cambridge: Cambridge University Press 2013) 46.
7 David Anderson, as quoted in Bernard Porter, "How Did They Get Away With It?" *London Review of Books* 27, no. 5 (2005): 3–6.

8 Harold Lasswell, *Propaganda Technique in the World War* (London: Kegan Paul, Trench and Trübner, 1927) 82.
9 MOD, file no. POL 42, "Overseas Press Comment on Kenya, 1954–56," AH/13/216, Kenya National Archives.
10 Paul Lowe, "Picturing the Perpetrator," in *Picturing Atrocity: Photography in Crisis*, ed. Geoffrey Batchen, Mick Gidley, Nancy Kay Miller, and Jay Prosser (London: Reaktion, 2012), 189–208.
11 Memo on "Public Relations – Kenya," prepared by C.Y. Carstairs for Lyttelton, 23 October 1952, CO 1027/40, National Archives (London).
12 Granville Roberts, *The Mau Mau in Kenya* (London: Hutchinson, 1954).
13 Ibid., 8.
14 Ibid., 44.
15 H. Dent, Secretary of Defence, to Commissioner of Police, Nairobi, memo on "Undesirable Photography," 12 November 1954, AH/13/220, Kenya National Archives.
16 Commissioner of Police to Secretary of Defence, memo, 26 November 1954, AH/13/220, Kenya National Archives.
17 G. Gribble for Commissioner of Police to Regional Commissioners of Police, memo, 9 December 1954, AH/13/220, Kenya National Archives.
18 RCMP Historical Collections Unit, file G838–8.
19 "Foremost Battler against Mau Maus Visiting Ottawa," *Ottawa Journal*, 24 July 1954, 3.
20 Bernard Dufresne, "Fighting Mau Mau: Terror Shadows His Kenya Beat," *Ottawa Citizen*, 26 July 1954.
21 Press Office file on "Police Publicity," April and May, 1955, Kenya National Archives.
22 L.H. Nicholson to Assistant Commissioner S. Bullard, 9 August 1954, RCMP Historical Collections Unit.
23 A.S. Band, Curator, RCMP Post Museum, "Depot" Divn., to Officer Commanding, 3 October 1963, file G. 838–8, RCMP Historical Collections Unit.
24 Dudley Thompson, *From Kingston to Kenya: The Making of a Pan-Africanist Lawyer* (Dover, MA: Majority Press, 1993), 86–7.
25 Bernard Allan Ogot as quoted by Yves Engler, "Canada's Contribution to Mass Murder and Torture in Kenya," https://yvesengler.com/2015/09/17/canadas-contribution-to-mass-murder-and-torture-in-kenya/.
26 Thompson, *From Kingston to Kenya*, 86.

9

The Meaning of the American Civil War in Southern Memory

Stephen J. Whitfield

This chapter explores the gap between national and regional public memory narratives, which influence how Americans build their individual identity. American history was taught in the eleventh grade in the public school that I attended in Jacksonville, Florida, a city just below the Georgia border. We sat in alphabetical order by surname; the student who was placed directly behind me was Terrell Wilson. When the lesson plan reached the topic of the Civil War, Terrell's normal geniality vanished: such was his indignation in denouncing the Union soldiers, whom he called "the Yankees," that I felt his spittle on the back of my neck. The war had ended nearly a century earlier, which – given the brevity of the nation's history – constitutes what Americans would deem *la longue durée*. Yet evidently the conflict had remained alive for my classmate – and perhaps also for the last living Confederate veteran, who died in that same year, 1959.[1]

Memory can be compelling in several ways. I can still recall the vividness with which a white Southern teenager could tap into a residue of outrage at what he remembered from family stories about what General William T. Sherman's army had done in marching through Georgia a century earlier. Those recollections may even have coincided with what a "Johnny Reb" might himself have summoned from the traumatic experience of warfare and invasion. But another way that the invocation of the past merits the special interest to historians is collective memory. It gets subtly entangled in the rest of the past, perhaps so much so that the relationship to the formal study of history can seem ambiguous and even problematic. For example, as late as the 1970s, one historian who was trained at the Sorbonne observed that she was taught how much scholarship required documents. Memory was regarded as the antinomy of history and as so suspect that a slogan continues to resonate for her: "pas d'archives, pas d'histoire."[2]

But this chapter reflects a trend that acknowledges the power of collective memory so that a socially constructed version of the past can affect the present. Indeed, sometimes the claims of remembrance can become too insistent, so that grievances become a scar that does not heal. They can distort and aggravate the conditions of the living generation; they can paralyze it. But the American danger is usually the opposite, which is to succumb to amnesia. The standard corrective is to quote the resonant 1905 dictum of the Harvard philosopher George Santayana: "Those who cannot remember the past are condemned to repeat it," to which the maverick historian James W. Loewen appended another warning: "Those who don't remember the past are condemned to repeat the eleventh grade."[3] And the past that an eleventh grader like Terrell Wilson "recalled" – the Civil War – continues to haunt the national (and not merely the sectional) imagination. The carnage of 1861–65 has bequeathed consequences, which is why it does not vanish into oblivion. Other historical grievances recede and come close to vanishing: the enmity between France and Germany is one impressive historical example. But somehow the consciousness of white Southerners will not let bygones be bygones.

Much of the explanation for this is the body count itself: the conflict extinguished more lives than all other American wars combined. The body count surely explains even more than the conflicts over federal jurisdiction, or over the separation of powers, that inaugurated constitutional dilemmas before and during the ordeal of the Union. Other wars have generated no major debate over nomenclature. But the conflict of 1861–65 is different. For the "fire-eaters" – the most ardent secessionists – of the mid-nineteenth-century South and their successors, it was the War of Northern Aggression, or the War for Southern Independence. For the broad swath of white Southerners who have believed that their region split from the rest of the Union over the issue of states' rights, over a dispute about federalism, it was the War between the States. In 1974, when a visiting scholar from Australia overheard a colleague of mine mentioning what he called the Civil War, the Australian intervened in the conversation by conjecturing: "I take it, sir, that you were not educated in Alabama." To which my colleague quickly retorted: "To the best of my knowledge, sir, *no one* has been educated in Alabama." But even in our own era, the majority of whites living in the South continue to believe that, however the conflict is named, it was fought over states' rights rather than over slavery. That difference was echoed half a century ago, when segregationists claimed that the crisis of civil rights was due to excessive federal authority rather than to the effort to maintain white supremacy.

No other feature of the national experience has proven to be so lingering, so inescapable, and still so divisive, due primarily to the tenacity of Southern white consciousness. This is the theme of my chapter, which is intended to contribute to the politics of memory. As late as the twenty-first century, the seemingly irrepressible conflict over the flying of Confederate flags and the preservation of statues of Confederate military heroes shows how ferociously the debate over the meaning of secession can be conducted.

The distinctiveness of Southern white memory has been a continuous phenomenon. It combines truculence with grievance, tenacity with defensiveness. But when did this sort of consciousness emerge? The historian Carl N. Degler dates it as far back as the crisis over how Missouri would be admitted into the Union. The Missouri Compromise (1820) registered a shock that awakened Jefferson, as he famously stated, "like a fire-bell in the night." He realized that, were slavery prohibited in Missouri, some Southern politicians were already primed to smash the Union. "All Southerners in Congress," Degler noted, "saw slavery as an institution with which they identified their region and one that ought not to be limited in the future." In their insistence that Missouri be admitted as a slave state, representative Southerners began to tie their regional interests to a narrow construction of the Constitution. The doctrine of states' rights was designed to limit the power of the federal government, especially were it to challenge the legitimacy and durability of "the peculiar institution" of slavery.[4]

The Southern states also diverged from others in the Union in intensity of commitment, in willingness to risk the very existence of the republic so that slavery might prevail. They were primed to go to the brink. By the winter of 1860–61, the heavily Democratic region reacted to the loss of a fair election by choosing "the rule-or-ruin mentality" of secession.[5] In his Second Inaugural address, Lincoln explained the distinction between the two sections at the origin of the conflict: "One of them would *make* war rather than let the nation survive and the other would *accept* war rather than let it perish. And the war came." But both sides knew, Lincoln added, that slavery had "somehow" caused the war.[6] The crowd listening to that speech included the Confederate sympathizer John Wilkes Booth, an actor born in the slave state of Maryland. As the military defeat of the South became certain, he feared that the president intended to install "nigger citizenship," a prospect that provoked Booth to join others in the plot to kidnap Lincoln. The plot was soon enlarged into assassination,[7] an "awful event," as the *New York Times* headline called it, that epitomized the lawless violence of the Southern animus against the ideal of equality.

FIN-DE-SIÈCLE

In the aftermath of the war, the myth of the Lost Cause, the need to cultivate and conserve the legendary gallantry of the Confederacy, became central to Southern white identity.[8] The formation of regional remembrance was neither accidental nor inadvertent. It emerged most spectacularly at the end of the nineteenth century and thus coincided with the "revolt of the rednecks" in politics, with the systematic installation of segregation in law, with the spread of an unashamed ideology of white supremacy, and with the frequency with which black men and women were tortured and lynched. These phenomena did not only occur simultaneously: they were related. The romance of the rebellion strengthened the will to impose subservience on blacks. The enforcement of Jim Crow helped to vindicate a cause that was fought to maintain regional autonomy in race relations.

Much of the work of promoting pro-Confederate monuments in the South was female. Miss Rosa Coldfield in Faulkner's *Absalom, Absalom!* (1936) belonged to this sorority, which the United Daughters of the Confederacy spearheaded. Founded in 1894, the UDC restricted its membership to descendants of the Confederate military or civil service.[9] The UDC and assorted ladies' auxiliaries in the cities and hamlets of the postbellum South promoted the Lost Cause so assiduously that even the victorious military came to accept the heroic and honourable mystique of the defeated foes. At least ten United States Army bases are named for Confederate generals – that is, for the commanders of troops whose purpose was to kill or wound American soldiers. Indeed, the Confederacy managed to kill more US military personnel than did the Axis in the Second World War.

The bases include Fort Lee in Virginia (Robert E. Lee); Fort Hood, Texas (John Bell Hood); Fort Benning, Georgia (Henry Benning, a subordinate of Lee's); Fort Gordon, Georgia (John B. Gordon, another of Lee's lieutenants as well as a probable Klansman during Reconstruction); Fort Bragg, North Carolina (Braxton Bragg); Fort Polk, Louisiana (Leonidas Polk, who owned several hundred slaves on his plantation); and Fort Pickett, Virginia (George Pickett). Consider the ironies. In the twenty-first century, blacks constitute about a fifth of the military, and they often serve on bases named for men who owned their ancestors. Those bases are named for generals who – to put it rather euphemistically – did not exactly uphold their oath to defend the United States.[10]

The United Daughters of the Confederacy did more than help alter the landscape, more than shape the *lieux de mémoire*. For example, the historian general of the UDC from 1911 until 1916 wanted to turn back the clock. Mildred Lewis Rutherford insisted upon a Constitutional right of secession, though no such clause can be found in the framing document. It does, however, describe the procedures for amendment, which the Radical Republicans followed in leading the fight to inject egalitarian ideals into the Constitution. Rutherford nevertheless considered the 14th and 15th Amendments illegal, and claimed that they were designed primarily to punish and humiliate the South.[11] This presumably invalidated them. Such recalcitrance and defiance of law did not augur well for sectional reconciliation – except, of course, on the fiercest neo-Confederate terms. But the UDC stance did reinforce the Southern belief that the region was under siege, under assault from the heirs of Abolitionism – a list that eventually included industrialism, unionism, Darwinism, feminism, socialism, communism, intellectualism, atheism, secularism, modernism, and liberalism. In 1925, when defence attorneys from Chicago and New York arrived in Dayton, Tennessee, to represent the biology teacher John T. Scopes, who had presumably taught Darwinism, Judge John Raulston welcomed to his courtroom the "foreign lawyers."[12]

Organizations like the UDC also sought to sniff out heresy, and the bar of what constituted heterodoxy was low. In 1911, a young historian at the University of Florida, Enoch Marvin Banks, quickly discovered the limits of intellectual tolerance when he published an article in the *Independent* exactly half a century after the formation of the Confederacy. Banks acknowledged that the cause of the Civil

War was slavery (the very claim that Confederate vice-president Alexander Stephens himself had made) rather than a political concept like "home rule." That put the South more in the wrong than the North. The incensed UDC and the United Confederate Veterans led the public criticism of the "false and dangerous" views of the Georgia-born Banks. Editorials in Jacksonville's *Florida Times-Union* piled on the hapless historian as well. To deny the Constitutional right of the states to secede, newspaper editor Willis M. Ball thundered, would give "moral consent to the subversion of government." Banks was forced to resign from the all-male, all-white University of Florida, which thus thwarted his "earnest desire," as he wrote in all innocence to the former president of the Gainesville campus, "to make some contribution toward promoting a liberal intellectual life ... in the South."[13] It regarded orthodoxy on the origins of the Civil War to be fully compatible with the scholarly ideal of objectivity. But any other stance, such as challenging the merits of the Lost Cause, was condemned as malice and mendacity. In 1906, the UDC persuaded the Kentucky state legislature to ban stage versions of *Uncle Tom's Cabin*.[14]

Ideological conformity was occasionally contested. The journalist Walter Hines Page urged the UDC to redirect its efforts towards building schools instead of erecting monuments to exalt the Confederate dead. What his native North Carolina needed, he added, were a "few first class funerals" so that the shame of ignorance and provincialism could be expunged. "There is not a man whose residence is in the state who is recognized by the world as an authority on anything," he lamented.[15] Page later served as ambassador to Great Britain in the administration of Woodrow Wilson, the first president since Zachary Taylor to be born below the Mason-Dixon Line. He had of course served as president of Princeton University and then as governor of New Jersey. But he identified as a Southerner. As the region struggled to defend the righteousness of the Lost Cause and to shape the collective memory of a nation that the Civil War had once divided, Wilson joined the culture wars. In 1896 he proclaimed that there was "nothing to apologize for in the past of the South – absolutely nothing to apologize for." That did not stop a Georgia demagogue, ex-congressman Thomas E. Watson, from accusing Wilson of being "ravenously fond of the negro."[16] As president he inaugurated the tradition of sending a wreath on Memorial Day to the Confederate Memorial, which the United Daughters of the Confederacy unveiled in 1914 and which is located in Arlington National Cemetery. (The tradition that Wilson began was amended when President Obama began sending a wreath to the African American Civil War Memorial, located in Washington.)[17] In 1913, when the fiftieth anniversary of the battle of Gettysburg was commemorated there, fifty-three thousand veterans showed up at the reunion, and all were white. At that ceremony the chief speaker was the commander-in-chief, Woodrow Wilson.[18] He bore responsibility that very year for the firing of black employees in the federal government and for the imposition of rigid rules of segregation on those public servants who remained.

Two years later Wilson hosted the first screening of a feature film in the White House, a movie that, when it opened in Los Angeles the following month, was still entitled *The Clansman*. This was the title of the Reverend Thomas Dixon's 1905 novel, dedicated to a former Klansman (the author's uncle). The stage version, which he adapted that same year, may have helped spark the 1906 race riot in Atlanta, where rampaging mobs killed two dozen African Americans. Several municipalities then tried to ban the play as incendiary. Another of the clergyman's works of fiction, *The Leopard's Spots* (1902), inspired the second half of the movie in particular.[19] Dixon had become Wilson's friend while studying political science with him in graduate school at Johns Hopkins University, and he would serve as a state legislator in North Carolina. Obsessed with the Southern past, Dixon felt a compulsion to justify it, and *The Leopard's Spots* and *The Clansman* were intended to rebut *Uncle Tom's Cabin*. (The publisher of these novels, incidentally, was Walter Hines Page.) Because controversy seemed bound to swirl around D.W. Griffith's film, even after he renamed it *The Birth of a Nation*, as the National Association for the Advancement of Colored People (NAACP) was soon calling for censorship, the novelist arranged for an endorsement from the executive mansion. On 15 February 1915, the famous presidential reaction to the film ("it is like writing history with lightning"), is praise that Dixon ascribed to Wilson but that he may not actually have uttered.[20]

In any case, the right question is: *Whose* history? Griffith's version was intended to be epic. He made a point to begin shooting the film on the fourth of July, and the release was timed for the half-century commemoration of the end of the Civil War. *The Birth of a Nation* quotes from Wilson's own *History of the American People* (1902), which attributes the emergence of the Klan to the abuses of Reconstruction. By 1915–16, history did not need to be written with lightning for the Southern perspective on the past to prevail. Much of the rest of the United States was "remembering" the Civil War and its immediate aftermath just as the UDC was. Dixon's fiction and Griffith's film exerted enormous influence in shaping a view of the Civil War as a heroic sacrifice that was independent of the moral issue of slavery (see part 1 of *The Birth of a Nation*). Reconstruction was depicted as a foolish transfer of political power from the self-rule of whites in the South to corrupt, rapacious, and "uppity" blacks (see part 2).[21] On the evening following the screening at the White House, the movie was shown to members of Congress as well as to the justices of the Supreme Court. One attendee, Chief Justice Edward White of Louisiana, was especially impressed, confiding to Dixon: "I was a member of the Klan, sir."[22]

Only two decades earlier the Lumière brothers had publicly projected a movie in Paris, and in that interregnum nothing would prepare audiences for the sweep and power of Griffith's three-hour epic. In 1915, very few viewers would have ever seen a "photoplay" that ran longer than twenty minutes. No audience would ever have heard a musical score specifically written for a movie, performed by a live orchestra. When the Klan rides to the rescue at the climax of *The Birth of a Nation*, orchestras were cued to play Wagner's stirring "Ride of the Valkyries";[23] and when white South and white North join together in unity, the title card hails their "defense of their Aryan birthright." The journalist Walter Lippmann regarded what he had watched on screen as "pernicious." But he doubted if "anyone who has seen the film" can ever suppress the frisson of "seeing those white horsemen."[24] The excitement of those galloping riders also helped inspire the rebirth of the Ku Klux Klan on Stone Mountain, about ten miles north of Atlanta, in 1915. Its members wore the regalia – the white robes and pointed headgear – that mimicked the movie, though members of the first Klan of the Reconstruction era rarely wore white hoods and robes. But those were the costumes in which the modern Klansmen paraded when *The Birth of a Nation* opened in their hometowns, giving the neo-Confederate film free publicity.[25]

Whatever the undeniable cinematic importance of *The Birth of a Nation*, it has remained vexingly irrepressible. It was playing, for example, in a downtown theatre in Little Rock, Arkansas, in 1957, during the riots that disrupted the efforts to achieve desegregation there. The opposition to the entry of nine black teenagers to Central High School required a military occupation, the first time that a Southern city had experienced such a fate since Reconstruction. Silent films were very rarely revived in such communities, especially when such movies were not comedies like Charlie Chaplin's or Buster Keaton's. The revival of *The Birth of a Nation* was no coincidence. A boulevard is named for Thomas Dixon in Shelby, North Carolina, and on that site state troopers arrested Dylann Roof for the murder of nine black churchgoers in Charleston in the summer of 2015. The bloodthirsty legacy of white supremacist views had returned with a vengeance.[26]

INTERWAR

Nothing demonstrated more tellingly the victory of the South in the contest over the past than a textbook: Samuel Eliot Morison and Henry Steele Commager's *The Growth of the American Republic*. Morison, a Harvard-educated patrician, was born and raised in Boston, the city that had been the ground zero of Abolitionism. Commager, an ardent liberal educated at the University of Chicago, had even published a biography of the Abolitionist Theodore Parker in 1936. But the Morrison-Commager textbook, which first appeared in 1930 and then went through several subsequent editions, treated slavery as benign and managed to claim that "Topsy and Tom Sawyer's nigger Jim were nearer to the average childlike, improvident, humorous, prevaricating, and superstitious negro than the unctuous Uncle Tom." The anguish of splitting families apart (as the Confederate general Nathan Bedford Forrest had done as an antebellum slave trader) constituted the aspect of slavery that "most appealed to human sympathy," the co-authors conceded. But, on the other hand, they added, "it was often asserted in defense that negroes had a very slight family attachment." Morison and Commager underscored the positive features of slavery, which might be understood as a transition "between barbarism and civilization." Besides, "Sambo" not only provided the South's economy with labour but also gave the nation's culture so much "rhythm and humor, for instance." Readers of this widely adopted textbook learned that "Sambo" probably "suffered less than any other class in the South from its 'peculiar institution.'"[27] If that were so, however, how could so civilizing and inoffensive a set of arrangements have caused as much carnage as did the Civil War?

A year earlier, an Indiana politician's best-selling version of Reconstruction, Claude G. Bowers's *The Tragic Era: The Revolution after Lincoln* (1929), described as a terrible blunder the postbellum occupation of Union troops to promote black citizenship. The defeated white Southerners, Bowers argued, had paid far too high a price for misguided Northern benevolence. This was the scholarly version of part 2 of *The Birth of a Nation*, and W.E.B. Du Bois sharply disagreed. Yet his *Black Reconstruction* (1935) was so eccentric in seeking to validate the egalitarian policy of the Radical Republicans that the nation's leading journal devoted to history, the *American Historical Review*, did not even bother to review the volume.[28] As early as 1899, Du Bois was expressing his exasperation with the notion that nothing could be done to satisfy the demands for racial equality. "The Negro problem ... [is] not more hopelessly complex than many others have been," he insisted. Even as "poverty [and] crime" among blacks needed to be addressed, he added, "ignorance" and "dislike of the stranger" among whites had to be confronted too. But the problem of the colour line, Du Bois added, was not beyond redress.[29] This view was not shared by the Harvard historian Paul H. Buck, whose own Pulitzer Prize-winning study of Reconstruction and its aftermath, *The Road to Reunion* (1937), glumly concluded that the national experience of the last third of the nineteenth century had shown the race problem to be "basically insoluble." But at least, Buck added, an "imperfect" reconciliation between North and South permitted "a degree of peace." He did not bother to tabulate the price that blacks were expected to pay for the South's "success in resisting an exclusively Northern formulation of Americanism."[30]

POSTWAR DECADES

The postwar crisis of civil rights in the South coincided with the centenary of the Civil War, and the synchrony revealed how closely the effort to maintain racial segregation was entwined with the sustained retrospective interpretation of the war that had crystallized in the late nineteenth century. In 1963, Alabama's governor, George C. Wallace, had ordered the Confederate flag to be flown atop the state capitol in Montgomery (the "Cradle of the Confederacy") to show the ferocity of his aim to block the desegregation of the University of Alabama. The flaunting of the Stars and Bars, which had flown over the Confederate capitol in Montgomery on the day that Lincoln had been inaugurated in 1861, had been intended to demonstrate particular disdain for the attorney general of the United States, Robert Kennedy, who was visiting the state. When he returned to Washington, he is reported to have described Alabama as "another country."[31] In Montgomery, earlier that year, Wallace had delivered a notorious gubernatorial inaugural address. He denounced the "tyranny" of the elected government in Washington, which Robert Kennedy's brother headed, and praised white Southerners as "the greatest people that have ever trod this earth."[32] (Most historians would find this claim dubious and have ranked the ancient Athenians higher.) But the raising of the Confederate battle flag, which the Army of Northern Virginia had adopted after it routed Union forces at First Manassas, was clearly designed to stir hostility towards civil rights and to mobilize passions against the prospect of desegregation. It cannot be accidental that Georgia adopted a new, neo-Confederate flag in 1956 or that the Confederate flag was first unfurled above the state capitol in Columbia, South Carolina, in 1962. Such symbolism thus tracks closely with opposition to desegregation, which is why the NAACP challenged (to no avail) the state sponsorship of the flag in Georgia and Mississippi. Athletic teams, such as those playing for the University of Mississippi, also adopted the Stars and Bars.[33]

The savviest Southern politicians fancied the postwar conflict over Jim Crow as a clash between overreaching federal authority and a heritage of states' rights, a claim that resonated with the political crisis of 1860–61. But a governor like Mississippi's crude Ross Barnett missed his cue and instead revealed the racial fears that animated the rise of massive resistance. In September 1962, three days after the Supreme Court upheld the right of James Meredith to be admitted to the University of Mississippi, Barnett defied anyone to find a "case in history where the Caucasian race has survived social integration," and therefore vowed that his state "will not drink from the cup of genocide." Such attitudes clouded the centenary of the Civil War, a commemoration that was supposed to rally around the ideal of national reconciliation. But when the anniversary took place in Charleston in 1961, the ubiquity of Confederate flags was evident, as though Lee's surrender at Appomattox had never happened. Slavery was unmentioned in the brochure, *Facts about the Civil War*, that the Civil War Centennial Commission published. Even the word "Negro" was omitted. When a black delegate from New Jersey tried to register at the headquarters hotel in Charleston, she was denied accommodations.

President Kennedy was forced to intervene. The compromise he worked out accepted the legality of segregation, and he switched the event to a local naval base, which provoked Southern delegates to secede from the federal celebration; instead, they held their own Confederate States Centennial Conference.[34] A journalist who asked the director of the Centennial Commission how the Emancipation Proclamation would later be honoured was told: "We're not emphasizing Emancipation."[35]

As president of the United States, Kennedy made civil rights, for the first time in the history of the White House, an issued of morality and, thus, repudiated his Democratic predecessor Woodrow Wilson. But as a historian, John F. Kennedy had helped perpetuate a view of Reconstruction that defined it as a "tragic era" that needed to be ended for the sake of national reunification. In 1957, Senator Kennedy had won the Pulitzer Prize for *Profiles in Courage*, which praised eight men who served in the Senate and the House of Representatives, and who defied, according to their own principles and consciences, their constituents – sometimes suffering defeat in the subsequent election. Among them was Lucius Quintus Cincinnatus Lamar, who had drafted Mississippi's ordinance of secession from the Union in 1861. After the war, however, he favoured sectional reconciliation. As a US Congressman, Lamar even praised the late Charles Sumner of Massachusetts, who had been the most radical egalitarian in the Senate. But Lamar's aim was the rapid withdrawal of Union troops and what the author called "the restitution of normal federal-state relations." In writing that "no state suffered more from carpetbag rule than Mississippi," Kennedy evinced no empathy for the freed slaves who had gained civil rights under that rule. He attributed to Lamar "a new kind of Southern statesmanship,"[36] without acknowledging how high a price the region's blacks were about to pay for the abrogation of that carpetbag rule. The historiographical indebtedness to Bowers is obvious. Less than a century later, liberal politicians like Kennedy would have to struggle to regain for Southern blacks some of the rights that Southern whites had systematically violated when they regained power in 1877.

In any case, Lamar was an odd choice to be included in *Profiles in Courage* because he did not pay any electoral price in Mississippi for his presumably brave views. Not only did he become a US senator but he also chaired the Senate Democratic Caucus. He held appointed offices too: secretary of the interior as well as associate justice on the Supreme Court. Far from directly challenging the restoration of white hegemony, Lamar benefited from it. That Senator Kennedy called him the South's "most gifted statesman ... from the close of the Civil War to the turn of the century,"[37] without noting the determination of such politicians to crush the hopes of Emancipation, is an irony worthy of note. In the antebellum period, no state harboured more forthright Abolitionists than Massachusetts; indeed, no other state even came close. Yet one measure of how smoothly the Southern historical sensibility had insinuated itself nationally is the work of John F. Kennedy, a successor to Charles Sumner in the upper chamber of the capitol a century later.

CONTEMPORARY CONFEDERATE MEMORY

Symbolic redress would nevertheless arrive as the new century turned. During Reconstruction, Jonathan J. Wright had served as the first black justice of South Carolina's highest appellate court. There only white justices had their portraits hung – until 1998, when a painting of Wright was finally unveiled. Three years later Georgia hurled some flags in the dust, when the Confederate battle flag was officially jettisoned.[38] In 1978, Ernest N. "Dutch" Morial, an attorney who grew up in a French-speaking home in New Orleans, became its first black mayor. In 1981, Morial tried to remove the Liberty Monument, an obelisk that is dedicated to a white supremacist group that tried to overthrow the city's biracial Reconstruction government. He failed to get rid of what he termed a "public nuisance." Four monuments celebrating the Confederacy were also constructed in New Orleans, including statues of Lee, Confederate president Jefferson Davis, and the Louisiana-born General P.G.T. Beauregard. In 2017, one of Morial's successors in City Hall succeeded in removing all four statues, despite a lawsuit from the Louisiana Landmarks Society and the New Orleans chapter of the Sons of Confederate Veterans.[39]

Though Tennessee was home to President Andrew Jackson, a slaveholder, he has been less represented on the statues, busts, or markers in that state than Nathan Bedford Forrest, who ordered the Fort Pillow massacre in Tennessee in 1864. Because black soldiers were killed even after they had surrendered to rebel forces, Forrest bore responsibility for an atrocity, a war crime. (Even before the Geneva Convention, prisoners of war, in uniform, had some reason to expect that they would not be murdered.) Forrest's postbellum career included pre-eminence in the Ku Klux Klan. He was, according to biographer Jack Hurst, "the Lost Cause's avenging angel, galvanizing a loose collection of boyish secret social clubs into a reactionary instrument of terror still feared today." Had my parents as well as Terrell Wilson's moved into a nearby neighbourhood in Jacksonville, we would have attended a public school named for this grand wizard. The name of Nathan Bedford Forrest High School was discreetly dropped in 2014.[40]

Not until 2015, which was more than four years after President Nicolas Sarkozy unveiled a monument in Paris dedicated to the remembrance of those held in bondage in the French colonies,[41] was the official flag of South Carolina removed and placed in the Confederate Relic Room and Military Museum. The Union had now decisively won the battle of ideas that was waged at so high a price over a century earlier. A stigma was finally and officially attached to the Stars and Bars. It "has always represented more than just ancestral pride," President Obama declared in eulogizing those whom Dylann Roof murdered in a church in Charleston. "For many, black and white, that flag was a reminder of systemic oppression and racial subjugation." As for the Confederate soldiers, he added, the removal of the flag "would simply be an acknowledgment that the cause for which they fought, the cause of slavery, was wrong." Lately, however, the momentum to repudiate the Confederate patrimony once so central to collective memory has stalled, either for political or legal reasons. But at least the nation's first black president had explicitly encouraged the Southern states to do the right thing – after they had evidently tried everything else first.[42]

Collective memory has effects, or at the very least it cannot be disentangled from the claims of sectional identity. Take the 2008 election, a punctuation mark in the political history of race relations. Only a minority of whites voted for Senator Obama. Nevertheless, over two out of five whites did cast their ballots for the Democratic nominee. But less than one in three Southern whites did so. The so-called Appalachian belt, which runs from New York through Mississippi (and is therefore not exclusively Southern), consists of 410 counties. Obama managed to win only forty-four of them, even as he was decisively defeating the Republican nominee, Senator John McCain, in the general election. With a recession crushing millions of families, and with two inconclusive wars raging in Afghanistan and Iraq, 2008 should have been considered a rather inhospitable year for *any*

Republican candidate. Yet in some areas of the South, McCain ran even better than President Bush had done against Senator John F. Kerry four years earlier.[43] In 2012, according to exit polls in Mississippi, the nation's poorest state, whites voted 89 per cent for Mitt Romney, a Mormon from cobalt-blue Massachusetts, as well as a multi-millionaire who got rich as a manager of a private equity firm. By contrast, President Obama garnered 96 per cent of the black vote in the state.[44] The racial divide could not be starker. Admittedly bigotry probably cannot account by itself for the lopsided margins that Southern whites gave Obama's two opponents, though it is hard to downplay the role of racism in explaining those two elections.

Some signs of public attitudes are dispiriting. In the election year of 2016, the single greatest measure to indicate allegiance to Donald J. Trump was the number of racist web searches by state. In the 2016 Republican primaries, no other correlation was statistically stronger.[45] In South Carolina, 70 per cent of Trump's Republican supporters wanted the Confederate flag to continue to fly over the state capitol in Columbia. Pollsters learned that 38 per cent of the white Republicans in South Carolina's presidential primary still wished that the Confederacy had won the Civil War, though countering that figure are the 24 per cent who were glad (or relieved) that the North did indeed emerge victorious. Thirty-eight per cent of Trump loyalists were undecided, and thus unsure about the political and moral consequences of the Confederate defeat. The overall figures for Republicans in South Carolina's primary may be a bit more reassuring: 36 per cent were pleased that the Union remained intact. But still, 30 per cent of the participants in the state's GOP primary would have liked secession to have succeeded.[46] The implication is clear. A significant bloc of South Carolina whites, by wishing that the Confederacy had won, in effect wanted slavery to be preserved. Even in the twenty-first century, the desire for racial supremacy is therefore not too far below the surface. No wonder, then, that Faulkner's small-town lawyer, Gavin Stevens, gets quoted so frequently: "The past is never dead. It's not even past."[47]

ACKNOWLEDGMENTS

Much appreciation is extended to Raymond O. Arsenault, David Engerman, Lawrence J. Friedman, Richard H. King, and Edward S. Shapiro, all of whom helped with this chapter.

NOTES

1 John Shelton Reed and Dale Volberg Reed, *1001 Things Everyone Should Know about the South* (New York: Doubleday, 1996), 97.
2 Sonia Combe, *Memory and History: Harmony or Dissonance* (William Phillips Lecture Series of The New School for Social Research), 17.
3 George Santayana, *The Life of Reason* (New York: Charles Scribner's Sons, 1905), 1:284; James W. Loewen, *Lies My Teacher Told Me*, 2nd ed. (New York: Simon and Schuster, 2007), 1.
4 Carl N. Degler, *Place over Time: The Continuity of Southern Distinctiveness* (Baton Rouge: Louisiana State University Press, 1977), 36–9, 41, 42; Thomas Jefferson to John Holmes, *The Portable Thomas Jefferson*, ed. Merrill D. Peterson (New York: Viking, 1975), 567–8.
5 John Burt, *Lincoln's Tragic Pragmatism: Lincoln, Douglas, and Moral Conflict* (Cambridge: Harvard University Press, 2013), 418.
6 Abraham Lincoln, "Second Inaugural Address," *Great Issues in American History: A Documentary Record*, ed. Richard Hofstadter (New York: Vintage, 1960), 1:416.
7 Burt, *Lincoln's Tragic Pragmatism*, 414.
8 Charles Reagan Wilson, *Baptized in Blood: The Religion of the Lost Cause, 1865–1920* (Athens: University of Georgia Press, 2009), 7.
9 Reed and Reed, *1001 Things*, 98.
10 Jamie Malanowski, "Misplaced Honor," *New York Times*, 26 May 2013, 5.
11 James C. Cobb, *Away Down South: A History of Southern Identity* (New York: Oxford University Press, 2005), 102.
12 Quoted in Angie Maxwell, *The Indicted South: Public Criticism, Southern Inferiority, and the Politics of Whiteness* (Chapel Hill: University of North Carolina Press, 2014), 58.
13 Quoted in Fred Arthur Bailey, "Free Speech at the University of Florida: The Enoch Marvin Banks Case," *Florida Historical Quarterly* 71 (July 1992): 8–9, 10, 11, 13; Cobb, *Away Down South*, 102.
14 M. Alison Kibler, *Censoring Racial Ridicule: Irish, Jewish, and African American Struggles over Race and Representation, 1890–1930* (Chapel Hill: University of North Carolina Press, 2015), 89.
15 Quoted in Cobb, *Away Down South*, 91.
16 Quoted in Maxwell, *Indicted South*, 16; and in C. Vann Woodward, *Tom Watson: Agrarian Rebel* (New York: Oxford University Press, 1963 [1938]), 426.
17 Jelani Cobb, "Last Battles," *New Yorker*, 91 (6 and 13 July 2015), 28; Blain Roberts and Ethan J. Kytle, "Remembering Our Original Sin," *New York Times*, 6 December 2015, 6.

18 David W. Blight, *Race and Reunion: The Civil War in American Memory* (Cambridge: Harvard University Press, 2001), 7–12, 383–89.
19 M. Alison Kibler, *Censoring Racial Ridicule: Irish, Jewish, and African American Struggles over Race and Representation, 1890–1930* (Chapel Hill: University of North Carolina Press, 2015), 93–6; Godfrey Cheshire, "Why No One Is Celebrating the 100th Anniversary of the Feature Film," *Southern Cultures* 21 (Winter 2015): 32.
20 Cheshire, "Why No One Is Celebrating," 30; Foner, "Selective Memory," *New York Times Book Review*, 4 March 2001.
21 Michael T. Gilmore, *The War on Words: Slavery, Race, and Free Speech in American Literature* (Chicago: University of Chicago Press, 2010), 277, 278; Cobb, *Away Down South*, 88.
22 Quoted in Thomas R. Cripps, "The Reaction of the Negro to the Motion Picture, *Birth of a Nation*," in *Focus on* The Birth of a Nation, ed. Fred Silva (Englewood Cliffs, NJ: Prentice-Hall, 1971), 115.
23 Cheshire, "Why No One Is Celebrating," 29, 30, 37.
24 Walter Lippmann, *Public Opinion* (New York: Free Press, 1965 [1922]), 61.
25 Cheshire, "Why No One Is Celebrating," 34; Eric Foner, *Who Owns History? Rethinking the Past in a Changing World* (New York: Hill and Wang, 2002), 193.
26 Melvyn Stokes, *D.W. Griffith's* The Birth of a Nation (New York: Oxford University Press, 2008), 249; Jason Morgan Ward, "The Cause Was Never Lost," *American Historian* 6 (November 2015): 26.
27 Samuel Eliot Morison and Henry Steele Commager, *The Growth of the American Republic* (New York: Oxford University Press, 1930), 415, 417, 418; Neil Jumonville, *Henry Steele Commager: Midcentury Liberalism and the History of the Present* (Chapel Hill: University of North Carolina Press, 1999), 26–31, 146–8.
28 John Hope Franklin, *Race and History: Selected Essays, 1938–1988* (Baton Rouge: Louisiana State University Press, 1989), 389.
29 W.E.B. Du Bois, *The Philadelphia Negro: A Social Study* (Philadelphia: University of Pennsylvania Press, 1996 [1899]), 385.
30 Paul H. Buck, *The Road to Reunion, 1865–1900* (Boston: Little, Brown, 1937), 297, 306.
31 Quoted in John P. Roche, *Sentenced to Life* (New York: Macmillan, 1974), 16, 38; Howell Raines, "The Dream World of the Southern Republicans," *New York Times*, 12 July 2015, 6.
32 Quoted in Marshall Frady, *Wallace* (New York: World, 1968), 142.
33 Cobb, *Away Down South*, 290–1, 297; Reed and Reed, *1001 Things*, 91–2.

34 Quoted in Jeffrey Lamar Coleman, "Civil Rights Movement Poetry," in *The Cambridge Companion to American Civil Rights Literature*, ed. Julie Buckner Armstrong (New York: Cambridge University Press, 2015), 146; Eric Foner, "Civil War in 'Postracial' America," *Nation*, 10 October 2011, 25; Michael Kazin and Maurice Isserman, *America Divided: The Civil War of the 1960s* (New York: Oxford University Press, 2000), 1–2.
35 Quoted in Kazin and Isserman, *America Divided*, 2.
36 John F. Kennedy, *Profiles in Courage* (New York: Harper and Brothers, 1956), 152–4, 161, 163.
37 Ibid., 176–7.
38 Philip Galanes, "The New 3 R's," *New York Times*, 26 July 2015, Sunday Styles, 8.
39 Daniel Victor, "Louisiana: New Orleans to Remove Monuments," *New York Times*, 18 December 2015; "Monuments' Removal Challenged," *New York Times*, 20 December 2015; Christopher Mele, "Dismantling a Monument, Under Guard," *New York Times*, 25 April 2017.
40 Brent Staples, "Confederate Memorials and Racial Terror," *New York Times*, 25 July 2015; Jack Hitt, "Set in Stone," *New York Times Magazine*, 18 October 2015; Emily Bazelon, "Sixteen Counties," *New York Times Magazine*, 28 August 2016.
41 Foner, "Civil War in 'Postracial' America," 26.
42 Quoted in Cobb, "Last Battles," 28; Editorial Board, "The Civil War Is Winding Down," *New York Times*, 6 July 2015; Alan Blinder, "Momentum to Remove Confederate Symbols Slows or Stops," *New York Times*, 13 March 2016; Cain Burdeau, "Backlash Halts Removal of Confederate Symbols in New Orleans," *Boston Globe*, 23 March 2016.
43 Adam Nossiter, "For South, a Waning Hold on National Politics," *New York Times*, 11 November 2008.
44 Toni Monkovic, "Why Donald Trump Has Done Worse in Mostly White States," *New York Times*, 8 March 2016.
45 Nate Cohn, "Trump's Secret Weapon: Blue-State Voters," *New York Times*, 27 March 2016.
46 "Trump, Clinton Still Have Big Leads," 16 February 2016, www.publicpolicypolling.com.
47 William Faulkner, "Requiem for a Nun" [1951], in *Novels 1942–1954* (New York: Library of America, 1994), 535.

10

Between Nigeria and Biafra: Locating Ethnic Minorities in Narratives of the Nigerian Civil War, 1967–70

Dominique Otigbah

The place of minorities in conflicts is not always made visible in war narratives, and this chapter aims to challenge existing bias by detailing the experience of minorities in what became known as "the Biafra story." On 30 May 1967, the Eastern Region of Nigeria declared its independence as the Republic of Biafra. This move followed months of political and social instability in Nigeria, which included two coups, retributive massacres carried out in the Northern region, and a large population displacement of Eastern migrants in the North moving back to the East. On 6 July 1967, after negotiations failed to come to a peaceful agreement, war broke out, with Nigerian troops advancing into Biafran territory. War lasted for two years and six months, ending on 15 January 1970. The conflict received widespread media attention, especially because of the blockade of Biafra and the resulting food crisis. The general discourse at the time of war and in its current historiography presents the conflict as an Igbo struggle for independence, a political struggle against the hegemony of the Yoruba and Hausa-Fulani, or a struggle anchored in ethnoregional rivalry between the North and the East. The war had two main belligerents, the Federal Military Government (FMG) and the Republic of Biafra, so to an extent it is possible to view the war in binary terms, Nigeria/Biafra or North/East.

Biafra was based on the colonial boundaries for the Eastern Region. However, the terms "Biafra," "Igbo," and "Easterner" are often conflated and have been used interchangeably. This area encompasses most of the Igbo homeland, but it is very ethnically diverse and includes a significant number of ethnic minorities who made up the populations of the peripheral regions. Minorities in this area include the Efik, Ijaw, Ibibio, Ekoi, Ikwerre, and Ogoni, among others. Working from the 1963 census, the total population of the Eastern region was 64.29 per cent Igbo, and the remaining 35.71 per cent consisted of Eastern region ethnic minorities, ethnic groups from other regions, other Africans, and non-Africans.[1] Thus, generalizing Biafra as solely Igbo neglects the true ethnic diversity of the region and neglects the experiences of these ethnic groups during the war and their interactions with the belligerents. Crucially, it does not allow for a thorough understanding of the situation during the civil war.

This situation could be best summed up as the result of an "imaginary bipolarity of belligerents focused mainly on one ethnic group rendering all other groups culturally and politically invisible."[2] One seeks to explore why the minorities have seemingly been left out of the Nigerian Civil War historiography by looking at how this binary image of the war, focusing on two belligerents, was initially created prior to the outbreak of the conflict and how it was later reinforced during the second half of the war. To give some context to the minority groups' response to the conflict I also include an overview of significant minority political activity throughout the period. In looking at these issues, it becomes clear that the ethnic minorities of Eastern Nigeria have an important place in the history of the civil war and that more nuanced views of the war are needed. In assessing several primary and secondary sources from all sides involved, including wartime reports by the British Foreign Commonwealth Office and press cuttings from the domestic and foreign media, it becomes clear that there is much more to be investigated than just the minority issue.

LITERATURE REVIEW

In terms of published scholarly work on the Nigerian Civil War, the minority groups' experience gets relatively little attention as compared to that of the Igbo majority. This is alarming, given the number of texts concerning the war. Most of the literature published in Nigeria and internationally takes a limited view of the war, with its being presented as between Nigeria and Biafra and/or the Igbo. A quick survey of several prominent texts written about the war by foreigners either at the time or more recently tend to fall into three categories: either they are pro-Biafra, fail to integrate the minority discourse into their writings, or focus on the military narrative rather than the social perspective.[3] For example, in *The Biafra Story,* Frederick Forsyth mentions foreign mercenaries in more detail than anything he says about the minorities, whereas in Suszanne Cronje's *The World and Nigeria* and Michael Gould's *The Struggle for Modern Nigeria and Biafra*, the conflict is presented in Hausa/Igbo/Yoruba terms and conflates the Eastern region as being all Igbo, all Biafrans.[4] In contrast, *The Nigerian Civil War* by John de St Jorre is more objective and provides a good but very brief overview of the minority issue prior to and after the outbreak of conflict.[5]

Unfortunately, general histories of the civil war by Nigerian authors tend to have the same issues as do some of the Western texts mentioned above, where Igbos are the only ethnic group mentioned and are associated with the Eastern region or where the minority issue is simply not addressed.[6] There are even a number of Nigerian texts that, ironically, fail to mention the minority narrative and focus only on the Igbo ethnic group, even though the minority narrative does appear in their sources.[7] In Chima Korieh's *The Nigeria-Biafra War* there is a chapter that includes a brief discussion of relations between minorities in the forty years after the war; however, the general theme of the book is dominated by a pro-Igbo and pro-Biafra narrative.[8]

This is not to say that the topic of the plight of Eastern minorities during the war has not been considered in published scholarly texts. Given Nigeria's and Nigerians' long-time fixation on ethnic origin, the issue of minorities in Biafra has been at least somewhat addressed. However, the majority of the texts that do so lack a clear argument and are mostly descriptive in nature,[9] whereas others are not in line with the main argument of this chapter.[10] The one exception to this is an article by Arua Omaka, which argues that minority groups who suffered during the conflict have been largely forgotten and excluded from the historiography of the Nigerian Civil War.[11] Based on a small number of interviews and archival sources, Omaka discusses these forgotten victims, highlighting instances of their suffering. In particular, he brings up the victimization of minority individuals in Biafra who were suspected of sabotage and of collaborating with the FMG.[12] However, he fails to discuss why exactly these minorities have become "forgotten victims," which it is as important to understand as is their victimhood.

Given the breadth of material on the Nigerian Civil War, one would expect a greater variety of academic texts relating to minorities. Yet, although the historiography of the war itself is quite extensive,[13] it is painfully limited regarding the minority experience, even though there is a wide range of primary sources concerning these minorities. In recent years, the topic of the war has reappeared in both the modern Nigerian consciousness and the Western consciousness, either through the international popularity of *Half of a Yellow Sun* by Chimamanda Adichie or through controversial Radio Biafra Broadcasts.[14] It is clear that a more comprehensive study of minorities during the Nigerian Civil War is needed. This may be beyond the scope of this text; however, it is possible to ascertain why the minority issue is seemingly forgotten in the historiography and modern images of the Nigerian war.

MINORITY POLITICS AND AGITATION, 1928–67: AN OVERVIEW

Before considering how and why the overall narrative of the Nigerian Civil War seemingly excludes detailed mention of minority involvement, it is important to look at minority political activity prior to the war. In the latter part of this period, minority politics were intricately tied to the looming conflict between Nigeria and Biafra. Though the ways minorities made political demands of the state differed over time, the overall goal was to gain self-governance for their areas while remaining within the state of Nigeria (with the one exception of the Niger Delta Republic, which attempted to secede in February 1966).

As a response to the expansion of colonial administration in Nigeria in the early 1900s, local movements began to arise throughout the colony. In the East, one of the earliest movements was the Ibibio Union, which was established in 1928 and which "sought to use collective representation to accommodate, resist, modify, and manipulate the colonial system."[15] Over time, other ethnic-based movements arose, like the Ijaw Rivers People's League or the Andoni Progressive Union. These movements spurred the creation of new provinces in 1947 and 1959, and many later became part of more political units, like the Calabar-Ogoja-Rivers (COR) State movement, which became a sub-political party in 1954.[16] This group, which feared political domination by Igbos, demanded that a separate COR state be carved out of the region in the hope that a separate state would afford their populace greater political representation. There were also other groups demanding a separate Rivers state, a separate Ogoja state, and a separate Calabar state.

In May 1957, minority demands at the Nigerian Constitutional Conference in London led, later that year, to the British government's setting up the Willink Commission, which was aimed at addressing minority fears throughout the colony.[17] The commission ultimately recommended that, for fear of delaying Nigeria's independence date, separate states should not be created. The key details to take away from the commission's report, however, are that, although there was an overall fear of political and economic domination by the Igbo ethnic group, there were also concerns about the relative underdevelopment of their regions. In Rivers Province, concerns were focused on the lack of development in the Niger Delta region and the government's lack of understanding of the difficulties experienced by the Indigenous peoples.[18]

As the report recommended against the creation of a separate state, it seems that, apart from the creation of the Niger Delta Development Board (NDDB) and the reshuffling of local council arrangements,[19] there was no sincere attempt made to grant the minorities their wishes. So, even though "all witnesses who came before us [the commission] were insistent that nothing but a separate state could meet their problems," ultimately the commission would leave the minorities to continue their campaign for separate statehoods.[20] Yet even the proposals made by the commission were not completely fulfilled, particularly with regard to Rivers Province, which, despite its abundance of crude oil,[21] remained highly underdeveloped until the outbreak of war.

The various minority groups of that region would continue to make their demands for their own states up to and following Nigeria's independence in 1960 and the outbreak of war in 1967.[22] These manifested in both peaceful and violent ways – for example, the attempted secession of the Niger Delta Republic led by

Isaac Adaka Boro and his Niger Delta Volunteer Force (NDVF) in February 1966 and renewed demands for state creation in August and September 1966.

Boro hailed from Oloibiri, and, as mentioned previously, the impetus of his revolt was the continued underdevelopment of the Niger Delta areas, an issue that had dominated the region since before the Willink Commission and that was not resolved by the NDDB.[23] Though the NDDB carried out initial projects to test the viability of the region for the establishment of fisheries, cash-crop production, and mineral extraction, only the latter resource was exploited, with crude oil production taking centre stage in the region. Most of the revenue generated from oil was not reinvested within the region.[24] There were no efforts to promote the large-scale development of agriculture and aquaculture, which could have brought economic improvement to the area's inhabitants.[25] In the case of social services and communications, the area was similarly deprived, with only a handful of schools, poor communication links, and three hospitals. And key industrial structures like fisheries and boatyards were situated far away from the riverine areas.[26] It is worth noting that the same underdevelopment argument that surfaced in 1957 resurfaced just ten years later with Boro's rebellion, and it has been a major grievance in the oil producing region ever since.

A month after the first coup, the NDVF launched its revolution in Yenagoa, leading to the looting of non-Ijaw businesses and the expulsion of non-Ijaw from the region.[27] The rebellion was quickly put down, and Boro and his associates were captured and sentenced to death for treason twelve days later. However, in 1967 he was pardoned and he served in the Nigerian Army, fighting against Biafra, and eventually dying in battle in 1969 fighting for the country from which he had initially wanted to secede.[28]

In 1966, more official (and less violent) demands were made for minority sovereignty. In August, Chief Harold Dappa-Biriye (a longtime advocate for the creation of separate states in the region) put forward his demands at the first Eastern Consultative Assembly meeting, with Ojukwu in attendance.[29] The assembly was meant to be a democratically chosen group of divisional and provincial representatives acting as advisors for the region; however, many of these representatives were alleged to have been handpicked by Ojukwu himself.[30] In September, calls were again made for the creation of a Rivers State in response to news that a conference to discuss

Nigeria's future would be held.[31] The main aim of the conference was to set up the broad outlines of a constitution.[32] In essence, it concerned what governance structure Nigeria should adopt: one based on a loose federation of autonomous regions, or one under a united central government (with the creation of more states).[33] According to Dappa-Biriye, Gowon was highly receptive to the latter option.[34] The conference stalled over this issue, as the Eastern delegation refused to agree to the creation of new states. Like the Willink Commission, the Eastern delegation stated that, although it supported the eventual creation of the new states, it would be inappropriate to do this during the current period of instability.[35]

In the following months, the issue of minorities continued to be a major topic throughout Nigeria, with the four regions being divided into twelve new states. Although brought up at the first ad hoc conference, the official proclamation that state creation would be considered came in April 1967 as part of the program for returning Nigeria to civilian rule.[36] Hence, minorities were brought back into the main narrative, with Gowon stating that if Ojukwu declared secession the COR state would be created by decree in order to protect the minority groups.[37] According to Gowon, he received petitions from COR peoples complaining about their poor treatment under the Eastern government, with minorities being imprisonment for "political reasons."[38] One must note that demands for state creation came from all over Nigeria, yet, arguably, Eastern minority demands had a greater bearing on the political situation and the lead-up to war in the region. The minorities eventually got the states for which they had so long campaigned on 27 May 1967, when Nigeria was split into twelve states, with the Eastern region becoming the Rivers, East-Central, and South-East states. The granting of statehood to the minority population in the Eastern region also highlights the importance of minority issues in relation to historiography as the quick declaration of Biafran secession three days later may be seen as a direct reaction to the creation of these states (to the dismay of pro-state creation minority peoples).[39]

Finally, it is interesting to note that the separatist agitation by minority groups in the Eastern region has a long history and that it clearly precedes the arguments for Biafra. In light of this, it could be said that a Biafran movement for a united Eastern Region lacked legitimacy as, in reality, a substantial part of its population demanded separation from the Eastern government unit rather than integration into it as part of Biafra.

COUPS, MINORITIES, AND ESTABLISHING THE NORTH/EAST BINARY

Crucial to the outbreak of war in 1967 were the coups of January and July 1966. This part of Nigerian history has been endlessly covered by academics and forms the basis of most Nigerian Civil War literature.[40] It is important to consider this time period, particularly with regard to understanding why the Nigerian Civil War has been presented in binary terms of North/East and why the Igbos seem to be associated solely with the East. This, in turn, links to why minorities seem to be relatively absent from this historiography. The political crises of 1966 are considered the definitive beginning of Nigeria's descent into civil conflict: most major texts on the war start with or include an account of the events of 1966. From the very beginning, the narrative has been posited as a North/East binary (with the Igbos often being singled out), and this has determined how the narrative developed over time.

To avoid repeating information about the lead-up to the war that can be found in great detail elsewhere, we can skip ahead and start with the alleged reasons for the 28 July 1966 counter-coup carried out by Northern officers and in which many Igbo officers were systematically targeted for killing. These were printed in an FMG pamphlet outlining the Nigerian crisis and included the 15 January coup, the failure to discipline the first coup plotters, and rumours of an "Igbo masterplan" to cleanse the army of Northern elements.[41] The first reason refers to Nigeria's first coup on 15 January 1966. Carried out by young army officers of mostly Igbo extraction, it involved the killing of a number of key political and military figures.[42] Given that the majority of the coup plotters were Igbo, and that there were relatively minimal Igbo losses, there were some suspicions that the coup was an Igbo plot against the North.[43] This was the case despite the fact that there were a number of Northern participants in the January coup and that,[44] from statements made by the coup leader Major Nzeogwu, there was little mention of regional or ethnic motivations.[45]

Even though the grievances mentioned above were specifically those held by the Northern counter-coup leaders, that contents of that same pamphlet, which represented the official federal position, clearly took on ethnic tones. The tone of this narrative of events appeared in other federal pamphlets as well. Igbos in high official positions either in the military or civil service were being singled out as orchestrating the first coup and were posited as key belligerents in Nigeria's civil war.[46] Despite the non-ethnic motivations of the first coup leaders, their political grievances against the old regime quickly became ethnicized. It is worth noting that many of these federal pamphlets were essentially propaganda pieces but were widely distributed and so helped to solidify the North/East narrative. Meanwhile, statements given by Gowon in the months prior to the first attack on Biafra mostly focused the blame on the reluctance of the Eastern regional government to come to agreement with other parts of the country. In these, ethnicity was not explicitly mentioned, yet there was frequent reference to those of "Eastern origin."[47]

During the counter-coup, Northern elements of the army decided to rectify the ethnic imbalance they perceived to exist. Initially, the violence carried out by Northern officers seemed to very systematic and clearly targeted Igbo soldiers.[48] But, by September, Northern civilians also began to take part in the violence, targeting Igbos and then non-Igbo Easterners of all backgrounds.[49] This volatile situation led to a mass population transfer between North and South, mostly consisting of Igbos returning to the East. For the non-Igbo Eastern minorities, it was a bit more mixed as they either stayed or migrated. It is estimated that 1.5 million returnees sought refuge in their Eastern homelands by the outbreak of war in 1967. Non-Easterners were expelled from the East in October 1966 on the grounds that their safety could no longer be assured. In a sense, these population movements demographically set the North/East binary.

The creation of this binary was not a one-sided affair. This North/East rivalry would crop up again and again in speeches, statements, and publications from the Biafran side. In Ojukwu's speech proclaiming his conviction regarding the agreements made at Aburi, he claimed that Lagos and Kaduna were acting together to dominate the Southern regions. The violence meted out against Easterners in the Northern riots of 1966 served as clear evidence that there was a Northern plan against the East. This idea was used extensively by the Eastern propaganda machine as Biafra further cemented its position against the FMG. The massacres were presented as a Muslim campaign aimed at wiping out the Christian South as Christianity was widespread among all the ethnic groups of the Eastern region. Whereas the exact demographics of the massacre victims are unclear, it is likely that the majority were Igbo (given their statistical presence). Yet, in order to rally support, the massacres were presented as an attack on all Easterners rather than just on Igbos, and minorities would figure greatly in early Biafran propaganda.

Ethnic stereotyping, prejudice, and discrimination are an unfortunate but recurring feature of Nigerian society.[50] It would be easy to blame the creation of the North/East binary on this notion of stereotyping and ethnic difference. However, one cannot exclude government involvement, and, as we have seen above, both the FMG and Biafra were involved in creating this binary.

MINORITIES FRONT AND CENTRE: PROPAGANDA AND SOLICITING MINORITY SUPPORT

Despite being relegated to the lesser explored parts of the history of the Nigerian Civil War, for a brief period following the second coup, the Eastern minorities were at the forefront of both FMG and Biafran propaganda. From the Biafran side, much of this propaganda was produced in the lead-up to the creation of new states in May 1967. Though the state creation decree was promulgated to satisfy the demand of ethnic minority groups and to allay fears of majority domination, it had a secondary effect of weakening the movement for Biafra. Given that Biafra was based on the Eastern region's boundaries, the creation of states broke up the territory even before Biafra was declared. It was also thought that those who were most demanding of state creation would become more likely to support the FMG, which was willing to give them their states, rather than the Biafrans, who proposed the provincial system. The creation of the Rivers and the South-East states meant that, from the outset, independent Biafra would lose its access to the sea and to the oil-producing areas that were in minority areas.

The most telling evidence highlighting the importance of the Eastern minorities in the lead-up to 27 May concerns the anti-state creation news articles from the Eastern region. Printed during this period, pro-Eastern government newspapers like the *Eastern Nigerian Outlook* and the *Eastern Nigerian Spotlight* carried articles with titles such as "COR State: Ogoja People Say No" (30 April), "Rivers Students Warn Yakubu Gowon" (26 April), and "Okrika Spurns State Plan" (11 May).[51] At the same time, there were a number of articles and photo-ops that also showed minority support for Ojukwu.[52]

10.1 "Together We Fight," *Eastern Nigerian Spotlight*, April 1967.

These articles are clearly propaganda pieces, yet they demonstrate that significant effort was spent to show that the Eastern government had widespread support from all the different minority areas. In light of the FMG's intention to divide the Eastern region, and the reality that many wanted their own state governments, it was crucial to

Rivers Students Warn Yakubu Gowon

The Rivers Students' Association in Lagos has warned Lt. Col. Yakubu Gowon to be prepared to bear the disastrous consequences of his action, if he dared create a COR State arbitrarily against the wishes of the "minorities" in Eastern Nigeria.

In a strongly worded three-page statement in Lagos yesterday deploring Lt. Col. Gowon's address to heads of diplomatic missions in Nigeria, that he would create a COR State in Eastern Nigeria militarily, the students said Gowon was living in a fool's paradise if he thought he would achieve his inordinate ambition to conquer the East, for the feudal North through the minorities in the East.

"Who are the minorities?" they asked, adding that Lt. Col. Gowon and his clique should come out openly if they were so ambitious and determined to invade the East instead of seeking the cover of the "minorities" in the region.

"They are soldiers and therefore behave as such and as comarades", the statement said.

The students also said they were not opposed to the creation of states in principle but felt that the present charged atmosphere was not conducive for the talk of a COR State in Nigeria.

"We sincerely believe such plan will bring more hardship and perpetual discomfort to the very minorities the Lagos authorities pretend to assist" they added.

The Rivers Students Association also remined Col. Gowon of the inhuman sufferings of the Tivs, and pleaded with him to save his own people from the series of oppression and intimidation meted against the Tivs by the Hausas and Fulanis of the far North before talking of the minorities in the East."

10.2 "Rivers Student Warn Yakubu Gowon," *Nigerian Outlook*, 26 April 1967.

present the image of unity. Most important, in order to legitimize their stance against the FMG, it was crucial to show that they were widely supported. This notion of a united Eastern region full of happily coexisting ethnic groups formed the Biafran narrative up until 1969.

On the federal side, too, minorities figured to a large degree in its anti-Eastern government argument. Since the state creation decree was posited as a way of protecting the ethnic minorities in that region, this notion was greatly expanded upon in FMG publications and statements on the crisis, with the population figure of 5 million non-Igbo Easterners and 7 million Igbo being repeatedly emphasized.[53] In another pamphlet, the idea of long-standing Igbo dominance throughout Nigerian history was also stressed.[54] The Igbos were put forward as the most "tribalistic," with the Hausas in the North being highly tolerant and patient with "the aggressiveness and insults of the Igbos."[55] The FMG simultaneously stressed the

OGOJA SPURNS GOWON'S BID

'It Is Wicked, Unsolicited'

The two-million people of the old Ogoja Province have debunked Gowon's proposal to carve out a state by decree in Eastern Nigeria.

Addressing a news conference in Enugu yesterday, Mr Mathew Mbu, acclaimed leader of the Ogoja people, described Gowon's programme on the creation of a COR State as "detached from reality."

"It reminds me of a red Queen in Alice in Wonderland who first demanded the verdict and then the evidence", he remarked.

Mr Mbu was surprised that Lt. Col. Gowon should plan to create states before ascertaining the wishes of the people.

"Is it being suggested?" he asked, "that states be now created without reference to the people or without ascertaining the wishes of the people through a plebiscite?"

Mr Mbu, who is also the Chairman of the Eastern Nigeria Public Service Commission said: "It is now obvious to us in Ogoja that Lagos and Kaduna have deliberately made up their minds to engage us as a sort of semantic people over the creation of states."

Ogoja people, he said,

(Continued on page 8)

AN EDITORIAL

OJUKWU RENEWS

10.3 "Ogoja Spurns Gowon's Bid," *Nigerian Outlook*, 27 April 1967 (note the cartoon depicting Gowon and a "Northerner" cutting off the East from the rest of Nigeria).

ethnic diversity of the region, focusing specifically on the minority groups and singling out the Igbos for their part in the conflict. This narrative would be maintained during most of the war, but, like on the Biafran side, the narrative based on the ethnic demography of the region would change in relation to how the war progressed.

As a final note, one must ponder the sincerity of either side with regard to the minority cause. Though it was important to show that they had the minorities' support, one can look at the situation more cynically. The peripheral areas of the Eastern region were regarded as highly valuable. The Niger Delta region, despite its long-standing underdevelopment, contained vast oil reserves. By May 1967, the Eastern region produced 404,000 barrels of crude oil per day, making up 65 per cent of Nigeria's total crude oil. Over 90 per cent of this oil was found in the Rivers area.[56] Not long after the outbreak of conflict, the Biafran government began to demand the oil

ENUGU, Sunday May 7, 1967

Three Pence

DENTS .S. MAN

epart- orary East- meri- rict- itive way nce of gos ire n- 's e 1

the Congress Government and people of the United States for the categorical assurance that the United States, mindful of the odious situation in Vietnam will not take sides in current Nigerian crisis."

The advertisement then quoted the speech as published in the congressional record.

The advertisement was signed by 15 chapters of the Eastern Nigeria Students' Association in the Americas.

GIFT FOR GOVERNOR

Lieutenant Colonel C. Odumegwu Ojukwu, Eastern Nigeria's Military Governor, recently visited Uyo.

The people of Uyo, known for their traditional African hospitality, took the opportunity to reaffirm their confidence in the Governor and his government.

They aso treated him to various types of cultural dances.

Picture above shows an Uyo young woman making a presentation to the Governor on behalf of her people. Other pictures at page 8.

TRIBUNAL HEARS MORE

On the second day of its two-day sitting in Ogoja, Mr Jacob Odida of Obudu, Ogoja Province, gave evidence to the Atrocities Tribunal of the tribulations of himself and his family in Northern Nigeria during the pogrom of last year.

Mr Odida, who lived as farmer and petty trader Moroa River, Benue Province, said that at first the Idomas natives of this area, pretended that the people of Ogoja were their brothers. However, they later attacked and

The Northern policemen assisted the soldiers in killing two of the policemen of Eastern origin, namely, Constable Ibeziako of Ogbunike and Sergeant Okani of Onitsha.

The remainder were taken away in an army

MILEAGE DETAILS

10.4 "Gift for Governor," *Nigerian Outlook*, 11 May 1967.

royalties of this region, now that it was part of Biafra.[57] Additionally, the Rivers and South-East states contained important commercial centres and sea links, like the towns of Calabar and Port Harcourt and their accompanying ports, the oil terminal at Bonny Island, and the refinery in Port Harcourt. In terms of agriculture, the minority areas held many rubber and palm oil plantations, and were key producers of staple crops such as yams and rice.[58] So in either case, these areas were crucial to secure, and in the Biafran case its economic viability as an independent state depended on these areas, which were the first to be captured by the advancing Nigerian troops.

LOSING THEIR PLACE: MINORITIES AND SHIFTS IN THE WAR NARRATIVE

As seen in the previous sections, minorities figured greatly in the lead-up to the war. Despite this, they rarely appear within the historiography of the Nigerian war, and, when they do, they are not afforded the recognition they deserve. I contend that this is due to how the war narrative has been formed over time, the first contributing factor being the articulation of a North/East Igbo-focused binary during the political crises of 1966–67. This section shows that changes in the dynamics of the conflict further solidified this binary and put even more focus on the Igbo experience, thus, in the long run, obscuring the minority experience. One may be tempted to say that the focus on the Igbo occurred simply because they are the majority ethnic group and so would logically dominate any narrative in the region. In reality, however, it has more to do with changes to the geographical scope and nature of the conflict than with demographics.

Between July 1967 and September 1968, the Nigerian advance into Biafran territory resulted in a drastic reduction in the size of the rebel state. After Port Harcourt was captured in May 1968, all the minority areas fell under Nigerian control. Biafra was now an ever-shrinking enclave located in the Igbo homeland. At the same time, the effects of Nigeria's blockade on Biafra and the disruption of seasonal agricultural patterns began to be fully felt, with malnutrition intensifying throughout Biafra. Yet Biafra persisted, and it would take another year for the FMG to finally achieve victory. These two facts received a great amount of attention in the foreign consciousness, helped along by pro-Biafran groups in Europe and North America and by the hiring of Markpress, a Swiss public relations firm that handled all of Biafra's foreign propaganda. Indeed, for many in the West, the word "Biafra" is more likely to conjure up images of sickly children suffering from Kwashiorkor rather than images of the Eastern minorities and their own strife.

In foreign press articles about the war published after 1968, the minorities were rarely mentioned, and, even when they were, they often took a backseat to the Igbos.[59] Nigeria, too, engaged a foreign public relations firm, but, arguably, the Biafrans were more successful in promoting and gaining sympathy and support for their struggle. This, in turn, has contributed to the large Biafran bias found in many Western texts about the war. Indeed, the efficacy of Biafran propaganda in the West has been discussed extensively.[60] Regarding the minorities and their relative obscurity in the war's historiography, changes in federal and Biafran approaches to the presentation of that war are important. Additionally, the images of the conflict in the foreign consciousness were dependent on the type of information and narratives being produced by the belligerents.

As discussed earlier, both Biafra and the FMG placed heavy emphasis on the ethnic diversity of the region (albeit for very different reasons). Yet, as minority areas were captured and Biafra receded into the Igbo heartland, this emphasis was no longer viable in terms of internal propaganda. In the lead-up to and immediately following the loss of minority areas, there were some mentions of minorities from the Biafran side,[61] but after that the notion of an ethnically diverse state that claimed support from all its minorities lost traction.[62] Rather, Biafra's struggle came to be articulated without any hint of ethnic diversity: those in Biafra were posited as one and the same, as if ethnic diversity had been transcended in the face of adversity.[63] In the Ahiara declaration, which was meant to represent the principles of Biafra's revolution, Biafran individuals are posited as one and the same, with the eastern minorities being mentioned only once (meanwhile, there is repeated reference to a Nigeria ruled by a "Hausa-Fulani oligarchy").[64]

Minorities also took a back seat on the federal side. Whereas the capture of minority areas and their support of the FMG was prominent news up until mid-1968, the FMG's focus on the war then turned towards what remained of Biafra. And, as Biafra was now concentrated in Igbo territory, the FMG began to focus on convincing the Igbo people to abandon their support for Biafra and for Ojukwu. One of the most notable ways it did this was by enlisting prominent Igbo politicians, some of whom had formerly been on the Biafran side.[65] Calls on Igbos to accept the FMG were made at conferences, in press releases, and in many government publications.[66]

Additionally, the FMG effort to gain the Igbos' trust also manifested itself in peace proposals and offerings. In these, emphasis is placed on plans for reconciliation with and rehabilitation of the Igbos in the war-afflicted Eastern region.[67] Crucially, the minorities are not mentioned in any of the offered peace settlements (which were based on the initial proposal at Kampala).[68] Hence, at another significant period of foreign media focus and of domestic hope for peace, the minorities and their own suffering during the war have been forgotten. At the same time, the narrative of the civil war is solidified as wholly Igbo-centric. In British Foreign Office texts and communiqués at the time and at the end of the war, there was a lot of concern about how the Igbos were to be rehabilitated and reintegrated into Nigeria. Entire folders were dedicated to this, but none were dedicated to the rehabilitation of minority areas.[69] And so, if the British government held this kind of bias, clearly focusing on the Igbos and not on the other inhabitants of the Eastern region, one can see how foreign images of the war would also be so aligned. Meanwhile, in Nigeria, if there was a focus on the Igbo, either as being the main agitators in the conflict or as needing extra attention at the end of the war, then it is understandable, albeit unfortunate, that most Nigerian texts on the war also follow this narrative.

CONCLUSION: BEYOND THE WAR

The Nigerian Civil War was essentially fought to gain control of the former eastern region, now known as Biafra, an area with a diverse range of ethnic groups and with many valuable natural resources. However, dominant narratives of the war do not reflect this ethnic diversity, and the conflict has been associated with only one of the many ethnic groups – the Igbo. Despite there being clear evidence of minority involvement prior to and throughout the war, from state agitation movements to attempts at secession and responses to the conflict, they are absent from the collective historiography of the Nigerian war. It is ironic that, during the war, both Biafra and Nigeria were desperate to win minority attention and support but that, afterwards, neither has really paid attention to these minorities.

Moving forward, it is clear that this situation needs to be changed and that it is possible to do so. The minorities were, and still are, some of the most vocal groups in Eastern Nigeria, like those in the Niger Delta, who have continuously challenged the Nigerian government and the international oil industry over the maltreatment of their communities and environments. In light of a Biafran resurgence in past years, minorities who had to navigate between state and rebel state during the war years also need to be brought back into focus, on a greater scale than heretofore. There is adequate source material to do this, and so it is possible to try toovercome the dominant and exclusionary binary narrative that has persisted since the 1960s. While the notion of a North/East binary is valid to a certain extent, the narratives and historiography of the war need to afford greater recognition to the complex experiences of the 5 million minority peoples of the former Biafra, all of whom suffered from the violent processes of the Nigerian Civil War. It is to be hoped that future studies and research on the war will follow this line, changing the historiography of this war from the outdated notion of a North/East binary to a fully representative narrative of civil conflict.

NOTES

1 *Population Census of Nigeria, 1963: Eastern Region* (Lagos: Federal Census Office, 1964), 1:vii.
2 Vilasnee Tampoe-Hautin, "Of Wars, Scars and Celluloid Memory: Representations of War in Sri Lankan Cinema (2000–2010)," paper presented at the International Conference "War Memories, Commemoration, Re-enactment, Writings of War in the English-Speaking World (18th – 21st Centuries)," 14–16 June 2016, Paris.
3 Suzanne Cronje and Auberon Waugh, *Biafra: Britain's Shame* (London: Michael Joseph, 1969); Frederick Forsythe, *The Biafra Story: The Making of An African Legend* (Barnsley: Pen and Sword, 2007, 1969); Suzanne Cronje, *The World and Nigeria: The Diplomatic History of the Biafran War, 1967–1970* (London: Sidgwick and Jackson, 1972); and Michael Gould, *The Struggle for Modern Nigeria: The Biafran War, 1966–1970* (London: IB Tauris, 2012).
4 Forsythe, *Biafra Story*; Cronje, *World and Nigeria*; Gould, *Struggle for Modern Nigeria*.
5 John De St. Jorre, *The Nigerian Civil War* (London: Hodder and Staughton, 1979).
6 These include: Ola Balogun, *The Tragic Years: Nigeria in Crisis, 1966–1970* (Benin: Ethiope Publishing, 1973); Ihediwa Nkemjika Chimee,

"The Nigerian-Biafran War: Armed conflicts and the rules of engagement," in *Warfare, Ethnicity and National Identity in Nigeria*, ed. Toyin Falola, Roy Doron, Opkeh O. Opkeh, and Opkeh Opkeh (Trenton, NJ Africa World Press, 2013), 111–38; and Tekena Tamuno and Samson Ukpabi, eds, *Nigeria since Independence: The First 25 Years*, vol. 6, *The Civil War Years* (Ibadan: Heinemann Educational Books, 1989).

7 See G.N. Uzoigwe, *Visions of Nationhood: Prelude to the Nigerian Civil War* (Trenton, NJ: Africa World Press, 2010); and Chima Korieh, ed., *The Nigeria-Biafra War: Genocide and Politics of Memory* (Amherst, NY: Cambria Press, 2012).

8 Philip U. Effiong, "40 Years Later the War Hasn't Ended," in Korieh, *The Nigeria-Biafra War*, 261–73 (Effiong is an Ibibio and the son of General Philip Effiong, Ojukwu's deputy).

9 For example, these pro-minority texts: Ebiegberi Alagoa, Tekena Nitonye Tamuno and John P. Clark, eds, *The Izon of the Niger Delta* (Port Harcourt: Onyoma Research Publications, 2009); Abiegberi Alagoa and Tekena Tamuno, eds, *Land and People of Nigeria: Rivers State* (Port Harcourt: Riverside Communications, 1989); Abiegberi Alagoa and Abi Derefaka, eds, *The Land and People of Rivers State: Eastern Niger Delta* (Nembe: Onyoma Research Publications, 2002).

10 Eghosa Osaghe, "Ethnic Minorities and Federalism in Nigeria," *African Affairs* 90, no. 359 (1991): 237–58, https://doi.org/10.1093/oxfordjournals.afraf.a098413; Kathryn Nwajiaku-Dahou, "Heroes and Villains: Ijaw Nationalist Narratives of the Nigerian Civil War," *African Development* 34, no.1 (2009): 47–67, https://doi.org/10.4314%2Fad.v34i1.57356.

11 Arua Oko Omaka, "The Forgotten Victims: Ethnic Minorities in the Nigeria-Biafra War, 1967–70," *Journal of Retracing Africa* 1, no.1 (2014): 25–40.

12 Ibid., 33–7.

13 For a relatively recent overview of the civil war historiography, see: Brian McNeil, "The Nigerian Civil War in History and Historiography," in *Africa, Empire and Globalization: Essays in Honor of A.G. Hopkins*, ed. Toyin Falola and Emily Brownell (Durham, NC: Carolina Academic Press, 2014), 539–51.

14 Chimamanda Ngozi Adichie, *Half of a Yellow Sun* (New York: Anchor Books, 2006). Radio Biafra is a station that calls for the independence and secession of Biafra from the Nigerian state once more. See Ibanga Isine, "Nigeria Arrests Radio Biafra Operators, Seizes Transmission Equipment," *Premium Times Nigeria*, 17 July 2015, http://www.premiumtimesng.com/news/186857-nigeria-arrests-radio-biafra-operators-seizes-transmission-equipment.html.

15 Monday Efiong Noah "The Ibibio Union 1928–1966," *Canadian Journal of African Studies* 21, no. 1 (1987): 38–53, https://doi.org/10.1080/00083968.1987.10803816.
16 Ben Naanen and A.I Pepple, "State Movements," in Alagoa and Tamuno, *Land and People of Nigeria*, 144; E.S. James, *Minorities in Biafra* (Stockton: P.L. Wood, 1969), 7; Richard L. Sklar, *Nigerian Political Parties: Power in an Emergent African Nation* (Trenton, NJ: Africa World Press, 2004), 137.
17 Sklar, *Nigerian Political Parties*, 138.
18 Lord Willink, *Report of the Commission Appointed to Enquire into the Fears of Minorities and the Means of Allaying Them* (London: HMSO, 1958), 42.
19 Ibid., 104.
20 Ibid., 97.
21 Ken Saro-Wiwa, *On a Darkling Plain* (Port Harcourt: Saros International, 1989), 51.
22 Cyril I. Obi, "The Changing Forms of Identity Politics in Nigeria under Economic Adjustment: The Case of the Oil Minorities Movement of the Niger Delta," *Research Report* 119 (Nordic African Institute: Uppsala University, 2001), 68.
23 Nwajiaku-Dahou, "Heroes and Villains," 56.
24 Isaac Boro, *The Twelve Day Revolution*, ed. Tony Tebekaemi (Benin City: Idodo Umeh Publisher, 1982), 64.
25 Ibid.
26 Ibid., 66.
27 Ibid., 127.
28 Nwajiaku-Dahou, "Heroes and Villains," 58–9.
29 Elechi Amadi, *Sunset in Biafra* (London: Heinemann Educational Books, 1973), 17.
30 Ibid., 16; Saro-Wiwa, *On a Darkling Plain*, 57.
31 Saro-Wiwa, *On a Darkling Plain*, 55.
32 "The Moment of Truth: Gowon opens the Ad-Hoc constitutional Conference in Lagos," 12 September 1966, in Ministry of Information, *Nigeria 1966* (Lagos: Ministry of Information, 1967), 7–40.
33 Ibid.
34 Harold Dappa-Biriye, "Democracy and Autocracy: My Testimony," *Guardian*, 27 May 1999.
35 "Statement by Dr Njoku on the Eastern Delegation's Stand on the Creation of States, Issued to the Press in Lagos," 19 September 1966, reproduced in A.H.M. Kirk-Greene, *Crisis and Conflict in Nigeria:*

A Documentary Source Book (London: Oxford University Press, 1971), 1:224.

36 "Programme for Return to Civilian Rule: Communiqué Issued by the Supreme Military Council, 22 April 1967," in Ministry of Information, *Blueprint for Nigerian Unity* (Lagos: Ministry of Information, 1967).

37 "Gowon Addresses Heads of Diplomatic missions in Lagos, 24 April 1967," in Kirk-Greene, *Crisis and Conflict in Nigeria*, 1:414.

38 Ibid.

39 "East's Reaction to the Creation of the States" and "Ojukwu Secedes and Declares the Republic of Biafra, 30 May 1967," in Kirk-Greene, *Crisis and Conflict in Nigeria*, 1:451.

40 For a highly descriptive narrative of events, see Max Siollun, *Oil, Politics and Violence: Nigeria's Military Coup Culture, 1966–1976* (New York: Algora Publishing, 2009), chaps. 1–12.

41 Ministry of Information, *Background Notes on the Nigerian Crisis* (Lagos: Ministry of Information, 1967), 3.

42 Siollun, *Oil, Politics and Violence*, app. 3: Casualties of 15 January 1966 Coup, 237.

43 Ibid., 34–5.

44 Ibid., 226.

45 "Nzeogwu Announces Martial Law in the North and Promulgates Ten Decrees," broadcast on 15 January 1966, in Kirk-Greene, *Crisis and Conflict in Nigeria*, 1:125–6.

46 Ministry of Information, *Nigeria – The 12 State Structure: The Only Way Out* (Lagos: Ministry of Information,1967), 6; and Ministry of Information, *Unity in Diversity* (Lagos: Ministry of Information, 1967), 9–10.

47 "Gowon's Private Address to Heads of African Diplomatic Missions in Lagos, 1 March 1967"; "Gowon's Broadcast to the Nation, 25 March 1967"; and "Gowon's Address to the First Civilian Members of the Federal Executive Council 12 June 1967." All in Kirk-Greene, *Crisis and Conflict in Nigeria*, 1:372, 405, 435, respectively.

48 Siollun, *Oil, Politics and Violence*, 115.

49 Ibid., 131–7.

50 See Ukoha Ukiwo, "The Study of Ethnicity in Nigeria," *Oxford Development Studies* 33, no. 1 (2005).

51 Nigerian Civil War Collection, News Material, April 1967, MS321463/08, and News Material, May 1967, MS321463/09, SOAS Archives (hereafter SA).

52 "Gift for Governor," *Eastern Nigerian Outlook*, 7 May 1967; and "Youths in Uyo Hail Ojukwu's Gesture," *Eastern Nigerian Outlook*, 12 May 1967, both in Nigerian Civil War Collection, News Material, MS321463/09, SA.
53 Ministry of Information, *Towards One Nigeria* (Lagos: Ministry of Information, 1967), 5 and 28; and Ministry of Information, *Challenge of Unity* (Lagos: Ministry of Information, 1967), 3 and 5. According to the 1963 census, the figures were closer to 4 million non-Igbo Easterners and 8 million Igbos.
54 Ministry of Information, *Unity in Diversity*, 9–10.
55 Ibid., 10.
56 Estrange to Davies, 2 August 1968, fol. 54 01/01/1968 – 31/12/1968, National Archives (hereafter NA); Shell-BP's Interest in Nigeria, and Rivers State, the Oil Rich Rivers State, Port Harcourt, 1967, FCO38/321, NA.
57 Permanent Secretary, Ministry of Finance, Enugu to Shell BP, 21 July 1967, "Enugu demanding oil royalties," FCO/38/112, NA.
58 UN Food and Agricultural Organization, *Agricultural Development in Nigeria 1965–1980* (Rome: UN Food and Agricultural Organization, 1966), 109 and 180.
59 Nigerian Civil War Collection, News Material, 18–23 January 1970, MS321463/44, SA; "The Tribal Tangle," *Sunday Times*, 18 January 1970.
60 Roy Doron, "Marketing Genocide: Biafran Propaganda Strategies during the Nigerian Civil War, 1967–70," *Journal of Genocide Research* 16 (2014): 227–46, https://doi.org/10.1080/14623528.2014.936702.
61 C. Odumegwu Ojukwu, "Response to address by the Efik community, Umuahia, 11 June 1968" and "Interview with John Horgan of the *Irish Times*, Umuahia, 20 February 1968," in *Biafra: Selected Speeches*, ed. C. Odumegwu Ojukwu (New York: Harper and Row, 1969), 2:132–3.
62 Although in a last-ditch effort to gain minority support, Biafra proposed a plebiscite for minorities to vote on whether they wished to stay in Biafra or not, see *Proposals for Plebiscites on the Future of Minority Tribes*, FCO65/266, NA.
63 "Statement by Lt. Col. Ojukwu, 7 October 1968, Biafran Overseas Press Division," Nigerian Propaganda, FCO65/445, fol. 2, NA; and "The Voice of Biafra, 22 December 1969," Civil War: Biafra Statements, FCO65/210, fol. 111, NA.
64 C. Odumegwu Ojukwu, *The Ahiara Declaration: The Principles of the Biafran Revolution* (Enugu: Biafra Information Service Corporation, 1969).

65 Like Dr Nnamdi Azikiwe, first president of Nigeria and former-Biafra spokesman, and Ralph Uwechwe, the former Biafran representative in Paris.

66 "Azikiwe's 14-Point Peace Proposal," 16 February 1969, 346; "Asika Looks Ahead to Reconstruction in East-Central State," March 1969, 355; "Azikiwe Appeals to his Fellow-Easterners," 8 October 1969, 422 (all in Kirk-Greene, *Crisis and Conflict in Nigeria*, vol. 2); Ministry of Information, *Ibos in a United Nigeria* (Lagos: Ministry of Information, 1968); and "Asika Appeals for a New Era," 11 November 1968, Nigeria: East-Central State, FCO65/213, fol. 1, NA.

67 "Agenda for Kampala Peace Talks" (May 1968), in Kirk-Greene, *Crisis and Conflict in Nigeria* 1:220.

68 "Enahoro Sets out Nigeria's Terms for a Settlement" (Addis Ababa, 7 August 1968), in Kirk-Greene, *Crisis and Conflict in Nigeria*, 1:272.

69 *The Position of the Ibo after the end of Civil War in Nigeria,* parts A and B, FCO65/735–736, NA.

3

Recollections of World Wars

11

Light and Not-So-Light Reflections in the *Wipers Times*' Trench Journal and in the Satirical Magazine *Punch or The London Charivari* (1939–45): What Narratives, What Recollections?

Renée Dickason

When pondering on memories of war, one can tackle the subject with an overtly biased representation (whether intimate, personal, collective, political, social, cultural, or merely ideological) or with balanced and cross-referenced argumentation. Moulding war memories is thus a subjective process that defies oversimplification and needs confrontations of viewpoints and opinions so as to grasp its full complexity. In this puzzle-like construction encompassing so many parameters, the added value of multifocal reflections on primary archives is undeniable.

As recollections of war events and testimonies can take several forms, I focus my attention here on two written media that emerged or evolved during the two world wars and that gave pride of place to humour and satire to transcend atrocities, apocalyptic events, absurd situations, and human incomprehension. One is the *Wipers Times*, a trench journal chosen for its direct evocation of First World War soldiers' states of mind, fears, apprehensions, dreams, and expectations; the other is the established satirical magazine *Punch or The London Charivari*, which aimed to sustain civilian morale on the "home front" during the Second World War. Despite their different contexts, the two media have common points: they are spontaneously reactive, there is little room for hindsight, and survival is perceived as an immediate and pressing preoccupation.

They also capture slices of life, insights into soldiers' or civilians' daily reality, and are representative of thoughts and anguish that shed light on individual but culturally determined illustrations of a universal phenomenon.

THE *WIPERS TIMES* OR NOT ALL (DIS)QUIET ON THE WESTERN FRONT

A few years back, while I was writing on *Punch* during the Great War and consulting various soldiers' narratives in the shape of letters, diaries, sketches, and drawings, I came across trench journals – living and breathing archives that encapsulate a sensitive and so palpable apprehension of life in a situation of extremes, between life and death, marked by repeated risk-taking, successive trauma (individual and collective), and an imminent need for survival in, to say the least, drastic conditions. The *Wipers Times* caught my eye and spurred my curiosity as it epitomizes so many facets of everyday themes composing *Britishness*, culturally, socially, and emotionally, along with broader universal and human preoccupations, and the fragile ephemera that sustained the will to survive catastrophic times in the desolate, bloody, and muddy trenches criss-crossing the north of France and Belgium. The *Wipers Times* constitutes a very valuable set of archives, documenting memories that tend not to be included in school history books and have somehow escaped mention in the narratives of wars. By commenting on this trench journal, my aim is also to do justice to this more discreet "monument" or "testimony" of war life, which is nonetheless so telling in insights into the frustrations and anxieties of a situation constantly bordering on the incomprehensible and the absurd.

In *The Soldiers' Press: Trench Journals in the First World War*, founder and convenor of the Australian Folklore Network historian Graham Seal extensively details the various trench journals that existed during the First World War and reminds us about the particular position the *Wipers Times* occupied. He also mentions recurring elements that are part and parcel of this type of literature: "Humour, satire, parody, cartoon and lampoon were the sanctioned modes. The stoically cheerful and the communal were generally preferred to the personal and the reflective."[1] It is this vein of reflections in the construction of war memories that I concentrate on, the fruits of the astonishingly fecund and vivid imaginations of those caught up in the conflict.

This chapter begins with the contextual understanding of the journal.[2] The *Wipers Times* is a masterpiece offering micro-analytical, often joyfully and ingeniously interwoven cultural references, impenetrable to the uninitiated and/or those living in a different period or different cultural reality. All these aspects are part of the (re)construction of what progressively comes out as the writing of war memories, a medley of real facts, fake news, illusions nourished by patriotically propagandist discourse, but also inventions of lives full of muffled expectations, fantasies, the recreation of intimacy, and the acceptance of the self and of others.

The Wipers Times*: Genesis, Context, and Aspirations*

The *Wipers Times* was born as a result of an unlikely combination of circumstances: the chance discovery in early 1916 in a cellar in the Belgian city of Ypres,[3] by members of the 12th (Pioneer) Battalion of the Sherwood Foresters Regiment, of a "slightly soiled" printing press; the presence of a sergeant, a printer in civilian life, who proposed to restore the machine to working condition; and the fertile mind of the company commander, Captain Frederick John (Fred) Roberts, who, along with Lieutenant John Hesketh (Jack) Pearson, seized the opportunity to launch an unofficial trench magazine intended for front-line troops. In all, twenty-three numbers were published between February 1916 and Christmas 1918, fifteen of them in the notorious Ypres Salient.

The moment was propitious for such a venture, relatively speaking. For some two years after the German advance of April/May 1915, the front line hardly moved, leaving space for "the long, grinding routine of trench warfare that was daily life."[4] During this time, the British lines and Ypres itself remained within easy range of the German artillery, which bombarded the ruined city night and day, leaving nobody, nothing, or nowhere ever entirely safe and establishing its unenviable reputation. As the war diary of the 1/8th Battalion of the Sherwood Foresters Regiment remarks: "The … description of [the Salient] as 'Bloody Ypres' … will be endorsed by all who ever had the misfortune to sample it at any period of the war. We have never met anyone who boasted of having found a 'cushy spot' in it."[5] Nor was the grim significance of the place lost on the enemy: it was to be the setting for Erich Maria Remarque's anti-war novel *All Quiet on the Western Front* (*Im Westen Nichts Neues*), first published in 1928 and burned by the Nazis in 1933, to which the title of this section alludes.

The journal makes no statement of its objectives: the opening editorial contents itself with the simple phrase "we have decided to produce a paper" (*WT* 12 February 1916). The circumstances of front-line service were such that the avoidance of death depended on discipline, concentration, and experience, and on a healthy measure of good luck or chance, aka a beneficent providence, although the latter is never evoked in the *Wipers Times*. Even if the paper could do nothing to alleviate the physical risks, its founders were aware of the role it could play in sustaining resolve and the will to survive.

This was achieved through a variety of complementary techniques, of which the paper's first number is fairly typical. Its ten pages contain an editorial and four pages of (spoof) advertisements, both presumed to be largely the work of Roberts himself, the reassuringly deadpan tone of the former contrasting sharply with the exuberant, sardonically humorous fictional detail in the latter. The remainder is a mixed bag of items from the pens of other (anonymous) contributors from the ranks of the PBI (Poor Bloody Infantry), notably the 24th Infantry Division to which the Sherwood Foresters Pioneers belonged. Most are in columns that were to be frequent features in future editions, offering a convenient framework into which short texts could be inserted for serious or comic effect (e.g., *People We Take Our Hats Off To*, *Things We Want to Know*, *Correspondence*, *Answers to Correspondence*, *Stop Press*, *Agony Column*), but there are also more lengthy individual prose and verse reflections and the start of a serial. Future editions were to extend this diversity with a wealth of linguistic devices (word play, puns, and deliberately inappropriate registers) and a range of genres and styles incongruously juxtaposed in the pages of the publication: one-liners, elaborate pastiches, irreverent parodies, cultured and doggerel verse, emotional reminiscences, or mere page fillers. The paper also had its catchphrase ("are we / am I being offensive enough" and its minor variants) playing on the double sense of the term "offensive," towards the enemy or towards higher authority.[6] Only two illustrations appeared in the paper before the Armistice was signed, but one of them contains this catchphrase (*NCT* 22 May 1916, see figure 11.1).

In view of the improvised nature of the journal and the "cacophonous quality of its different offerings,"[7] the following discussion cannot refer to any overarching strategy but, rather, to a variety of

QUESTIONS A

PLATOON COMMANDER

SHOULD ASK HIMSELF.

ENGRAVED BY SAPPER COUZENS R.E.

1: Am I as offensive
as I might be.

Hauptmann Van Horner,
in trench traverse corner,
Once heard what he thought
was a "goer";
But he was mistaken
Said Fritz Carl Von Haken,
"I'll write to his widow, I know
her."

11.1 *Wipers Times*, column from NCT, 22 May 1916, containing one of the only two pre-Armistice illustrations and a piece of doggerel verse with echoes of the children's nursery rhyme "Little Jack Horner."

aids to survival, to the fragile cocoons of shreds of human reflections in a situation of extremes. It invites the reader to linger over the bewildering journey traversed by the imaginations of the diverse contributors to the publication, which are both complementary and simultaneous. Among the elements emerging from this improvised pot-pourri are denial, downplaying the existence or the importance of the war, the extensive use of humour as an enduring source of relief, not to mention the journal's role both as a mouthpiece for common grievances and as a vehicle for acknowledging the sacrifices of others.

"Don't Mention the War":[8] *A Declension of Denial*

The pretence of studiously ignoring the "elephant in the room," or at least of denying the war's significance, was sustained in the early months of publication. The first editorial (*WT* 12 February 1916) sets the tone, wryly expressing the need, for reasons of censorship, to "refer to the war, which we hear is taking place in Europe, in a cautious manner," which duly occurs: the pages of the paper are devoid of details of the fighting, about which the PBI needed no reminders. The Battle of Verdun appears only three times, twice in the *People We Take Our Hats Off To* column expressing admiration for "The French of Verdun" (*WT* 6 March 1916) and simply "The French" (*WT* 20 March 1916), and in an editorial mentioning "the international show at Verdun" (*NCT* 1 May 1916), while the Battle of the Somme is reduced, in a masterly euphemism, to "a disturbance which is taking place down South" (*BEFT* 1 December 1916, see figure 11.2).

In contrast, the publication itself is often centre stage. Fictional details mimic a real newspaper's narcissistic concerns with its own survival: its price, its print runs, competition from a local rival "Messrs. Hun and Co.," its expensive special correspondents, and summer and Christmas double numbers. All this serves to sustain two typically tongue-in-cheek editorial arguments, namely, that "war is all very well in its way, but when it interferes with the publication of a journal it's a ___ ___" (*BEFT* 15 August 1917) and that producing the paper fills "all the spare time which we have on our hands during these pleasant winter months" (*BEFT* 20 January 1917). Ironical tone aside, the underlying message remains that the journal is vital to survival for, as Roberts's "Foreword" to the 1918

THE

B. E. F. TIMES.

WITH WHICH ARE INCORPORATED

The Wipers Times, The "New Church" Times, The Kemmel Times & The Somme-Times.

No 1. Vol 1. | Friday. 1st December, 1916. | Price 1 Franc.

EDITORIAL.

A LONG time has elapsed since our last number. This was unavoidable, in fact at one time it seemed that our tenth number would also be our last, as the press was marooned in the midst of a disturbance which is taking place down South. However the outfit is once more safely housed, and our new premises, although draughty, are at least in a quieter situation where the street calls and other noises are not so persistent. Many cheery faces are missing from the Division; and it seems we must get a new lot of contributors. For reasons over which we have no control we are compelled to alter the title of our journal, and so we now appear under the all-embracing name of "The B. E. F. Times." May we here and now beg everybody to send along every incident which might be adapted, humorous or otherwise. We have great pleasure in announcing that, for an enormous fee, we have secured another effort from the pen of the author of "Herlock Shomes." Owing to the popular demand some characters of the story have been resuscitated, we will leave to the reader's imagination how they got over the little difficulties in which the end of the last serial saw them. The first instalment appears in this number. We are fortunate in having another graphic article from our special correspondent Mr. Teech Bomas. Also we have opened a new branch in "Our Matrimonial Column," and any candidates for a suitable partner can probably be accommodated from the splendid selection of charming young ladies we have on our books. For this branch we have secured the services of "the brunette

11.2 *Wipers Times*, a typical title page with heading, illustration, and the start of an editorial.

published edition cryptically observed, its "hilarity was more often hysterical than natural,"[9] suggesting that the paper was as therapeutic for its contributors as it must have been cathartic for its readers.[10]

Roberts further claimed in "How it Happened" that "little incidents of daily life in the Salient were turned into adverts or small paragraphs," but this is far from the whole unvarnished truth. Many of the shorter items, notably in the *Correspondence* or *Stop Press* columns, are pure invention, while others, *Things We (Really) Want to Know*, are mere trivia: anecdotes, rumours, innuendos, or gossip designed to divert attention from unpalatable day-to-day realities while creating a sense of complicity between contributors and readers,[11] who share dubious secrets others thought concealed: certain officers' ready acquisition of whisky, the decoration of their dugouts with titillating illustrations or graffiti (respectively termed the "Kir(s) chner collection" or the "Munque art gallery"), or suggestions of clandestine dalliance (e.g., between an "incarnadined" [i.e., staff] major and a "dark-haired Belgique"). Along with such attempts to mask the realities of war and fill the absurd vacuum of endless waiting, the journal's contributors employ the comforting evasion offered by comic devices and derision. Let us now examine how the paper strove to sustain spirits to cope with the harshness of war.

"Through Dread of Crying You Will Laugh Instead": The Use of Humour and Cultural References

As suggested in this line from the poem "With the Usual Apologies" (BEFT 1 November 1917), a parody of Rudyard Kipling's famous "If," each edition of the paper makes use of a wide diversity of humour (frivolous, absurd, ironical, dark, or derisory) to help create a "free-floating circuit of logic independent of the outside world,"[12] which sustains a detached attitude to the conflict and thereby preserves the will to survive in a hostile foreign environment.

Very little in the paper is without several layers of meaning, but a few items intend nothing more than amusing casual diversion from the daily routine. One-liners (e.g., "Who discovered the salient. / Why?" WT 26 February 1916, and "What does an 'A' frame?" BEFT 25 December 1917) may attract a brief smile, and the numerous inventive parodies and pastiches are equally distracting through their implausible transposition of the originals to a different time and place: for instance, the brilliant detective Herlock Shomes, his sidekick Dr Hotsam,

Restoration diarist Lieutenant Samuel Pepys, and, in the *As of Old* stories, Old Testament figures, all find themselves in the trenches. All these familiar cultural references evoke facets of British common popular culture and collective memory. They also reflect levels of education through examples appealing to the literary knowledge of the social classes involved in the battlefield *huis clos*, where quoting Kipling or Pepys's *Diaries* has a particular pertinence for a desperate need to feed the imagination so as to escape from the horrifying reality of war. These, like the transparent and acknowledged allusions of the "Warlord and the Chancellor" (*BEFT* 1 December 1916) and "A Dweller in Wipers' Elegy to That Town" (*WT* 6 March 1916),[13] would have been familiar to a large percentage of the PBI. On the other hand, only the best educated of readers would have been able to appreciate the humour of such parodies as the "Rub*á*iy*á*t of William Hohenzollern" (*BEFT* 10 April 1917).

While appealing similarly to memories of a kinder, familiar world far from the reality of the trenches, many of the paper's numerous spoof advertisements have a harsher comic edge,[14] for very few of them are without potentially disturbing ironical overtones. The announcements for fictional Music Hall venues are a good example. Each edition of the *Wipers Times* promotes its own variety theatre, with names such as Opera House, Electric Palace, or Hippodrome, associated with actual places where the Pioneers were stationed (see figure 11.3).

On the face of it, they are a reminder of happier times, evoking what Robert L. Nelson calls a "unifying cross-class popular culture" common to officers and ORs.[15] Just below the surface lies a darker reality, however. The Cloth Hall Ypres/Cloth Hall Wipers, a fictional venue on no fewer than nine occasions, is praised for the quality of its "ventilation," a grim joke as this gem of thirteenth-century Flemish architecture was no more than a roofless shell by 1916. Many of the "acts" also carry their share of premonitory word play. Placed in an incongruous Music Hall setting and called by familiar nicknames, shells and explosives might seem to lose some of their power to intimidate, but they remained a constant, unpredictable threat to life as the word play in the subtexts demonstrates: "The Great Silent Percy Brings the House Down" (*WT* 26 February 1916); "'Big Guns.' ... Including that versatile and popular Artiste 'Harry Howitser as the Housebreaker.'" "'A thundering good show'" (*NCT* 29 May 1916); "Miss Minnie Werfer. Always Meets with a Thunderous Reception"

(BEFT 25 December 1916). In other adverts the irony is manifest with troops themselves becoming the performers, executing unlikely dance-steps such as the "Duck-board Dangle" and the "Whizz-Bang Hop" (NCT 22 May 1916) or appearing, on the eve of the Battle of the Somme, as "Professor Scrapper's Performing Troupe of Highly-trained Animals" (KT 3 July 1916).[16]

To alleviate tensions in a world of uncertainty, the paper makes full use of the potential of absurdity as a source of derision or (self-)mockery. "The Sybarite's Soliloquy" to his "Dearest, at break of Dawn, I need you most," who, it transpires, is none other than a "Stick of Superfatted Shaving Soap" (BEFT 25 December 1916), is a classic case of self-deprecation, albeit with a sidelong glance at front-line privations. Nevertheless, the force of much of the paper's ridicule depends on point of view. Domestic visions of "normality" come off worse when they collide with the reality of the trenches. Lengthy fictional correspondence triggered by a reader's claim to have heard a cuckoo while rambling at night along the Menin Road (WT 12 and 26 February, 20 March 1916) or letters complaining about the state of the roads and the street lighting in Wipers enable serving troops to mock the futile, petty preoccupations of newspapers at home. Similarly, they were sure to see the pure absurdity of the idea of falling into a police (speed) trap on the Menin Road between Hell Fire Corner and the Culvert (WT 12 February 1916) or of buying the Salient Estate, despite its attractions of perfect shooting and good fishing (WT 20 March 1916).[17] Some advertisements mock the presumed ignorance of readers at home by making ludicrous, highly alarming suggestions. One in NCT 17 April 1916 recommends sending a soldier friend an "Improved Pattern Combination Umbrella & Wire Cutter," promising "no more colds cutting the wire," while another promotes an equally hazardous purchase: "Velveteen Corduroy Plush Breeches to enable a man to Look Like a Soldier when Going over the Top."[18]

Absurd army procedures are another sitting target for derision in the WT. The language of quartermasters (QMs) ("the tribe of Kew … speaking a strange language whereof the last words do appear to be the first" [BT November 1918]) includes suitably ridiculous minor examples ("pencil, ink, copying, one" [BEFT 10 April 1917]) and the ubiquitous "gum boots thigh." But copious scorn is reserved, sometimes in verse, for the burdensome bureaucracy of QMs' repeated demands for futile form-filling at critical moments. *An Ode to Q* sums up the mood: "The foe comes on in countless thousands /

THE

B. E. F. TIMES.

WITH WHICH ARE INCORPORATED

The Wipers Times, The "New Church" Times, The Kemmel Times & The Somme-Times.

No 1. Vol 2. Wednesday, August 15th, 1917. Price 1 Franc.

CLOTH HALL, WIPERS.

Under Entirely New Management.

THE VENTILATION OF THIS THEATRE HAS BEEN ENTIRELY OVERHAULED DURING THE SUMMER MONTHS.

SPECIAL ATTRACTIONS.

—o—o—o—o—

Haig's Company in a Stirring Drama, Entitled:

PILKEM'S PROGRESS.

—o—o—o—o—

WILLIAM'S TROUPE:—
"THE COCKCHAFERS"
IN A HUMOUROUS KNOCKABOUT SCENE.

—o—o—o—o—

MANY OTHER ATTRACTIONS.

—o—o—o—o—

PRICES AS USUAL. BOOK EARLY.
WE HAVE A BOMB PROOF CELLAR IN THE EVENT OF AIR-RAIDS.

11.3 *Wipers Times*: advertisement from *BEFT*, 15 August 1917.

Bearing down with savage cry. / Jones receives the frantic message / 'Indent now for gum boots thigh'" (*WT* 26 February 1916). In a similar vein, "Letter R of the BEF *Alphabet*" (*BEFT* 5 March 1917) waxes lyrical about "Returns to be rendered by noon / Of the number of men who have seen a blue moon / Speak Japanese or who have been to Rangoon, / Before they came out to the trenches." Inefficiency or incompetence are a further target for mockery. "*A Story without a Moral*" (*BEFT* 10 April 1917) tells a tale of woe whereby, "in due course," after an administrative paper chase between high-ranking officers known only as AA, DAA, QMG, and DAQMG and in accordance with CRO 758,[19] fifty men who have refused to be inoculated are mistakenly awarded the Military Medal instead. What remains unsaid is that those who should have been decorated find themselves in line for a second injection. This is cause enough for derision, for outsiders, but what happens in "Ours or Theirs" (*BEFT* 5 March 1917) is well beyond laughter, the account of a full-blown military cock-up in which, thanks to poor communications, carelessness, and defective equipment, a test firing organized by a Heavy TM officer and a Medium TM officer ends with a "toffee apple" (aka trench mortar) destroying the battalion's own cook house and latrines. The scene is set in "Any Battalion Headquarters in Line," which suggests that such mishaps are no rarity, which can easily prompt the short step from derision via complaints to anger and resentment.

The Virtues of Truculent Cheerfulness:[20] *A Cathartic Safety Valve?*

Life in the Salient, and the other theatres of war where the 24th Division was engaged, was not only risky and frightening but routinely uncomfortable. In such circumstances, grumbling easily became a way of life. Contributions to the journal voice a wide range of more or less serious grievances, which help to engender a feeling both of shared suffering and of resolution in the face of adversity, while paradoxically ultimately maintaining a sense of unity and of common purpose.

Complaints could provoke resigned humour or varying degrees of grumpiness or exasperation, depending on the frequency and the seriousness of the problems encountered. At one end of the annoyance scale come the persistent rain, the resulting mud ("Belgium's finest") and flooded trenches, which the paper describes light-heartedly as suitable for fishing in the absence of "the U-boat menace" (*BEFT*

15 August 1917). Other frequently voiced but comparatively minor grievances concern poor and repetitious food, the least appreciated being "tickler" (i.e., jam, almost invariably plum and apple) and maconochie (a tinned meat and vegetable stew named after its manufacturer), and the unreliable supply of alcohol (rum undersupplied by quartermasters or purloined on the way to its destination). Grumbles about unsuitable materiel include the comparatively trivial, the nail and hammer in the Pioneer's lament (illustrated in *WT* 6 March 1916), with which he is expected to build a dugout; the consistently annoying, cumbersome A frames, intended to shore up parapets and walls, which Pioneers transport to the front at night, either at high risk above ground or in the trenches, while balancing precariously on unstable duckboards; and the downright dangerous, defective shells that fail to explode ("duds") or do so in the wrong place ("short ones"), putting friendly forces at risk. All these laments are notable for their regularity rather than their originality.

These examples are, however, far from exhausting the litany of woes echoed in the pages of the paper. Many concern favoured individuals or, more generally, those who fail in their patriotic duty. It gives short shrift to those who refuse to join up, caricatured by a certain S. Lacker (*WT* 26 February 1916) and "Wata Funk, The Conscientious Objector," appearing in the film *He Didn't Want to Do It* at the "Dead Cow Farm Cinema" (*NCT* 17 April 1916). On the real war front, the PBI found other recalcitrants to grumble about, "those 'back' who have never been through, / Or stood their whack in the trenches" (*BEFT* 5 March 1917), such as an anonymous master-general of ordnance (MGO) who has the onerous task of commanding a camp at Pop(eringe), with a personnel of three NCOs and one private (*WT* 6 March 1916). The general complaint is that such staff officers are pampered individuals who enjoy unmerited privileges, a sentiment cogently expressed in lines from "Rhymes without Reason, by PBI" (*BEFT* 20 January 1917): "Realising Men must laugh, / Some wise man devised the Staff: / Dressed them up in little dabs / Of rich variegated tabs / ... Often, often gave them leave, / Decorations, too, galore." In a similar vein, "Late News from the Ration Dump" (*BEFT* 1 November 1917) makes the exasperating, but fortunately apocryphal, announcement that "pay for men at the base is going to be doubled owing to the increased cost of living, and halved for those in the line as they are not in a position to spend it."

These extended "flights of outrage" indicate that truculence must have been second nature at various levels in the military, and contributions in the *Wipers Times* serve both as a mouthpiece for grievances and as a safety valve for anger and frustration. Their effect is not necessarily negative, however, for they do not imply disloyalty, disobedience, or dereliction of duty as, along with a common language[21] and shared experiences, good and bad, they help create a sense of unity in the face of adversity. By the time the journal was born, any hope of rapid victory had long since disappeared, but survival remained a necessity if one wished to remain loyal to living comrades or show due respect to the fallen. These sentiments are summarized, as well as anywhere, in the Kipling-esque "With the Usual Apologies," already quoted (*BEFT* 1 November 1917), which ends with "If you can ... / think a muddy ditch a bed of clover, / You'll be a soldier one day, then, my son." Such serious messages of the positive value of resilience and endurance appear with increasing regularity in the later pages of the journal, many of them phrased in the simplest of terms, which give an impression of both sincerity and commitment: "never mind" (*NCT* 1 May 1916); "cut the grousing and for God's sake wear a grin" (*BEFT* 10 April 1917); "are we downhearted?" "NA POO!" (*BEFT* 26 February 1918). In this surprising world of new-found harmony, disagreements give way to expressions of unity between ranks, as in the "Song of Any Infantry Brigadier to His Men" (*BEFT* 15 August 1917): "I know the stuff of which you're made, / I know the old 'Umpteenth' Brigade. / ... Just understand (it's nothing new) / How proud I am of all of you," or between units. "Their Union Our Strength" (*BEFT* 10 April 1917) concludes that the trust between the three rival branches – gunners, sappers, and PBI – "has stood the test, / ... Each at his job has proved the best."

More than anywhere, however, it is in the respect for the dead that such sense of common purpose prevails, when the paper abandons its editorial stance of ignoring the war. *BEFT* 1 December 1916, the first number published after the Battle of the Somme, laments the loss of "many cheery faces ... missing from the Division," and similar understated respect is expressed in *BEFT* 15 August 1917 for the loss of "many old chums," including three lieutenant colonels mentioned by name in what amounts to a short, black-lined obituary column. *Mort pour la France* (*BEFT* 1 November 1917) extends the idea of common sacrifice by offering a timely reminder of losses suffered by Britain's principal ally, but the most striking of all such messages is the simple poem "To My Chum" (*WT* 16 March 1916). In this poignant anonymous tribute, the author summarizes the will to survive, evoking the good and bad experiences he shared with his now dead pal, whose only memorial is a wayside cross, but whose sacrifice will inspire others to carry on "till the end of the game." Such examples serve to reinforce the destructive horrors of war and the need and duty to remember those who valiantly gave their lives to defend their country in order to ensure the future of their compatriots.

The Wipers Times, *an Archive for Posterity…*

The *Wipers Times* was created as a vehicle to entertain, sustain, and unite troops in the Ypres Salient, but it was never intended as a vehicle for memory: it remained throughout a publication with a life expectancy potentially as short as those of many of its readers and contributors. Nevertheless, it was a resounding success: the first print run was of only one hundred copies, but later editions doubled or trebled that number, and the reasons for this popularity are not hard to find. It was written by and for front-line troops, officers and men, it focused on their own concerns and irritations, and contributions could be read collectively and/or aloud, which had the advantage of bringing the sometimes excruciating puns and word play to life. It therefore had an authenticity that publications destined for home consumption but that found their way into the trenches did not.[22] It also met another unexpected need: the desire of families at home for information they could not receive,[23] because of censorship of soldiers' mail and, no doubt, because men on leave were often reluctant to shock their loved ones by telling the horrible truth about life at the front.[24] The frontier between this *huis clos* and Blighty was nevertheless sufficiently porous for news of the *Wipers Times* to seep out, and interest was strong enough to justify the first publication of the journal in (truncated) book form by Herbert Jenkins in 1918.[25]

One further question remains concerning the wartime survival of the paper: its relationship with censorship and higher authority. The former is broached in the opening editorial, but signs of direct intervention from above are rare.[26] Elsewhere, references are clearly light-hearted: the words "censored ed." appear occasionally and the entry for "E" in the "BEF Alphabet" (*BEFT* 5 March 1917) suggests that the editor would "blue pencil this if I said all I knew." How much control Roberts himself exercised over the content of the paper must remain a matter of conjecture, but his touch does seem to have been light, in part no doubt because the paper's survival depended on the willingness of volunteers to provide copy.

Nevertheless, if the paper contained little that would have been of use to the enemy, it was often unquestionably insubordinate, to say the least. Criticism of staff officers, quartermasters, and mocking pastiches of military procedures were more than occasional features of the paper, and some of its columns (e.g., *Things We Want to Know*, *News from the Ration Dump*, *Military Definitions/Terms Defined*, and *Correspondence*) offered a ready framework for such ridicule. That these remained uncensored may be due to a variety of factors: a particularly fraught military context, which encouraged the authorities to tread lightly; the anonymity of the contributions and the absence of specific or identifiable targets; and the fragmented nature of the paper, with potentially offensive material interwoven with the simply diverting. It is nevertheless surprising that certain items got through unscathed, most notably three advertisements. The gallows humour of the first might seem to have been detrimental to morale at the key period of the Battle of the Somme: "Are You a Victim to Optimism?" (*ST* 31 July 1916) proposes a remedy for the dread disease of optimism, "We Can Cure You. Two Days Spent in our Establishment will Effectually Eradicate all Traces of It from Your System." The second appears sceptical about the conduct of the 3rd Battle of Ypres (*BEFT* 8 September 1917) through a reference to the Western Advance Co., General Manager D. Haig, a fictional financial institution capable of making advances at short notice requiring (or should it be guaranteeing?) "no security." The following number of the paper (*BEFT* 1 November 1917) comes close to making a damaging declaration of war weariness with a comically framed but nevertheless earnest plea for the fighting to end by promoting a Great Anti-War Demonstration at Passchendaele organized by the Great Army Peace Movement presided over by D. Haig, Esq and with speakers R. Tillery and T. Atkins. The messages here are unequivocal and are ominously representative of thoughts and opinions that were current in fragmented form at the time and that now resurface in other media and other contexts.

The war and its sacrifices were omnipresent in hearts and minds for much of the 1920s. Survivors and their stories were very much alive, and the decade was marked by commemorations, the establishment of the minute's silence at 11:00 a.m. on 11 November, the unveiling of national and local war memorials, and the creation of war cemeteries and of the Menin Gate Memorial in Ypres, which bears the names of more than fifty-four thousand soldiers who have no known grave. The publication of the first complete version of the paper, *The Wipers Times: Including for the first Time in One Volume a Facsimile Reproduction of the Complete Series of the Famous Wartime Trench Magazines* (London: Eveleigh, Nash and Grayson, 1930), demonstrated both that original copies had survived and that the conflict had lost none of its interest.

It was, however, over forty years before the *Wipers Times* was republished. The Second World War, the "people's war," in which around sixty thousand British civilians perished, brought its own quota of stories and memories. Of course, the Great War was never completely forgotten, and the 1960s saw it return to centre stage as the fiftieth anniversary of the conflict approached, commemorated notably by an authoritative BBC documentary of the same name screened between May and November 1964, which included a substantial amount of footage from the above-mentioned 1916 film *The Battle of the Somme*. More or less simultaneously in this irreverent and imaginative period, alongside historian Alan Clark's vehement criticism of First World War generalship (*The Donkeys*, 1961), came Joan Littlewood's *Oh! What a Lovely War* (stage production 1963, film 1969), which combined scorn for the military hierarchy with a sympathetic evocation of the culture and resilience of those in the trenches, a vision smacking strongly of the pages of the *WT*. Moreover, the decade spawned new forms of comedy, satirical, witty, and exuberantly anarchic,[27] strikingly similar to the techniques in the journal, with which a new generation of readers could therefore identify.

Indeed, it was not long before a new edition appeared, *The Wipers Times – A Complete Facsimile of the Famous World War I Trench Newspaper, Incorporating the "New Church" Times, the Kemmel Times, the Somme-Times, the BEF Times and the "Better Times"* (London: Peter Davies 1973), which cast the journal in a new role. The publication included copious notes and an extensive glossary by Patrick Beaver, whose introduction stated the aim of making the *Wipers Times* a vehicle of commemoration through which Ypres could be celebrated as "a symbol of tenacity and victory" for the British, just as were Verdun and Vimy Ridge for the French and the

Canadians, respectively. To this end, Beaver called upon the first-hand recollections of the diminishing number of ageing survivors in order to explain "the more puzzling 'advertisements' and other pieces" in the journal while the opportunity to solve such "riddles" still existed.[28] The new edition was therefore intended to connect the memories of veterans with younger readers and inspire the latter to try to explain unsolved mysteries for themselves.

The imminence of the centenary of the Great War sparked substantial fresh interest in the *Wipers Times*. In 2006, Little Books published a new edition (with introduction by historian Malcolm Brown and foreword by Ian Hislop, editor of the satirical magazine *Private Eye*), reissued in 2013 to coincide with the broadcasting of a TV play, *The Wipers Times*, by Ian Hislop and Nick Newman, which was screened by the BBC in September 2013 and subsequently travelled as a stage performance. These latter ventures, widely acclaimed for their faithfulness to the spirit and often to the letter of the original, played a substantial role in bringing the paper to the attention of a wider public and secured its place as a monument to one of the most traumatic events in recent times.

A hundred years after the end of the Great War, the *Wipers Times* lives on. Thanks to the recent anniversary commemorations, it is more widely known now than at any point in its history. Like trench journals in general, which have benefited from its longevity and notoriety, the paper is now enjoying a new lease on life as a valuable, "authentic," uncensored primary document to be exploited by current and future generations of historians and researchers in other fields. More than all else, however, it stands as a memorial both to those it helped to survive the horrors of the trenches, in the Ypres Salient in particular, and to those who perished. It offers an insight that other sources cannot match into the individual emotions and the collective mind-set of those who fought and died for their country, and it is an enduring reminder of the remarkable courage, the unshakeable determination, and the sheer resilience typical of all those, on both sides of the barbed wire and in whatever theatre of war, who were willing or unwilling participants in the "war to end all wars."

The atmosphere conveyed and the wealth of details sometimes lavishly displayed in the various examples extracted from the *Wipers Times* give a particular human flavour to war memories, which official restitutions have, all too often, failed to reconstitute or recreate. In the cultural history cornucopia of memories, sharing of experiences, and recollections of things personal and collective, the medium of communication now to be examined is at once dignified, comic, and satirical: the cartoons published between 1939 and 1945 in *Punch*, a magazine already renowned for the quality of its art work, and one whose pages, by the time of the Second World War, had already discussed and depicted major conflicts involving Great Britain: the Crimean War, the Boer Wars, and the Great War. Its Second World War cartooning was initially strongly reminiscent of the methods and techniques employed during the Great War and reflected the mentality that had prevailed in 1914–18. This was, however, merely a starting point for the Second World War, after which, since the context was so radically different, *Punch* contributors found themselves faced with the task of discovering and developing new approaches suitable for what was no longer a distant military adventure but a "people's war" to be won on the home front. The content of both publications examined in this chapter was conditioned by immediate events and experiences, and both, implicitly or explicitly, acknowledged that war was too serious to always be taken seriously or, indeed, at face value and that humour was a suitable antidote to unrelenting doom and gloom. Although the diverse texts in the *Wipers Times* and the cartoons in *Punch* were intended for immediate consumption by a specific (narrower or wider) audience, they carried messages that were multifaceted, with layers of meaning that have enabled them to reinforce the recollections and incite the curiosity of readers well beyond those originally targeted.

A CARTOONING MISSION: *PUNCH OR THE LONDON CHARIVARI* (1939–45)

Different wars, in changing times, places, and circumstances, each bring their particular challenges for the media, to which each publication has to find its own appropriate solution. During the Great War, the London newspaper of record, the *Times*, employed its resources and experience to produce, alongside its extensive coverage of ongoing events, *The Times History of the War* (22 volumes), in weekly instalments from 25 August 1914 to 27 July 1920, asserting, in the words of the preface, that it would be "written by men of great experience in political, military, and naval matters" and claiming that it would have additional value as a living archive, containing, as it did, "a great deal of first-hand material that [would] be really valuable to historians of the future." Needless to say, such narratives, even if written by "experts," follow the flow of events and offer immediate insights/reactions without the benefit of hindsight, which guides the reflections of later analysts.

Another valuable, if apparently less serious, technique among the so many facets available when remembering wars, is the revisiting of facts and events in times of conflict, with a potentially lighter touch offering measured proportions of amusement and (self-) mockery. The Second World War is a case in point: it has long been a subject of fascination for the British media, which are traditionally inclined to use humour, derision, sometimes even sarcasm and cynicism, to deal with either past or current events. Indeed, by the 1930s, cartoons had become an essential feature of a flourishing daily press and were proving to be an effective means of casting an unexpected, divergent glance at events. The weekly magazine *Punch* is a particularly pertinent example for the study of the Second World War, for practical as well as for historical reasons. The wide availability of *Punch* cartoons in anthologies and on websites (notably the World War II Gallery and those of individual artists at World Warw.punch.co.uk) far exceeds that of other illustrations.[29] By 1939, the publication was well established, having been founded in 1841, and was justly proud of its reputation and, in particular, of its artwork for, as Mark Bryant and Simon Heneage – respectively honorary secretary and later vice-president of the British Cartoonists' Association, and co-founder of the London Cartoon Art Trust – explain, it was thanks to the magazine's first comic illustrator, John Leech, that "the word 'cartoon' began to acquire its modern meaning of 'funny picture,'" distinguishable from "'caricature' [which] came to stand for distorted portraiture emphasizing characteristic traits."[30] By the outbreak of the Second World War, its position as a comic magazine was secure for, along with substantial advertising revenue, it enjoyed the loyalty of a mostly middle-class readership who appreciated its cartoons, with their "humour of the gentler sort."[31] Moreover, like the *Times History*, albeit in a very different vein, *Punch* proclaimed itself as a publication with a Janus-like mission, as a guardian of what might be termed archive value, connecting past, present, and, as it turned out, future in unequivocal terms in an advertisement placed in the magazine itself on 20 March 1940 intended to attract new readers or to persuade existing readers to renew their subscriptions. It runs:

> PUNCH in War Time
> In 1914–1918, PUNCH was regarded as essential to the Allies' War effort.
> Again now, it is the endeavour of PUNCH to do everything possible to maintain the good spirit of the NATION and her ALLIES in the present conflict.
> PUNCH is considered a faithful historian and a humorous recorder of the way in which the general public are "carrying on" in Wartime. Do not deny yourself the pleasure of its stimulating cartoons, humorous drawings, poems and articles.

To reinforce *Punch*'s view of itself as a serious publication with both a historical vision as a preserver of Great War memories and a contemporary role, the advertisement was completed by two thematically similar illustrations reproduced from its own pages, scaled-down versions of whole-page cartoons (known to *Punch* as "big cuts"): Bernard Partridge's "Unconquerable" (21 October 1914), in praise of Belgian resistance to German invasion, and E.H. Shepard's "Holding the Bridge" (24 January 1940), featuring Finland as a latter-day Scandinavian Horatio standing firm against Russian aggression.

In fact, *Punch* in 1914 had been much more hesitant than the above advertisement might seem to suggest. "At the outbreak of war, [editor Owen] Seaman had been doubtful whether there was any place for *Punch* at all in wartime. His natural tendency was to see public affairs dramatically and this made him unable to conceive of humour as emerging from the conflict."[32] In 1939, however, the magazine was more obviously at ease with its role. The Great War had shown that humour in various guises, and cartoons in particular, had a part to play in sustaining civilian morale with the cathartic effect of banishing, however temporarily, the pain and grief occasioned by the fighting and its inevitable consequences.[33] It was soon to become clear, as *Punch*'s advertisement implied, that the battle front/home front dichotomy of the Great War no longer existed: "carrying on" was a task for military and civilians alike, as the latter themselves were also liable to become war casualties.

This discussion of *Punch*'s role in the process of commemoration begins with a brief outline of the favourable but testing situation in which the magazine found itself in 1939, before moving on to consider the development of its new vision of heroism, more appropriate to the context. There follows an examination of the essential role played by its cartoons, throughout the conflict, in defusing the persistent tensions of life on the home front by playfully evoking common everyday experiences that were to become the stuff of popular memory. Finally, it considers how *Punch* used historical parallels and a dose of humour to instil realistic expectations of what peace might bring.

Opportunities and Challenges

Opportunities for *Punch* were plentiful. The magazine was well placed to meet the demands of wartime illustration. Its art editor, appointed in 1937, was the cartoonist Fougasse,[34] who had at his disposal a large and diverse team of talented, younger contributors to complement the work of the magazine's staff illustrators: Bernard Partridge (chief cartoonist since 1910), second cartoonist E.H. Shepard, and Leslie Illingworth, alongside veterans such as George Stampa, George Belcher, Frank Reynolds, and George Morrow. The latter two were both sometime art editors. Partridge, Shepard, and Illingworth were responsible for drawing the big cuts, whose subject was decided at the weekly lunch-time meeting of the Punch Table (i.e., the magazine's editorial board), but the other contributors enjoyed considerable liberty over the choice of their subject matter and the manner of handling it. Thus, *Punch*, a weekly publication free from the constraint of daily papers to be up to the minute, had the great advantage of being able to treat the ongoing themes of the war from various, more or less conventional, angles or to publish cartoon series such as Pont's *British Character* (begun in 1936 but with content blending seamlessly into wartime numbers under the headings of *Popular Misconceptions* and *Wartime Weaknesses*) and Fougasse's *Changing Face of Britain* and its successors.

By early 1939, after Germany's occupation of Czechoslovakia in breach of the Munich Agreement, war appeared inevitable, and the British government began imposing regulations that were to mark the entire conflict. They were also to provide an ongoing array of subjects for cartoonists and to loom large in individual and collective recollections: rationing (petrol, food, and later clothes and household goods); the evacuation of children from towns and cities likely to suffer aerial bombardment; the need to carry gas-masks; air raid precautions (ARP) and the blackout; and military conscription. Britain's declaration of war aroused none of the popular enthusiasm seen in 1914, and the lack of immediate conflict directly involving the United Kingdom during the lengthy period of the Phoney War brought with it anxiety and presented cartoonists with the dilemma of how to strike the right chords to sustain national morale, a challenge that their reliance on the established practices of traditional war cartooning often initially left them unable to meet.

Punch did publish a few light-hearted, early wartime cartoons. Pont's "A.R.P. Dept." (21 December 1938), showing a well-dressed lady incongruously declaring, "I feel sure we [by which she meant I] could drive a fire-engine," was simply funny, but many others shunned all frivolity, unashamedly attempting not merely to reflect opinion but to mould it along lines that now seem utterly predictable. Illingworth's "Call to Training," 1 February 1939 (see figure 11.4 below) and Shepard's "Registered" (27 September 1939), both big cuts, struck an unreservedly patriotic note, evoking the archetypical personification of Great Britain – John Bull – to instil a national sense of destiny and to remind current readers that they should do their duty just as their ancestors had done. Another recycled, tried and tested theme that had been a mainstay of wartime cartooning from as early as the Napoleonic Wars was denigration of or derision for the enemy. Hitler was systematically caricatured, but initially inspired fear, contempt emerging only much later, once it was clear that Germany's military fortunes were on the wane. Other Axis leaders enjoyed no such period of grace. Mussolini appeared, from the start, as an untrustworthy, inconsistent figure, "Humpty Dumpty on the Roman Wall," unable to decide whether to join the Allies or Germany (Shepard, 9 May 1940). The Italian declaration of war on Britain and France, only days before the latter's capitulation, prompted Illingworth to reduce the *Duce* to bestial status of the worst kind in "The Hour of the Hyena" (17 June 1940), and the subsequent consistently unimpressive performance of Italian armed forces made him even easier to lampoon. Goering was another figure of scorn, his rotund figure allowing him to be caricatured as a barrage balloon in Illingworth's ironical depiction of him as the "Great

THE CALL TO TRAINING

11.4 Illingworth, *Call to Training*, 1 February 1939.

Protector" (31 January 1940), confiscating all private property, but it took the failures of the Luftwaffe under his command to complete his transformation into a figure of ridicule.

Other big cuts expressed an understandable, but soon to be dashed, enthusiasm by giving a uniformly positive image of what were, from the British point of view, still remote events of the war, while adopting a symbolism and an exaggeratedly heroic tone recalling memories of 1914–18. Among the more implausible scenarios evoked in the autumn of 1939 were direct allied military assistance to Poland in her hour of need (Illingworth, "Three as One," 30 August 1939), successful Polish resistance to the German invasion (Partridge, "Salute to the Brave," 20 September 1939), and imminent joint Anglo-French combat operations (Shepard, "Vive l'entente. 'I think our fathers fought together,'" 20 September 1939). The trend continued into the following year. Partridge's "British Blockade," 3 April 1940 and "The Lion at Bay," 20 May 1940, along with Shepard's "Sea-King's Brothers from over the Sea" (1 May 1940, referring to the Norway campaign) were examples of similar wildly optimistic estimations of British and Franco-British military capacity. If that were not enough, cartoonists' imaginations even extended to discerning straws in the wind for a successful long-term outcome to the conflict through untimely suggestions that foreign (American) military aid might be coming (Shepard, "Wings from the West," 15 November 1939), that Hitler might already have overreached himself and neglected the lessons of history by starting the war (Partridge, "Crystal Gazer," 6 September 1939), or that Russian-German cooperation would be short-lived (Partridge, "The Two Constrictors," 8 November 1939). Of course, the constraints of censorship were such that cartoonists were in no position to know the real truth of the military or international situation and, no doubt, had no desire to mislead, but it would seem difficult, even without hindsight, for them or their readers to have turned a blind eye to the discrepancy between the over-optimistic content and messages of these illustrations and the ever more pervasive government regulations designed to ensure the safety of the population in the event of bombardments or invasion.

Revisiting Heroism

An abrupt reality check was about to become crucial. The sudden end of the Phoney War with the German airborne invasion of the Low Countries, prophesied with uncanny accuracy in Illingworth's "The Combat" (6 November 1939), was rapidly followed by the evacuation of British and Allied troops from Dunkirk (Operation Dynamo), and the ensuing fall of France put an end to any lingering optimism.[35] *Punch*'s weekly publication schedule allowed a brief, but necessary, breathing space for editorialists and cartoonists to find new perspectives from which to view events. The small vessels that participated in the rescue from the beaches of Dunkirk first appeared in a cartoon only on 4 September 1940, in Shepard's "The Little Ships Sail On." By then, other media, notably Churchill's speech to Parliament on 4 June 1940, novelist J.B. Priestley's radio broadcast *Postscript* the following day, and selective footage on Pathé news had firmly established the successful evacuation of British troops from Dunkirk as a miraculous escape. *Punch*'s role was therefore to sustain rather than to create the myth that Shepard's illustration exploited with recourse to an exaggerated historical parallel, the victory of Sir Francis Drake over the Spanish Armada in 1588, Churchill's speech to Parliament having already, more realistically, warned that "wars [we]re not won by evacuations." A little closer to reality, the Battle of Britain, with the defeat of the Luftwaffe and the postponement of Operation Sealion, the planned German invasion of England, was illustrated in the most favourable light by powerful maritime imagery of a different kind, with Britain being the rock on which the storm of Goering's air forces pounded in vain (Shepard, "The Rock and the Storm," 25 September 1940).

Most important, Dunkirk and the Battle of Britain offered cartoonists what had thus far been lacking in the war: the opportunity to concentrate not only on the smaller picture, the daily concerns and occupations of ordinary citizens, but also on the vision of a new kind of war hero, not the political and military leaders who were to feature later, Churchill and Montgomery being prime examples, but the civilians risking their lives on the ships at Dunkirk and the small number of mostly anonymous RAF Battle of Britain pilots, dubbed "the few" by Churchill. Airmen were to appear regularly in subsequent cartoons, typically depicted as the prime minister had portrayed them: brave but calm, unassuming, and with a ready wit (e.g., Pont's, "But you must remember that I outnumbered them by one to three," 19 June 1940; and Lawrence Siggs's imperturbable bridge players, "Dummy pops off and attacks the docks at Brest – OK," 27 August 1941). Other unsung and unassuming heroes appeared in *Punch* cartoons: civilian "Watchers of the Sea" in a church tower, on the lookout for invaders and ready to sound the alarm by ringing the bells (Shepard, 18 September 1940), or firefighters "On the Front Line" after bombing raids (Partridge, 25 September 1940). All served to boost national morale, and all exemplified the qualities that would enable other civilians in real life to be heroes on the home front, "carrying on" in the face of the rigours of this people's war.[36]

Smiling Through[37]

The majority of the cartoons discussed thus far have been big cuts offering serious treatment of military events and issues of the war, and leaving little room for amusement. Nevertheless, it was no secret that comedy had played a significant role in sustaining morale in the Great War, and it was to do so to an even greater extent in the Second World War. In *Punch* illustrations dealing with the vagaries of life on the home front, this humour was often conveyed in subtle, understated, almost unobtrusive ways, snapshots of everyday life relying on superimposed touches, minor verbal or visual details, implicit comparisons between what they showed or told (i.e., between the images and the words in the captions and/or titles), or by the gentle evocation of lateral thinking bringing an original viewpoint. A smile rather than a laugh was thus the mood of the times, and humour was to prove all the more essential as war dragged on … and on. The evacuation from Dunkirk and the Battle of Britain were, in fact, only the prelude to worse things to come. The Blitz, the bombing of London and other major cities in 1940 and 1941, and other later bombardments left some sixty thousand civilians dead, to say nothing of collateral material damage. Admittedly, the start of Operation Barbarossa, the German invasion of Russia in June 1941, meant that Britain was no longer alone but,[38] rather, had powerful, if distant, support from the USSR; and the Japanese attack on Pearl Harbor finally brought the world's other great military power into the war on the Allied side. But neither of these developments had much immediate impact on the home front. Indeed, even after the successful Normandy landings, the war still had a sting in its tail, as London was attacked once more by the Germans' new superweapons, the V1 flying bombs and the V2 rockets.

Throughout this long haul, civilians were subject to privations and the rigours of austerity, and they were constantly assailed by a plethora of well-intentioned but sometimes contradictory regulations and instructions, leaving them understandably confused or perplexed, as suggested most graphically in Fougasse's "If only they'd tell us all what to do" (see figure 11.5).

It was, of course, far from the magazine's mission to criticize government policy, and it had no wish to suffer the same fate as the Communist *Daily Worker*, which was closed on the orders of the home secretary from January 1941 to September 1942 for

"If only they'd tell us all what to do."

11.5 Fougasse, "If only they'd tell us all what to do," 25 March 1942.

undermining the war effort.[39] Rather, *Punch* endeavoured to maintain morale by proposing a model of calm behaviour and good citizenship as advocated in the ubiquitous government posters. It further sought to defuse the tensions of wartime life by the safety valve of humour, inviting readers to make light of the restrictions brought by the war, to smile at the foibles of their neighbours, and to view with more or less gentle mockery the failings of officialdom.

This latter phenomenon had made a first timid appearance in cartoons published towards the end of the Great War,[40] and it was renewed with greater vigour in 1939 by the BBC radio broadcast *It's That Man Again*, usually shortened to *ITMA*, whose caustic tone contrasted markedly with *Punch*'s more restrained approach. This program, which attracted an audience of 20 million, featured comedian Tommy Handley encountering a range of odd or incompetent individuals, most revealingly when working as Minister for Aggravations and Mysteries in the Office of Twerps, a barbed and scarcely concealed reference to the Ministry of Information, responsible, among other things, for government advertising campaigns.

Among *Punch*'s principal exponents of morale-boosting techniques were Pont and Fougasse. The former specialized in the understated portrayal of the British self-image, more particularly of the "self-assured upper and middle classes in a state of unruffled calm,"[41] which was to be an abiding self-congratulatory wartime memory. A typical Pont cartoon, from 25 September 1940 at the height of the Battle of Britain, showed a family seated in deckchairs taking a leisurely afternoon tea in the garden, with the husband/father training his binoculars on the sky, calmly counting enemy aircraft, "Three hundred and seventy-six, three hundred and seventy-seven, three hundred and seventy-eight. Swastikas as plain as pikestaffs." Nevertheless, many of Pont's drawings went beyond this narrow brief. One significant example, published on 14 August 1940, reminded readers of the continued existence of the propaganda war between Britain and Germany by depicting two elderly working-class men calmly and silently drinking beer in a pub, in total contrast with and in contradiction of the words of Josef Goebbels booming from the radio: "meanwhile, in Britain, faced by the threat of invasion, the entire population has been flung into a state of complete panic."

Fougasse, on the other hand, was notable for his sharp eye for the ridiculous, the "incongruity and faulty proportion" that needed to be corrected,[42] and for his economical style. His cartoons in the series *The Changing Face of Britain* (started in August 1939) and its successors *Still Changing*, *Changing Face Again* … are good examples. They employed the simple method of juxtaposing almost identical diptychs, or occasionally triptychs, and allowing the reader to interpret the differences between the original and the changed versions. Some were simply funny (e.g., the triptych "XXIX Newsstand," in which the illustrated covers of a line of publications on sale showed, in turn, sporting activities, military training, and finally a number of barely clothed pin-ups); others brought a reminder of the need to adjust one's habits and expectations (e.g., "III The Public Park," with the transformation of a flowerbed into an exit from an emergency shelter). In many cases, the change was positive. In "IV The Town Hall," the previously underused municipal building had become the venue for myriad indispensable wartime activities – National Service enrolment, first aid lectures, or serving as an ARP headquarters – while "XVI Residential Area" showed neighbours actually talking to each other, suggesting that the growth of the new, much-vaunted community spirit would bring a sense of common purpose that could only be beneficial to the nation at war. Not the least of Fougasse's qualities was his awareness of the dangers of "information fatigue," a public aversion to the endless repetition of the same, sometimes out-of-date, instructions, and in "Growth of Obedience to Official Instructions" (30 September 1942) he seized the opportunity to celebrate the end of the need for gas masks by changing the wording of a wall poster from the familiar "Hitler will send no warning. Always carry your gas mask" to "Conservation of Rubber. Don't carry your gas mask." The eight posters Fougasse drew, free of charge, in 1940 for the Ministry of Information's Careless Talk Costs Lives campaign, the work for which he is most frequently remembered, reflected his approach to cartooning. They employed a number of comically improbable scenarios to gently stress the same anti-gossip message: Hitler and Goering seated behind two women in a bus (captioned "You never know who's listening") and a man telling a woman in a restaurant "Of course, there's no harm in your knowing," blissfully unaware that the Führer concealed under the table was noting his every word.

Home Front Problems

The obstacles and privations citizens had to confront on a daily basis throughout the war gave the large number of *Punch* contributors abundant scope to develop an evolving variety of more or less overtly comic cartoons evoking incidents disconcertingly close to real life, which helped to sustain national morale by reminding readers that others shared the same difficulties and which were, ultimately, to be the stuff of postwar memories.

Shortages were a perennial and vital issue, and government intervention was inevitable. *Punch* gave the introduction of food rationing a somewhat cautious welcome in Illingworth's big cut, "Fairly Happy Families" (17 January 1940). This typically detailed illustration had a ludic, light-hearted touch. It presented, as central figure, Minister for Food William Shepherd Morrison in the guise of Mr Grits, the grocer from the popular card game Happy Families, originally created for the Great Exhibition in 1851. His appearance and that of other characters in his shop recalled the original drawings made for the game by John Tenniel, appointed *Punch*'s chief cartoonist in 1864. The government claimed that rationing would be beneficial by bringing "fair shares for all," an argument only partly endorsed by the illustration's title. In fact, restrictions became more and more severe as the war developed, prompting later cartoons to adopt a tone emphasizing the importance of lower expectations or of the benefits of improvisation. In Mervyn Wilson's cartoon of 27 May 1942, a couple watching a luxurious banquet in a film at the cinema could take little more than crumbs of comfort from the husband's whispered aside, "By the way, I managed to get some corned beef for dinner." But the incongruous juxtaposition of the two meals could still raise a resigned smile. Making do had become an officially encouraged way of life, but so, unofficially, had bending the rules. Of course, *Punch*'s status could not allow it to advocate recourse to the widespread (but illegal) black market. Nevertheless, an Anton cartoon comprising six untitled, uncaptioned vignettes (25 November 1942), which intimated that there was always room for ingenuity, came close to doing so and implicitly suggested readers might like to draw their own conclusions. Two slim young women solved the problem of their lack of clothing coupons by persuading a portly acquaintance to purchase a dress, which they then transformed into one for each of themselves.

Rationing was only one of the headaches citizens encountered. Many initially justified regulations could rapidly appear excessive or ridiculous, making them ripe for comic treatment. The decision to remove all road signs in order to impede the progress of any invading German army was perfectly logical, but, as Pont pointed out on 24 July 1940, it was not only putative enemy forces that could be frustrated. His cartoon showed a small boy, determined to obey to the letter instructions not to divulge any information, refusing help to a lost British military convoy with the words "I'll tell nobody where anywhere is." George Stampa had already given a humorous take on the same subject the previous month. A man knocked down by a car sits up in the road and asks the question "Where am I?"

only to hear the unhelpful reply, "I'm afraid we're not allowed to tell you" (12 June 1940). But one of the more incongruously funny sidelights on the subject appeared in an (untitled and uncaptioned) Harold Hewitt drawing of 28 August 1940. A housewife is seen rushing indoors clutching the name-plate of her house, Ivy Cottage, thereby avoiding its removal by workers demolishing road signs in the area. In the light of the role of cartoons in prompting popular memory, the depiction of such trivial, plausibly everyday, incidents raises one significant issue concerning the interaction between the artist's vision and real life. Is the cartoon pure invention, the fruit of his/her imagination, or a version of something seen or (over)heard and therefore already in the public domain and thus within the personal experience of readers? How will it later be remembered, as a personal recollection or as a cartoonist's vision, or both?

Officialdom in all its guises was the butt of much of *Punch*'s home front humour. Impenetrable paperwork was mocked in "Come and help me to fill in this form for Six Easy Lessons on Filling in Government Forms" (Stampa, 1 July 1942), as were out-of-touch civil servants in their own ivory tower, like the two bespectacled figures identically dressed in pin-striped trousers crossing a college quadrangle in an Acanthus cartoon dated 19 October 1942. With due gravitas, one informs the other that "Balliol may be a bit earlier but this is one of the oldest Ministries in the University." Ill-conceived government posters could also raise a smile, but cartoons took the liberty of highlighting potential absurdity for comic effect by creating their own versions of official pronouncements. As early as 18 October 1938, Lawrence Siggs showed a masked burglar, jemmy in hand, pondering exactly how he could "Follow [his] trade in the fighting services," while airmen rushing to take off on a night bombing raid were no doubt left wondering whether a notice reading "Fires are caused by careless unthinking people" should really apply to them (David Langdon, 16 February 1944). Other illustrations played to the popular dislike of the blackout and the sometimes officious ARP wardens who over-enthusiastically enforced it. In William Sillince's 27 August 1941 cartoon, the excessive zeal of a colleague provoked the observation, "It's gone to old Brown's head a little – catching the first chink of the season," behaviour which readers found it only too easy to recognize. Langdon's cartoon (17 July 1940) had already gone much further. A bumptious official had apparently come to warn a searchlight battery that they were committing an offence, a patently preposterous suggestion worthy only of a petty bureaucrat, but the artist made full capital of the scenario. The battery commander did not deign to talk to the warden, telling a subordinate to transmit the message couched in ironically formal terms: "Give him my compliments and tell him that, while we admire the subtlety of his point, we prefer to assume that the black-out regulations do NOT apply to searchlights."

The illustrations just discussed, with the possible exception of the last, were all liable to ring true with daily experiences and grievances, but many of the most effective humorous cartoons were no doubt those that simply helped readers to cope by offering them a snapshot of the ridiculous lying just below the surface of many wartime situations. Some examples concerned personal preferences: a couple quarrelling as they look in vain for their cat during the blackout (John G. Walter, 15 November 1939, "Well, it was your idea to have a black cat"), or another, elderly, couple deciding to "take a walk through the park and 'ave a look at the vegetables" (Sillince, 24 July 1940). Others exploited incongruous situations caused by shortages: a customer in a completely unstocked shop being addressed: "Good morning, madam – and what can't I show you today?" (Fougasse, 12 January 1944), or a visitor telling the owner of a stately home, "You'll have to do something to this room to match your Utility furniture" (Anton, 15 March 1944). Finally, the question asked in Sillince's illustration of a family eating sandwiches in a Morrison shelter (a rudimentary refuge consisting of nothing more than a table with wire-netting fixed on its sides) offered a clear opportunity for lateral thinking in the face of absurdity, "By the way, did you remember to feed the canary?" (25 August 1941). Meanwhile, Rowland Emett's typically surreal drawing of a huge American tank drawn up outside a quaint olde worlde "tea nooke" to request "two hundred and eighty-seven dainty afternoon teas" was a striking visual metaphor both of a culture clash and of the blatant disproportion between Britain and its transatlantic ally (9 February 1944).

Facing the (Uncertain) Future

As we have seen, *Punch* endeavoured in a variety of ways to sustain morale by keeping its readers cheerful throughout the war. The imminence of peace in 1945, however long awaited, sparked other, less reassuring, reflections. Thoughts began to focus on the future, while, paradoxically recalling the lessons and disappointments of the past, the significance of historical comparisons, already familiar to *Punch* readers, was reinforced by the interweaving of illustrations echoing those of earlier times.

Thus, the sentiments of two Tommies traversing a scene of total destruction in 1945, remarking "the old place hasn't changed a bit from 1917" (Sillince, 10 January 1945), may have evoked memories of the two Contemptibles,[43] who expressed similar views in Fougasse's cartoon of 30 October 1918, which cast doubt on the value of any war. Similarly, the child asking his father, "What was the post-war world like after the last war?" (Siggs, 2 May 1945) offered a less than comforting reminder both of the Great War recruitment poster "What Did You Do in the Great War, Daddy?" and of the broken promises made in 1918.

Moreover, the eagerly awaited peace in 1945 would, *Punch* illustrations rightly warned, bring a substantial number of challenges. In Illingworth's "The Price of Flags" (25 April 1945), John Bull reappeared to warn that the war still had to be paid for, and at a high price, while the same artist's "Britannia in Wonderland" (8 August 1945) pointed to an impending housing crisis (already present in 1939 but which bomb destruction had done nothing to improve). For his part, Frank Reynolds's shopkeeper wishing a customer "Happy new ration book, madam" (20 June 1945) was doing little more than stating the obvious unpalatable fact that the end of austerity was not yet in sight. For readers who could remember what happened after the Great War, the fear of unemployment was a renewed unspoken threat as some jobs might well no longer exist. Hewitt's cartoon captioned "So I wrote to my MP and finally got back the job I was doing before the war" (11 April 1945) raised this serious point, but with a touch of humour, for the speaker is painting railings once more, but now only the stumps remain, the rest having been removed to provide metal for more useful wartime purposes. Finally, uncertainty prevailed about what might become of the wartime sense of unity now that party politics had returned, the coalition government having been replaced by Britain's first majority Labour government. In short, people would have to adjust to living in peace again but in a changed world, as shown in the ironic juxtaposition of words and image in Leslie Starke's paradoxical vision ("As far as I'm concerned I don't care if I never see another uniform") in the cartoon below (see figure

"As far as I'm concerned I don't care if I never see another uniform"

11.6 Starke, "As far as I'm concerned I don't care if I never see another uniform," 21 November 1945.

11.6) from 21 November 1945. The war was over, the wheel had come full circle from the call to service with which this section opened, but would the world ever be the same again?

Punch's predictions were not unfounded. The postwar world was as challenging as it had imagined, but the magazine itself emerged strengthened from the war with increased circulation figures. It had fulfilled its mission, done its self-appointed historic duty, and helped its readers to do theirs and left them memories, collective, public, or official, of major events, or, perhaps more important, intimate, private, and personal of day-to-day life to prompt tales to sustain them, to delight or weary future generations in the years to come, and to fascinate those aspiring to fathom the intricacies of endlessly intertwined stories in the best storytelling tradition.

A FEW WORDS AS A WAY OF CONCLUSION

Through these two examples of war memories in the written media, in light and not-so-light versions, we catch glimpses of what makes the richness of the ideas employed for the purpose of morale boosting, or of pinpointing by an ironic word or a mocking gaze the absurdities of human and material destruction or of social, cultural, and political preoccupations. The abundance of clichés, miscomprehensions, understatements, and biased visions conveyed by the palimpsestic nature of the documents studied suggests the perpetual reinterpretation of "truths" amid a multifarious circulation of ideas that prompt spontaneous recollection of forgotten memories, one's own or those of others, and form a solid foundation for the propagation of real or not so real news. Presentations of facts and events help to deepen our reflections on the several layers of perceptions and possible interpretations inevitably present in the memorialization process. The wealth of haphazard details displayed in the narratives and images under consideration highlights the complexity of trying to define cultural traits and tropes in a world at war, and unceasingly complements the ongoing challenge of understanding the shaping of war memories.

NOTES

1 Graham Seal, *The Soldiers Press: Trench Journals in the First World War* (Basingstoke: Palgrave Macmillan, 2013), 8.

2 In this chapter, unless otherwise indicated, the name the *Wipers Times* (aka the *WT*, the journal, the paper, the publication) refers generically to all twenty-three published numbers: in chronological order, four of the *Wipers Times or Salient News*, four of the "*New Church*" *Times*, one each

of the *Kemmel Times* and the *Somme-Times*, eleven of the *B.E.F. Times*, and two of the "*Better Times*." They are identified below by their initials, *WT*, *NCT*, *KT*, *ST*, *BEFT*, and *BT*, along with publication dates.

3 Although the city lay within the Flemish-speaking part of Belgium, British forces (along with troops from the Empire and the dominions) made almost universal use of the French name (Ypres), systematically mispronounced as "Wipers."

4 Commonwealth War Graves Commission, https//issuu.com/wargravescommission/docs/ypres_salient/ (page no longer active).

5 Capt. W.C.C. Weetman, MC, Croix de Guerre, *The Sherwood Foresters in the Great War, 1/8th Battalion*, chap. 3, "*The Salient*" (June–October 1915) (Nottingham: Thos Forman and Sons, 1920), n.p., https://archive.org/stream/thesherwoodfores20527gut/pg20527.txt.

6 J.G. Fuller's *Troop Morale and Popular Culture in the British and Dominion Armies, 1914–1918* (Oxford: Clarendon 1991) observes that the "aggressive policy of trench raiding [was] likely to inspire [soldiers'] antagonism towards their own staff" (Introduction, 2).

7 Mary Reid Kelley, "But Kultur's Nar-Poo in the Trenches," *Art in America*, June/July 2014, 120.

8 Catch-phrase taken from the BBC TV series *Fawlty Towers*, season 1, no. 6, 1975, "The Germans," spoken by Basil Fawlty alias John Cleese, a member of the Monty Python troop. See also note 27.

9 Frederick Roberts, "Foreword, How It Happened," *The Wipers Times a Facsimile Reprint of the Trench Magazines: – the Wipers Times – the New Church Times – the Kemmel Times – the Somme Times – the B.E F. Times* (London: Herbert Jenkins 1918), vii.

10 On this subject, John Ivelaw-Chapman suggests that "most of the readers and correspondents of *The Wipers Times* were suffering from shellshock to some degree." See John Ivelaw-Chapman, *The Riddles of Wipers* (Barnsley: Pen and Sword, 2010 [1997]), 53.

11 Opinions differ as to the status of the authors of the paper: John Ivelaw-Chapman (*Riddles*) and Patrick Beaver ("Introduction" to the 1973 Peter Davies edition) assume that most were ex-public-school subalterns, while Esther MacCallum-Stewart claims that "most of the submissions were by working-class soldiers." See "Satirical Magazines of the First World War: Punch and the Wipers Times," feature article 2009, n.p., http://www.firstworldwar.com/features/satirical.htm. In either case, readers were troops in the front area for whom the key distinction was between themselves and more privileged individuals out of harm's way at the rear.

12 Kelley, "Kultur's Nar-Poo," 124.

13 Evoking, respectively, Lewis Carroll's "The Walrus and the Carpenter" (from *Alice through the Looking Glass*) and Thomas Gray's "Elegy Written in a Country Churchyard." (Having no reference works to hand, the *WT* contributors misspell both authors' names.)

14 All the advertisements are fakes, with the single exception of "the only pukka advert we've ever had" (*BEFT* 25 December 1917), announcing the imminent publication of the *Wipers Times* by Herbert Jenkins and Co.

15 Robert L. Nelson, "Soldier Newspapers," *1914–1918 online. International Encyclopaedia of the First World War* (Berlin: Freie Universität 8 October 2014), 13.

16 Major-General J.E. Capper, familiarly known as "the Professor," commanded the 24th Infantry Division until May 1917. PBI troops were their own beasts of burden when going into action.

17 Hell-Fire Corner, as its name suggests, was a notoriously hazardous spot in the Salient, where the once poplar-lined Menin Road was crossed by a railway track, two prominent features that allowed German artillery to target it with the greatest precision. All vehicles passed at full speed and foot soldiers literally ran for their lives. The implications of shooting in the second advertisement are obvious, the fishing being a jibe at the standing water in many of the trenches.

18 In truth, anything attracting attention in no man's land at night could draw potentially fatal enemy gunfire, and any visible or distinguishing feature made an advancing officer an equally easy target. Shiny badges, leather puttees, and what Paul Fussell calls "melodramatically cut riding breeches" all came into this category. See Paul Fussell, *The Great War and Modern Memory* (New York: OUP, 2000), 42.

19 Acronyms for Assistant Adjutant, Deputy Assistant Adjutant, Quartermaster General, Deputy Assistant Quartermaster General, and Corps Routine Orders.

20 A phrase recalling the observation in Sidney Rogerson's *Twelve Days on the Somme: A Memoir of the Trenches* (London: Greenhill Books 2006 [1933]), 130: "the British soldier … when by every law he should have been utterly weary and 'fed up' … invariably managed to be almost truculently cheerful."

21 This includes military jargon, armaments known only by their nicknames, trenches named after London streets, and pidgin French, the commonest example being "Nar Poo / Na poo / narpoo," a distortion of *il n'y a plus* (= there is / are none left), with extended meanings of "no good" or "no way."

22 One notable exception was the cartoons drawn by Bruce Bairnsfather for the *Bystander* magazine, a publication mentioned in the *WT* both by its real name and by the anagram *Stybander*. After being wounded in the 2nd Battle of Ypres (April–May 1915), Captain Bairnsfather returned to England, where he spent the rest of the war training new recruits. The grumpy "heroes" of his cartoons, based on troops he encountered in the trenches, have the same survival mentality as the contributors to the *WT* and air many of the same grievances.
23 The official documentary propaganda film *The Battle of the Somme*, released in August 1916, at an early stage of the battle attempted to satisfy this demand and was a runaway success, attracting a cinema audience of 20 million.
24 On censorship, see, for example, Anthony Richards, "Letter Censorship on the Front Line," *Telegraph*, 30 May 2014.
25 As early as 22 March 1916, little over a month after the birth of the journal, the *Nottingham Evening Post* carried news of its launch in an article entitled "Sherwood Pioneers Turn out a Newspaper," quoting extensively from a letter sent to his parents in West Bridgford by QMS Leslie W.L. Tyler of the Pioneer Battalion. Tyler's letter promises to send his parents a copy of the paper "under another cover." See http://blog.britishnewspaperarchive.co.uk/2013/09/11/the-wipers-times-a-review-of-the-soldiers-newspaper-march-2016/.
26 The only examples seem to be the enforced change in the paper's title when it became the *BEF Times* and the blank spaces replacing the names of the departing GOC and one of the original members of the Division (*BEFT* 8 September 1917).
27 Obvious examples include *Private Eye* magazine (founded in 1961 and still in print), *That Was the Week That Was* (TV, 1962–63), *Round the Horne* and *I'm Sorry I'll Read That Again* (radio 1964–68 and 1964–73), and, most notably, *Monty Python's Flying Circus* (TV 1969–74), whose continuity catch-phrase "and now for something completely different" could have been inspired by the structure and content of the *WT*.
28 Patrick Beaver, "Introduction," *The Wipers Times – A Complete Facsimile of the Famous World War I Trench Newspaper, Incorporating the "New Church" Times, the Kemmel Times, the Somme-Times, the BEF Times and the "Better Times"* (London: Peter Davies, 1973), xv, xx.
29 "Many British history books – if they use cartoons at all – tend to concentrate on the artists of the weekly *Punch* and the work of David Low in the *London Evening Standard*." See Mark Bryant, "Introduction," *Illingworth's War in Cartoons* (London: Grub Street, 2009), 12. Each of

the drawings quoted in this chapter is identified by the name of the artist, the date of publication, and words from the caption or title, if any, thus enabling readers to discover the illustrations for themselves.

30 Mark Bryant and Simon Heneage, "Preface," *Dictionary of British Cartoonists and Caricaturists, 1730–1980* (Aldershot: Scolar Press, 1994), viii.

31 Joseph Darracott, *A Cartoon War: World War II in Cartoons* (London: Leo Cooper, 1989), 74. According to Mark Bryant, the magazine's weekly print-run in 1939 was 116,000 copies (*Illingworth's War*, 12n1). As early as 1925, the magazine had increased in size from eighteen to twenty-eight pages in order to avoid the editorial content being swamped by advertisements. The wartime seasonal numbers (spring, summer, etc.) and annual almanacs could afford the luxury of illustrations in full colour.

32 R.G.G. Price, *A History of Punch* (London: Collins, 1957), 218.

33 Apart from the illustrations in *Punch*, notable examples included W.K. Haselden's series of derisive portrayals of the Kaiser and his son the Crown Prince as "Big and Little Willie" in the *Daily Mirror* (1914–18, partly reprinted in *The Sad Experiences of Big and Little Willie*, 1915) and Bruce Bairnsfather's down-to-earth depiction of life in the trenches, published in the *Bystander* (from 1915) and subsequently in book form as *Fragments from France* and *Billets and Bullets*.

34 Real name Cyril Kenneth Bird. Fougasse had been severely wounded at Gallipoli in 1915 and first drew for *Punch* the following year. Several other wartime artists are also best known by their pseudonyms: Acanthus (Harold Hoar), Paul Crum (Roger Pettiward, deceased 1942), Pont (Graham Laidler, deceased 1940) and Anton (brother and sister Harold Thompson and Beryl Antonia Yeoman). Thompson's service in the Royal Navy meant that the majority of Anton's wartime drawings were the work of Antonia, *Punch*'s first woman cartoonist.

35 Partridge's *Homage to France*, published on 19 June 1940, almost on the eve of the armistice that ended French resistance, was the sole, highly symbolic, cartoon dealing immediately with the Battle of France.

36 Illingworth's *The Common Task* (19 March 1941) epitomized the new spirit of united purpose. It showed an elderly man "digging for victory" (planting vegetables) in his garden while glancing seawards at a passing naval flotilla.

37 Words from a line of "We'll Meet Again," the celebrated wartime song recorded by Vera Lynn in 1939.

38 Despite the defiantly patriotic message in David Low's "Very well, alone" (*Evening Standard*, 18 June 1940), arguably the most famous cartoon of

the war, Britain was never completely alone. Two subsequent *Punch* illustrations, "Gangways of Empire" (Shepard, 3 July 1940) and "So our poor old Empire is alone in the world. Aye, we are – the whole five hundred million of us" (Fougasse, 10 July 1940), rightly observed that the nation could still count on the support of the Empire.

39 The *Daily Mirror* was threatened with similar treatment in March 1942 after publishing a cartoon by Philip Zec that was interpreted as suggesting that sailors' lives were being put at stake to enhance petrol companies' profits.

40 For more details, see, for instance, Renée Dickason, "The Nuanced Comic Perspectives of the Cartoons in Mr. Punch's History of the Great War," in *Humor, Entertainment and Popular Culture during World War I*, ed. C. Tholas-Disset and K. Ritzenhoff (New York: Palgrave Macmillan, 2015), 123–34.

41 Darracott, *Cartoon War*, 74.

42 "The function of humour is essentially corrective: it is a corrective, for instance, of incongruity and of faulty proportion." See Fougasse, *A School of Purposes* (1946), quoted in Darracott, *Cartoon War*, 74.

43 That is, members of the British Expeditionary Force in 1914, dubbed a "contemptible little army" by the Kaiser.

12

Writing the Blitz, Listening to the Nation: Personal Narratives of the Blitz and the Construction of a Collective Aural Identity in British Cinema of the Second World War

Anita Jorge

"The whole of the warring nations are engaged, not only soldiers, but the entire population, men, women and children. The fronts are everywhere. The trenches are dug in the towns and streets ... The front line runs through the factories. The workmen are soldiers with different weapons but the same courage," Winston Churchill declared in his August 1940 speech – commonly referred to as "The Few" – to the House of Commons. Images of a continuity between all sections of society – soldiers on the front line and civilians on the home front; men, women, and children; rich and poor; young and old; city-dwellers and countryfolk – formed the basis of what was later to be called "The People's War," the most emblematic period of which was the Blitz, which spanned a period of eight months (September 1940 to May 1941) during which London and other cities, such as Coventry, Glasgow, Liverpool, and Birmingham, were subjected to systematic bombings by enemy aircraft. The fantasy of a united people "pulling together" beyond social and geographical divisions, as well as the alleged determination and heroic stoicism of the British people, pursuing their "business as usual" in the face of adversity – debunked though they were by the historiographical trend initiated by Angus Calder – still have currency for contemporary representations and understanding of the period. For Angus Calder, these representations were part and parcel of what he named the "myth of the Blitz,"[1] created and disseminated by official propaganda in order to conceal deep social divisions and unheroic stories of opportunistic looting and rape. This "mythology," however, appears to permeate individual writings of the period – in the form of letters, diaries, memoirs, and autobiographies as well as interviews and accounts written by Mass-Observation enquirers.[2]

As historian Penny Summerfield states in her book *Contesting Home Defence*, "Personal testimony is not simply a window on the past ... Memories are formed through a complex process of interaction between an individual's experiences and publicly available constructs, including prior accounts of similar experiences."[3] Wartime personal writings can be considered, to a certain extent, as mirrors of the dominant ideology, immersed as they were in the political and cultural discourse conveyed through the press, literature, radio, and cinema of the period, but also transmitted by their relatives and neighbours.[4] These writings then became, in turn, channels of the collective discourse.

This chapter deals with the principles underlying this "cultural circuit" – a concept developed by Richard Johnson[5] – as far as the aural perception of the Blitz is concerned. In other words, it explores the way in which "aural testimonies" of the Blitz, as they appeared in personal writings, responded to and interacted with an official discourse on sound, which emerged in cinematographic works of the period. Prior to these considerations, it also addresses the fundamental conditions of the emergence of an official collective discourse on sound during the Second World War in Great Britain. In order to do so, excerpts taken from the literature of the Blitz – diaries, memoirs, and letters chiefly – will be examined in conjunction with specific examples of documentaries and " semi-documentaries" that were screened in cinemas during the Blitz.[6]

"LONDON IS THE BIG NOISE NOWADAYS":[7] SONIC MEMORY AND THE HISTORY OF THE BLITZ

The rationale behind the emergence of an official discourse hinging on an aural sense of identity lies in two main factors that led, between the two world wars, to the emergence of a new "culture of aurality," itself conditioned by what was considered by some as a reconfiguration of the sensorial hierarchy. Accounting for this cultural shift was, on the one hand, the modernization of listening practices, chiefly related to the technical innovations of the period in the domain of field recording, and the spread of radio in the domestic space. Indeed, owing to technical developments and lower manufacturing costs, three-quarters of British homes possessed a wireless set in 1939. On the other hand, the upheavals brought about by the two world wars – namely, unstable identity frameworks as well as the fragmentation of reality and isolation of individuals – led to what was deemed by such theorists and artists as the Italian Futurists to be a sensorial shift whereby the sense of sight was superseded by the sense of hearing. As early as the 1910s, hearing was considered by Luigi Russolo, whose experiments in sound echoed in many respects those of the British documentary movement,[8] to be the only sense that could enable people to grasp the fragmented complexity of the real, owing to the essentially multi-directional and fleeting quality of sound. In the Futurist manifesto, published in 1913, Russolo claimed: "In modern warfare, mechanical and metallic, the element of sight is almost zero. The sense, the significance, and the expressiveness of noise, however, are infinite."[9]

For Londoners, who spent a considerable amount of time in collective underground air-raid shelters plunged into complete darkness, hearing was almost the only sense that they could rely on to gauge their surroundings during enemy attacks, a fact that was upheld by anthropologist Tom Harrisson, one of the founding members of Mass-Observation: "For nearly everyone, hearing was the most important sense used."[10] Consequently, what could be labelled as "aural hypotyposes" abound in the literature of the Blitz, testifying to the acquisition of a new "sonic culture." By way of example, war memoirist Harvey Klemmer offers a list of sounds, such as "the never-ending whine of approaching machines; the roar of guns; the crack of exploding shells high in the heavens; the clang of fire engines and the occasional swift rush of an ambulance; the high-noted siren of a 'suicide squad' hurrying to deal with a time-bomb; shouts, and the sounds of struggle; curses and exhortations; the *psing* of falling splinters; the tinkle of broken glass; and always, through all and over all, that deadly crump of bombs."[11] The diversity of noise-related substantives ("whine," "roar," "crack," "clang") and the occasional use of onomatopoeia ("psing," "crump") are representative of what is to be found in personal narratives of the period, bearing witness to the remarkable acuteness of the Londoners' sense of hearing. A great many of them felt that they were capable of differentiating between RAF fighters and Luftwaffe bombers, and even between specific types of armament, just by the sound they made. Certain bombs were even personified and given reassuring pet names, such as "Weeping Willie," "Cracking Ronnie," or "Humpty Dumpty."

The literary retranscription of this sonic environment emerges, in some cases, as very idiosyncratic, relying as it does on a profusion of personal images and similes. But it also appears to hinge on certain tenets of the official discourse on sound, as it was conveyed in cinema.

METAPHORICAL REPRESENTATIONS OF THE REAL

According to historian of the Blitz Amy Helen Bell, civilians had a strong sense of their own historical importance and felt it their duty to record the details of their personal lives and their perceptions of the national crisis in Britain in order to rectify the fallacious and monolithic official discourse formulated by the authorities. The diversity of personal accounts could therefore be seen as a foil to this uniformity.[12] Such diversity is seen, for instance, in the variety of descriptions regarding civilians' perceptions of the permanent din of enemy attacks, which seemed to have had varied and mitigated impacts. In some accounts, the sound of British aircraft and armament is described in laudatory terms, whereas the sound of enemy bombers and rockets is universally execrated. For instance, English journalist and author Fryniwyd Tennyson Jesse wrote in a letter dated 15 September 1940, "the guns, besides being reassuring to hear, are really not a very unpleasant sound,"[13] and a young man interviewed by Tom Harrisson claimed that he found the sound of guns extremely satisfying.[14] However, some diaries and memoirs display a generalized aversion to all the sounds of war, be they British or German, especially as they were a source of insomnia, night terrors, and psychological disorders. Contrary to what was maintained by some diarists, the sound of the all-clear siren was far from a soothing one. As noted by American war correspondent Edward Murrow, "Many [people] don't like the sound of sirens. It is loud, penetrating, and can't very well be ignored. The sirens seem to be about as disturbing and upsetting as the distant crump of bombs."[15] The appropriation and internalization of these sounds, retranscribed in a personal and unique manner, bears resemblance to what sound historian Jonathan

Sterne labels the "commodification" of radio waves,[16] a phenomenon that coincided with the mass commercialization of domestic wireless sets, starting in the 1930s. In the case of the London Blitz, this "commodification" came in the form of what a Mass-Observation enquirer called the "my bomb syndrome," referring to the commonly shared feeling that each bomb that fell on the city was aimed at someone in particular. Phrases such as "I could have sworn it fell absolutely in the back garden here," "when the explosion went off everybody … thought it was right here," "it had my number on it,"[17] are typical of this paranoid feeling. A similar phenomenon commands the famous and oft-quoted passage from Graham Greene's war novel *The Ministry of Fear* (1943), whose narrator fancies hearing the continuous "mumble" of the enemy bomber chasing him down and repeating, "Where are you, where are you?"[18]

Resorting to personal images, similes, and metaphors was a common way of appropriating and domesticating these alien sounds. The following examples exhibit an astonishing degree of inventiveness as to the choice of similes to refer to the sounds of bombings: "there was a series of horrible whistles overhead … Sometimes it seemed to be drumming out tunes, like an enormous Boys Brigade, drum and fife band," "there was a sound which I can describe only as like thousands of pebbles dropping on corrugated-iron sheets," "the sound was as if someone was scratching the sky with a broken finger nail," "[the bomb] makes a very loud whistling sound – exactly like a car in first gear that is going too fast."[19] The richness and variety of the figurative language displayed in these examples epitomizes the Londoners' sense of their own historical importance within the collective experience.

To James Olney, metaphor is grounded in a psychological process of domestication. He defines the principle underlying the metaphor as follows: "To a wholly new sensational or emotional experience, one can give sufficient organization only by relating it to the already known … This is the psychological basis of the metaphorizing process: to grasp the unknown through the known, or to let the known stand for the unknown and thereby fit that into an organized, patterned body of experiential knowledge."[20] The mechanism underlying the "metaphorization" of sonic perception therefore consists in transfiguring an unknown and intangible reality into corporeal and familiar objects, a process aided by the agency of the cinematic medium.

A CINEMATIC EXPERIENCE OF SOUND

The generalized use of figurative language is also a test case for the subtle conflation of idiosyncratic discourses on sound and the collective discourse, itself influenced by that of the Ministry of Information. The memoirs, diaries, and letters of the period feature a series of recurring stock metaphors, or similes having undergone a certain degree of lexicalization. The comparison of the roar of aircraft engines with the buzz of a swarm of bees, for instance, is extremely commonplace. "The bombers came back and back like urgent swarms of jungle bees,"[21] described Fryniwyd Tennyson Jesse in a letter, and memoirists Harvey Klemmer and Basil Woon, referring to Luftwaffe bombers, spoke of a "dreadful, angry sound like the buzzing of millions of bees" or a "double-waspish hum."[22] The comparison of airplanes with swarms of bees ominously roaming the skies is redolent with biblical meaning and conjures up Old Testament imagery of disease and invasion, such as the Plagues of Egypt (the swarm of flies and the invasion of locusts) or enemy menace – as read in Isaiah 7:18: "The Lord will whistle for ... bees from the land of Assyria" (the Assyrians being the inveterate enemies of Israel). But taken conversely as the embodiment of British and Allied bombers, the bees could be seen as yet another avatar, of the *miles christianus* (or Christian soldier) busily going about his work with single-minded dedication, while the beehive, traditionally taken to represent the Church, would stand for the British nation itself.

In a similar fashion, enemy sounds are very often correlated with the grunt of wild beasts – "the bestial grunt of bombs," "the tiger's purr,"[23] for instance – while, on the other hand, British bombers and guns are placed on a par with noble and virtuous animals – "at last, the All Clear sounded, triumphant as the crowing of a celestial rooster."[24] Most striking of all, perhaps, is the customary perception of the roar of British bombers as a reassuring and vivifying sound, and a metonymy of Allied victory.

But the conflation of individual and collective discourses is first and foremost made manifest by the constant comparison between scenes witnessed and experienced by Londoners during the Blitz, and their cinematographic representations. Londoners' descriptions of air raids are very often accompanied by comments highlighting the cinematic quality of their experience, such as "it's like the films, only we never thought it was true, on the films,"

"We never thought we'd see it like in the cinema," "This reminds me of a horror film," or "I couldn't believe it, it was like a film being shown before our eyes."[25] The same process recurs here: the unfathomable – the extreme violence of an air raid – is made sense of through the prism of familiar and common objects, in this case, popular films.

Finally, that diarists should sometimes resort to shared and hackneyed signifiers and images, rather than personal and original ones, also appears as a consequence of what one of them referred to as a "suspension of the imagination," a form of psychological disorder induced by the terror of aerial attacks. "I find one practises a sort of suspension of the imagination. I do not think that the drone in the sky means death to many people at any moment,"[26] diarist Harold Nicolson confesses. Cinematographic representations of traumatic episodes therefore served as surrogates for the Londoners' momentarily suspended imaginations.

One has to bear in mind, however, that the memoirs, diaries, letters, and interviews that have been handed down to the modern-day researcher were all scanned for potentially seditious material and that many of them – letters in particular – were at least partially altered by censorship so as not to reveal information that might be of use to the enemy. Hundreds of diaries written during the Blitz were destroyed or never made it to publication. It is not surprising, therefore, that a majority of those first-hand accounts should mirror the official discourse.

The frequent recourse to and reliance on metaphors and images testifies to the highly traumatic and alienating experience of the Blitz for Londoners. Relocating the soundscape of the Blitz within the frame of a collectively shared experience on the screen was, therefore, a way of tightening the bonds of a dislocated community.

SOUND REAPPROPRIATION: DOMESTICATING THE ALIEN

During the Blitz and throughout the war period, Britons became increasingly acquainted with the documentary form – a cinematographic genre heretofore associated with an educated and intellectual elite. The Crown Film Unit – previously known as the GPO Film Unit, before it came under the aegis of the Ministry of Information – the Army Film Unit, or the illustrious Ealing

Studios began producing documentaries and semi-documentaries (also known as "drama documentaries"), which were screened in cinemas, generally between the newsreels and the main feature, or at the top of the bill, in the case of feature-length documentaries. More than one thousand documentary films were produced during the war, many of them under the Ministry of Information's "five-minute films" scheme, whereby propaganda messages were distributed free to cinemas to play in their main program. These documentary or semi-documentary films, although not as popular as feature films – with some notable exceptions – could be perceived by civilians as mirrors of their experience of war and served as artistic and pragmatic tools enabling them to label and domesticate the sounds that surrounded them. Some of the five-minute films focused on very practical details, such as *Ack-Ack* (1941), minutely delineating the different types of armament and the sounds they produced, or *If War Should Come* (1939), a "catalogue" of the different sounds that Britons would need to get used to, and know how to react to, in the case of bombings. Being exposed to these informative documentary films several times a week enabled civilians to hone their listening skills and become increasingly "sound-savvy." As suggested above, diarists of the Blitz frequently mentioned a newly acquired ability to pinpoint and discriminate between different wartime sounds. As war chronicler Basil Woon remarks, "It is amusing to note how bomb-minded we have all become ... We can distinguish a Dornier from a Savoia and a Heinkel from a Messerschmitt (or we think we can) by the sound of its motors ... It's elementary, now, to distinguish between the sound of a bomb and the sound of a gun ... We know when our own guns are in action."[27] The reservation he momentarily expresses – "or we think we can" – is a tell-tale sign of how cogent collective discourse, official or unofficial, was in civilians' minds. That there should be an unmistakable distinction between the sound of Spitfires and that of Messerschmitts was itself the product of a deep-rooted belief, the former (steady and reassuring) reflecting the moderate, phlegmatic, and sophisticated temper traditionally attributed to Britons, and the latter (cacophonous and unpleasant) mirroring the choleric and coarse German character. "The bombers have a heavy massive hum, quite different from the lighter, more casual-sounding British machines. All the difference between the Teutonic and Anglo-Saxon temperaments seems to lie in those two familiar noises,"[28] writer Vera Brittain explains in her famous autobiography *England's Hour* (1941).

Second, rather than aggressive and bellicose, these sounds are made to appear banal and homely, as though they were an intrinsic characteristic of the home front or indeed the whole nation. *Our Country*, a forty-minute semi-documentary made in 1944 by John Eldridge, epitomizes the assimilation of the "business as usual" ethics, a recurring motif of wartime propaganda, into civilian everyday life. A sequence taking place in Sheffield immerses us in the domestic life of a family of miners. The soundtrack consists only of the diegetic sounds of the bombardments, while enemy aircraft are never seen, and the images are those of the interior of a family home, whose sounds are never heard. The effect of this syntagmatic manipulation is striking since it gives the apocalyptic sounds of bombs a domestic dimension, underlining the internalization of the war effort. Thus, while the father of the family turns on the wireless, the sound we hear – and whose source is revealed shortly afterwards – is that of the anti-aircraft guns. Later on, a young woman places the kettle on the stove and, thanks to a subtle cross-fade, the kettle's nozzle becomes the beam of the anti-aircraft gun. When the all-clear siren is finally heard, a close-up shows, instead of the siren, the kettle starting to whistle. The use of contrapuntal sound editing is not a refusal to confront the reality of war, or a negation of the psychological effects of the Blitz, as the principle of "business as usual" denoting the "stiff upper lip" attitude that the British must adopt in all circumstances might suggest. Rather, the sequence illustrates the daily survival of an ordinary working-class family who manage, despite the constant terror, to keep hope and carry on.

LISTENING TOGETHER THROUGH UNIVERSAL SOUNDMARKS

For Canadian composer and inventor of the term "soundscape" Raymond Murray Schafer, any given community possesses a certain number of aural symbolical icons that he calls "soundmarks" and that he defines as "community sound[s] which [are] unique or possess qualities which make [them] specially regarded or noticed by the people in that community."[29] These soundmarks "make the acoustic life of the community unique" and, for this reason, "once a soundmark has been identified, it deserves to be protected."[30] Such noticeable aural icons are made salient in almost every wartime British film as symbols of the vivacity of the nation that, precisely, "deserve to be protected" and cherished. The aforementioned roar of Spitfires – "that steady hum night after night, that noise which is oil for the burning fire of our hearts," as member of the Dutch resistance movement Jo de Vries fervidly states in Powell and Pressburger's *One of Our Aircraft Is Missing* (1942) – stands as one of the most fundamental of those soundmarks. Equally outstanding were the striking of Big Ben; the ringing of church bells, which had been silenced by decree in the early years of the war and reserved for sounding a warning if the German invasion should begin; the wail of the all-clear siren; the national anthem "God Save the King"; and the illustrious "fate knocking at the door" motif consisting of the first four notes of Beethoven's Fifth Symphony, whose recognizable rhythm – three short notes followed by a long one – was Morse code for the letter "V," for Victory, and which came to be used as the aural symbol of resistance shared by all Allied or occupied countries. However, these symbolical sounds of resistance were also buttressed, in a great number of films, by ordinary prosaic sounds that, combined, formed a "symphony of war" (*Listen to Britain*). The resilience of these ordinary sounds, which cannot be drowned out by the sounds of the enemy, was considered the resounding proof of the vivacity of the British nation itself. The most "symphonic" of the wartime films is indisputably Humphrey Jennings's masterpiece *Listen to Britain* (1942), a collage of everyday sounds ("the sound of [Britain's] life by day and by night") that, "blended together in one great symphony," constitute "the music of Britain at war."[31] Be it the clank of factory machinery, the whistles and sounds of shunting trains, BBC broadcasts blending with the chime of Big Ben, Flanagan and Allen singing "Round the Back of the Arches" at lunchbreak in a factory canteen and pianist Myra Hess performing Mozart's Concerto in G Major during one of her lunchtime concerts at the National Gallery, Canadian soldiers singing "Home on the Range" accompanied by the guitar, the chatter of children at play, the clap of horses' hooves and the chirping of birds, or the roar of tanks and Spitfires, these familiar sounds together testified to the British nation's resilience and fighting spirit. But beyond embodying the vigour of the nation, they emerged as so many landmarks of cross-geographical identification and as a means of bridging the gap between soldiers and civilians, urban and rural areas, Southerners and Northerners, and even between Great Britain and its Allies. "Sound ... creates common ground and identification, perhaps more powerfully even than other senses," as Greg Goodale states.[32] Wartime cinema thus provided the means for the emergence of an "imagined community" (Benedict Anderson),[33] cohered by a communal experience of sound.

A "SYMPHONY OF WAR": HARMONY IN CHAOS

The term "symphony," employed in a film such as *Listen to Britain*, but also in several documentary films of the period, is far from fortuitous. Postulating the intrinsic musicality of a discrete collection of sounds was yet another way of rationalizing the unknown and controlling the fears of civilians through a newly found sense of purpose and harmony. Indeed, such phrases as the "melody" or the "voice" of the guns, the various "tones" or "ranges" of cannons as well as their fluctuating "tempi," or else the "score" or the "overture" played by the British aircraft, are commonplace not only in experimental films such as those produced by members of the documentary film movement but also in films that did not possess any specific aesthetic qualities. This dialectical opposition between noise-chaos and music-positive order is a secular one. All founding myths praise the figure of the musician as a bulwark against chaos and political dissent, characterized by noise. Orpheus's death, for instance, as it is related in Book XI of Ovid's *Metamorphoses*, is described as a combat between the cacophonous yells of the Bacchanals and the soothing strains of his lyre: "In louder strains their hideous yellings rise, / And squeaking horn-pipes echo thro' the skies, / Which, in hoarse consort with the drum, confound / The moving lyre, and ev'ry gentle sound." [34]According to Jacques Attali, noise channelling and its transformation into music is the condition for the emergence of positive society and political power: "Everywhere codes analyze, mark, restrain, train, repress, and channel the primitive sounds of language, of the body, of tools, of objects, of the relations to self and others. All music, any organization of sounds is then a tool for the creation or consolidation of a community, of a totality."[35] He distinguishes three cases in which music is strategically used as a form of submission by the established power, among them "music ... used and produced ... in an attempt to make people forget the general violence," and "music ... employed to make people believe in the harmony of the world." This is precisely what underlies the process of "musicalization" of wartime sounds in the cinema of the Second World War: it would help people psychologically escape the violence of daily bombings, while maintaining the illusion that there existed organized and systematic retaliation to enemy attacks.

But the term "symphony" bears a yet deeper significance. It harks back to the cinematographic genre of the "city symphony," whose most prominent forerunners were German filmmaker Walter Ruttmann, who directed his silent movie *Berlin: Symphony of a Great City* (1927) in collaboration with Ealing Studios documentarist Alberto Cavalcanti, and Soviet filmmaker Dziga Vertov, whose masterpieces *The Man with a Movie Camera* (1929) and *Enthusiasm: Symphony of the Donbass* (1931) both screened at the London Film Society. A certain number of British films of the period were claimed to have been made to fit a symphonic form. Although there is no fixed formal structure to the symphony, it is traditionally agreed that it should comprise four (or five) parts, the first of which would follow the sonata form (ABA'). Basil Wright's innovative documentary *Song of Ceylon* (1934), whose soundtrack was composed by Walter Leigh, is arguably one of the first British documentaries that was purposely made with the symphonic form in mind,[36] and it served as a blueprint not only for many later documentaries, such as Jennings's *Listen to Britain, Ordinary People* (1942) by J.B. Holmes, and *London 1942* (1943) by Ken Annakin, but also for feature films, among them *49th Parallel* (1941) by Michael Powell and Emeric Pressburger. The artistic trend that consisted in circumscribing the soundscape of the Blitz within a pre-existing cinematographic tradition was, on the one hand, prompted by a desire to objectify it and abstract it from reality, thus making it paradoxically more comprehensible. But perhaps more characteristically so, the process of musicalization of wartime sounds partakes of a whole discourse seeking to restore the fundamental musicality of Great Britain, mockingly dubbed "the land without music" (das Land ohne Musik) by its German neighbour since the end of the nineteenth century.[37] The conviction that Britain was not only a nation of accomplished musicians and composers who could vie with the greatest European composers of the time but also a land of music-lovers whose intrinsic musicality permeated its "music of the everyday" was, throughout the first half of the twentieth century, a nodal point of the official discourse channelled through the cinema. The diaries, memoirs, and letters that were published during the London Blitz are invaluable tokens of the pre-eminence of the collective discourse on the way civilians formulated their sonic experience of the conflict, and, as such, they continuously fuel the mythology of the Blitz and condition our perception and comprehension of those events.

CONCLUSION

On 29 December 2015, the streets of London echoed again with the infamous wail of the air raid siren, to mark the seventy-fifth anniversary of the "Second Great Fire of London" – one of the most destructive bombing raids that took place during the London Blitz, on the night of 29 December 1940, killing more than 160 civilians. That such a soundmark should have been chosen as the symbol of that historical event bears witness to the fundamentally aural quality of the Blitz, whose resilience up to this day originates in the construction of an official discourse on "the sounds of a country at war" (Michael Powell) that was popularized by cinematographic productions of the period. As cinema historians Tony Aldgate and Jeffrey Richards, writing about Humphrey Jennings's films, argue in their book *Britain Can Take It*, "Together, they decisively shaped and defined the image of Britain at war that was to be circulated round the world and handed on to the generations to come."[38] The soundscape of the Blitz could therefore be regarded as what historian Pierre Nora calls a "memory space" (*lieu de mémoire*),[39] for it emerges as a symbol of British national identity, whose memory is perpetually fuelled by the added discourses of those who experienced it, permeated as they were by the widely circulated collective discourse. Nevertheless, these testimonies should not be dismissed as sheer mirrors of a dubious "official myth," the dissemination of which would be willingly carried out by individual filmmakers. One needs only to watch Humphrey Jennings's films to grasp how immersed in humanist social ideals were his beliefs in the cohesive power of sound and his rejection of a mortifying fragmentation as he strove to transmute an individual, alienating, and fragmented aural experience of war into one to be collectively savoured.

NOTES

1 See Angus Calder, *The Myth of The Blitz* (London: Random House, 1992).
2 The Mass-Observation movement, founded in 1937 by anthropologist Tom Harrisson, poet Charles Madge, and filmmaker Humphrey Jennings, was a sociological movement that aimed at collecting and compiling the habits of Britons and reactions to their political and social environment by means of interviews, journalistic reports, and polls. Although it retained its independence throughout the Second World War, the data it collected were occasionally used by the Ministry of Information.
3 Penny Summerfield and C.M. Peniston-Bird, *Contesting Home Defence: Men, Women and the Home Guard in the Second World War* (Manchester: Manchester University Press, 2007), 14.
4 The following excerpt from Vera Brittain's autobiography *England's Hour* (1941) is an apt illustration of that phenomenon: "In less than a second, my blood is chilled by a descending scream – a scream so often described by friends in war-books that I recognise it immediately, though I have never heard it before." See Vera Brittain, *England's Hour* (London: Continuum, 2005), 117.
5 The concept of "cultural circuit," coined by Richard Johnson, draws on the assumption that public discourses are taken up and reworked in personal narratives. These accounts become, in turn, incorporated into the collective discourses, thereby achieving wider distribution, which makes them readily available for a new mass audience. See Richard Johnson, "The Story So Far and Other Transformations," in *Introduction to Contemporary Cultural Studies*, ed. David Punter (London: Longman 1986), 277–313. See also Maurice Halbwachs's notion of "hegemonic collective memory" (Maurice Halbwachs, *On Collective Memory* (Chicago: University of Chicago Press, 1992).
6 "Drama documentaries" or "semi-documentaries" were a documentary form that emerged during the Second World War, characterized by a mixture of fictional elements (i.e., the use of a scripted story and dialogue, and an organized chain of events) and documentary-style elements (e.g., location shooting, the use of non-professional actors, and a truthful representation of actual incidents or events). Such films as *Men of the Lightship* (1940), *Merchant Seamen* (1941), *Target for Tonight* (1941), *Coastal Command* (1943), *Fires Were Started* (1943), and *Western Approaches* (1944) are typical of this aesthetics.

7 "Letter from the wife of a painter to an American friend," London, 7 October 1940, quoted in Diana Forbes-Robertson and Roger W. Straus, *War Letters from Britain* (New York: G.P. Putnam's Sons, 1941), 119.
8 For the connection between the Futurists and the British Documentary movement, see early experimental documentary films such as *Coal Face* (1935, dir. Alberto Cavalcanti) or *Night Mail* (1936, dir. Harry Watt), which were quite similar in their approach to sound to what Russolo advocated in his *Manifesto*. However, as opposed to the Futurists, who celebrated the advent of the machines and their characteristic sounds, British documentary filmmakers emphasized the harmony between humans and their environment, and never shared the former's fascination with modern warfare.
9 Luigi Russolo, *The Art of Noises: Futurist Manifesto* (New York: Pendragon Press, 1986), 49.
10 Tom Harrisson, *Living through the Blitz* (London: Collins, 1976), 102.
11 Harvey Klemmer, *They'll Never Quit* (New York: W. Funk, Inc., 1941), 157.
12 See Amy Helen Bell, *London Was Ours: Diaries and Memoirs of the London Blitz* (London: IB Tauris 2011).
13 "Letter 24, 15th September 1940," in *While London Burns: Letters Written to America, July 1940–June 1941*, ed. Fryniwyd Tennyson Jesse and Harold Marsh Harwood (London: Constable and Co. Ltd, 1942), 82.
14 "I love the sound of the guns!" quoted in Harrisson, *Living through*, 101.
15 Edward R. Murrow, *This Is London* (London: Cassell, 1941), 156.
16 See Jonathan Sterne, *The Audible Past: Cultural Origins of Sound Reproduction* (Durham, NC: Duke University Press, 2003).
17 Quoted in Harrisson, *Living through*, 76–7.
18 Graham Greene, *The Ministry of Fear: An Entertainment* (London: William Heinemann Ltd, 1950), 22.
19 Quoted in Harrisson, *Living through*, 94, 79; Basil Woon, *Hell Came to London* (London: P. Davies, 1941), 71; Forbes-Robertson and Strauss, *War Letters from Britain*, 109 ("Letter from Quentin Reynolds, American author, living in London, to his brother in New York").
20 James Olney, *Metaphors of Self: The Meaning of Autobiography* (Princeton, NJ: Princeton University Press, 1972), 30–1.
21 "Letter 26, near Manchester, October 1940," Jesse, *While London Burns*, 126.
22 Klemmer, *They'll Never Quit*, 2; Woon, *Hell Came to London*, 8.
23 Murrow, *This Is London*, 183; Vincent Sheean, *Between the Thunder and the Sun* (New York: Random House, 1943), 177.

24 Jesse, *While London Burns*, 131.
25 M-O A, TC Air Raids, 9/T, "Attitude of Non-evacuees," quoted in Katherine E. Fisher, "Writing (in) the Spaces of the Blitz: Spatial Myths and Memory in Wartime British Literature" (PhD diss. University of Michigan, 2014), 138; Bernard Kops, quoted in Juliet Gardiner, *The Blitz: The British under Attack* (London: Harper Press, 2010), 22.
26 Harold Nicolson, *Diaries*, ed. Nigel Nicolson (London: Collins, 1967), 129.
27 Woon, *Hell Came to London*, 201.
28 Brittain, *England's Hour*, 61.
29 Raymond Murray Schafer, *The Soundscape: The Tuning of the World* (New York: Knopf, 1977), 10.
30 Ibid.
31 The opening commentary of *Listen to Britain*, added to the film for the American release, was spoken by famous Canadian journalist Leonard Brockington.
32 Greg Goodale, *Sonic Persuasion: Reading Sound in the Recorded Age* (Champaign: University of Illinois Press, 2011), 108.
33 See Benedict Anderson, *Imagined Communities: Reflections on the Origin and Spread of Nationalism* (London: Verso, 2006).
34 Ovid, *Metamorphoses*, Book XI, trans. Dryden Garth et al. (London: Printed for J. Tonson in the *Strand*, 1727).
35 Jacques Attali, *Noise: The Political Economy of Music* (Manchester: Manchester University Press, 1985), 6.
36 About *Song of Ceylon*, Walter Leigh declared "The film has, in fact, been cut throughout with an eye to the sound-score. Its form is musically conceived; an analysis of its four movements would read like that of a symphony." Quoted in Roger Manvell and John Huntley, *The Technique of Film Music* (London: Focal Press, 1975), 48.
37 The origin of the expression is still uncertain, but it appears to have been first used by the London-based German musicologist Carl Engel who, in his *Introduction to the Study of National Music* (1866), not only questioned the existence of a British national musical tradition but also criticized the country's musical philistinism. The expression was later popularized by the German journalist Oscar Adolf Hermann Schmitz in his book *Das Land ohne Musik: Englische Gesellschaftsprobleme* (The land without music: Problems of English society), published in Munich in 1914, in which he stated that the British nation was the only one that did not have a national musical tradition.

38 Anthony Aldgate and Jeffrey Richards, *Britain Can Take It: British Cinema in the Second World War* (London: I. B. Tauris, 2007), 220.

39 See Pierre Nora, *Rethinking France: Les lieux de mémoire* (Chicago: University of Chicago Press, 2001).

13

The Literature of Intervention: US Participation in the Second World War

Tatiana Prorokova-Konrad

Although the Second World War remains one of the bloodiest conflicts, the US government skilfully and successfully played on feelings of patriotism to make it known to subsequent generations as a "good war" – or even *The* Good War. The government and the powerful elite played a large role in persuading the nation – and the soldiers in particular – that the war was worth fighting, that the goals were just, and that America had the truth on its side. Therefore, such attitudes as "a patriotic fervor, a single-minded devotion to cause and country that crossed all class and ethnic barriers" were naturally heavily propagated.[1] However, deployed military men were not necessarily inspired by high-minded patriotic appeals; they were fighting for each other and for the families that were waiting for them far away in their homeland, not necessarily out of a feeling of patriotism in its purest form.[2] These ideas widely found their expression in war literature, which implicitly questions the necessity of the intervention, the aims that were pursued, and the means used to achieve them.

American literary fiction about the Second World War includes many outstanding novels and short stories. Among the most well known are Norman Mailer's *The Naked and the Dead* (1948), Joseph Heller's *Catch-22* (1961), Kurt Vonnegut's *Slaughterhouse-Five* (1969), and William Wharton's *A Midnight Clear* (1982). These novels describe the events of the war and the soldiers' experiences in the Good War, yet the most fascinating question is perhaps how these writings shape and influence the understanding of US participation in the Second World War. All the aforementioned books proffer insight into the inner thoughts of fighting soldiers, plumbing their attitudes towards both local combat and their role in the war. They do so through an exploration of American exceptionalism, an ideology that was both essential to justify US military intervention in the Second World War on national and international levels and helpful to understand the military failures and frustration of those who were (in)voluntarily involved in the conflict.[3]

CATCH-22: THE "BUREAUCRATIC WAR"

Perhaps the most radical standpoint can be found in Heller's *Catch-22*, in which the author distances himself far from the customary praise of the US Army and patriotic reasons for soldiers' participation in the war. In doing so, he turns the stereotypical justification of a good fight on its ear to some degree. Quite impressively, Heller achieves this effect without devoting himself to a meditation on the fighting as a savage, brutal, and eternal condition that achieves its expression in war but, rather, by focusing on the US Army and stripping the phenomenon of its military dimension, thereby revealing the grotesque deformities of the institution itself. Kenneth D. Rose writes that, in the beginning, the novel is "less about the horror of war than the horror of bureaucracy."[4] From the very first pages Heller questions American values as he reveals the soldiers' inability to formulate their personal reasons for fighting in Europe, apart from stating sometimes meaningless icons that are unique to US culture. Yossarian, the protagonist of the novel, is the first to articulate the shallowness of their motivation: "The hot dog, the Brooklyn Dodgers. Mom's apple pie. That's what everyone's fighting for. But who's fighting for the decent folk? Who's fighting for more votes for the decent folk? There's no patriotism, that's what it is. And no matriotism, either."[5] The description of another soldier, Appleby, explicitly identifies this superficial rationale: "Appleby was a fair-haired boy from Iowa who believed in God, Motherhood and the American Way of Life, without ever thinking about any of them."[6] The toxicity of national propaganda is underlined too: "Men went mad and were rewarded with medals. All over the world, boys on every side of the bomb line were laying down their lives for what they had been told was their country, and no one seemed to mind, least of all the boys who were laying down their young lives."[7]

There is one character in the novel who does mind, however, and that is Yossarian, a rebellious soldier who does not want to accept the conditions in which he finds himself or, more specifically, the conditions that the war and the US Army in particular have forced him into. The war seems wrong and illogical to him. His commanders also fail to provide him with a proper explanation of why he must fight. To them, it is Yossarian's seemingly anti-patriotic objections that are illogical and wrong:

> Strangers he [Yossarian] didn't know shot at him with cannons every time he flew up into the air to drop bombs on them, and it wasn't funny at all. And if that wasn't funny, there were lots of things that weren't even funnier. There was nothing funny about living like a bum in a tent in Pianosa between fat mountains behind him and a placid blue sea in front that could gulp down a person with a cramp in the twinkling of an eye and ship him back to shore three days later, all charges paid, bloated, blue and putrescent, water draining out through both cold nostrils.[8]

This "aggressively, even belligerently, antiheroic" character should not be regarded entirely as a rebel,[9] a nihilist, or a traitor. On the contrary, one can view him as a hero for he is the only one who succeeds at a straightforward interpretation of the system under which the soldiers operate. He does not want to glamorize the war (or perhaps simply does not see a solid reason for doing so) in his mind or in his words and thus refutes the nation's general perception of the Second World War as a good war and the soldiers' participation in it as an honour. At the same time, Yossarian's refusal to fight and risk his life is not to be perceived as an act that subverts the nation in general and US interests in particular. Quite the contrary, it illustrates how misguided and disingenuous the system as well as its representatives – that is, the military commanders – are since they do not appear to value the individual lives of their soldiers. This idea is persistently stressed in the novel by the quota of sorties that becomes a moving target for the soldiers, for they can never reach the final number and therefore cannot "quit" fighting. The war thus becomes not a finite and well-planned military activity but a daunting, hopeless, Sisyphean obligation, the most probable end of which will only come with the soldier's death. Yossarian constantly tries to connect with the sympathies of his comrades: for example, he clarifies his refusal to go on missions because the enemy is "trying to kill" *him*, and when met with the response that "they're [the enemy] trying to kill everyone," he asks in return, "And what difference does that make?"[10] More than that, Yossarian openly rejects Clevinger and ex-P.F.C. Wintergreen's belief that it is his "job" to be killed on a mission in Bologna: "Am I supposed to get my ass shot off just because the colonel wants to be a general?"[11] This important yet controversial issue often becomes a stumbling block as we consider the moral responsibility of officers toward their subordinates. For some, the American military intervention is perceived as a career-making opportunity to attain medals and higher rank, but soldiers under their command will have to die in action in order to make that happen.

Catch-22 is not steeped in spiritual and moralistic claims in terms of one's national obligation to take up arms as part of the war effort; rather, it proclaims a unique wartime viewpoint – namely, that dying is not an option: "He [Yossarian] had decided to live forever or die in the attempt, and his only mission each time he went up was to come down alive."[12] Heller's protagonist demonstrates that there is nothing disgraceful in attempting to survive; no soldier would choose dying. The controversy lies in the impossibility – or self-defeating unpopularity – of expressing that notion as in the army soldiers are trained and expected to be fearless and prepared to sacrifice their lives. The novel addresses the issue head-on in the scene where Major Major claims, "One of the biggest jobs we all face in combat is to overcome our fear," but Yossarian acidly disagrees: "Oh, come on, Major. Can't we do without that horseshit?"[13] A soldier's life, according to *Catch-22*, no longer belongs to the individual: it becomes the property of the armed service that then orders the soldier to risk his life by fighting. Additionally, the novel articulates that the army not only asserts a metaphysical possession over the individual but a physical one too. Nurse Cramer points out that Yossarian is not in command of his own body: "It certainly is not your leg! ... That leg belongs to the US government. It's no different than a gear or a bedpan. The Army has invested a lot of money to make you an airplane pilot, and you've no right to disobey the doctor's orders."[14] Yossarian refuses to abide by this convention, advocating his right to his own life, which makes him morally the strongest character in the novel. His inquiry, "But why must I be one of them?" – meaning people who have to "suffer" in war[15] – is crucial in the novel as it demonstrates the widespread hopelessness that prevails in the wartime army along with the futility of escape, as going absent without leave (AWOL) would be ignominious for a soldier who is placed, willingly or not, in the role of a protector. This gives the novel enough material to play with the notions of sanity and insanity that are obviously central to *Catch-22* and are linked with Yossarian's behaviour. Because Yossarian does not want to obey the orders of his commanding officers and because he does not want to be part of the war at all, he is considered crazy. To demonstrate his disagreement with the army's orders, he refuses to wear his military uniform, which is to be understood as a visibly symbolic protest against being part of the US Army. Contrary to expectation, his disobedience and unwillingness to fly on dangerous missions is noted affirmatively by Dr Stubbs: "That crazy bastard may be the only sane one left."[16] The quote aptly illustrates the paradox on which the novel is built: Yossarian is considered insane because he challenges the regulations of the army, yet this could be the most defining indication of his sanity.

The issue of the absence of choice is further raised in the episode in which a military doctor says to Yossarian, "You've always got to do what your commanding officer tells you to … But they don't say you have to go home. And regulations do say you have to obey every order."[17] However, whether commanders really "know what they're doing" is questioned when the unit gets an order to bomb an Italian village.[18] Some soldiers refuse to do it, but Colonel Korn underlines the importance of the mission: "Would it be any less cruel to let those two German divisions down to fight with our troops? American lives are at stake, too, you know. Would you rather see American blood spilled?"[19] Thus, the rationale behind intervention is called into question. If the reason for US involvement in Europe was to fight against evil and free the oppressed, why did the US choose to bomb civilians? And why did Captain Aarfy later rape and kill an innocent Italian servant girl? What gave these American soldiers in Italy the right to iniquity? Was it because they were sure that in the chaos of war no one would pay attention to yet another corpse and, further still, that no one would force American soldiers to stand trial for these crimes because they were on a greater and more noble mission?

The most remarkable aspect of all this is to what extent the idea of American exceptionalism found its reflection in *Catch-22*. When discussing the outcome of the war with an old Italian, the young soldier Nately proclaims that "America is not going to be destroyed!"[20] However, the old man insists that even such great countries and empires as Rome, Greece, Persia, and Spain collapsed and that America is fated to reach a similar end. Such speculation prompts the following comment from Nately: "I suppose we can't last forever if the world itself is going to be destroyed someday. But I do know that we're going to survive and triumph for a long, long time."[21] The role of God as a divine basis for US intervention is not evaded in the novel either, although the author questions the idea that God is on the side of Americans. Whereas Colonel Cathcart conceives of God as an "immortal, omnipotent, omniscient, humane, universal, anthropomorphic, English-speaking, Anglo-Saxon, pro-American" deity who nevertheless "had begun to waver,"[22] Yossarian calls God "a country bumpkin, a clumsy, bungling, brainless, conceited, uncouth hayseed,"[23] blaming Him for all the misfortunes people must experience.

For these reasons *Catch-22* is an important literary work that provides a solid basis for how the idea of American exceptionalism was and continues to be applied to US military intervention. The novel also takes two stances that could be read as two core principles: it claims that American soldiers fought in the Second World War because theirs was the only nation that could stop evil; but it also suggests that the same ideology could be used merely as a powerful propaganda tool to encourage people to go and fight, even though good American men were dying in the war – despite supposedly having God on their side – and there was nothing graceful about it.

SLAUGHTERHOUSE-FIVE: THE PILGRIM SOLDIERS

The idea of American exceptionalism is even more pronounced in Kurt Vonnegut's *Slaughterhouse-Five, or The Children's Crusade: A Duty-Dance with Death*. In calling his main character Billy Pilgrim, the author clearly embraces symbolization; in fact, he makes it the novel's core technique. Possible origins for the main character's name, of course, come from the plot itself as it focuses on Pilgrim's travelling in time. Since Pilgrim's adventure partially takes place in Germany, where he fights against Nazis alongside other US soldiers, *Slaughterhouse-Five* inevitably focuses on the aspects of US intervention in the Second World War. Pilgrim's name creates a figurative parallel between the first Europeans who came to the United States as pilgrims and Billy, who, conversely, is an American who has arrived in Europe. The full title of the book also implicitly hints at the perceived divinity of the mission as it calls it a "crusade," a term that historically has always been associated with religious warfare. Therefore, my further proposition is that Vonnegut created a character, a soldier, whose journey to Europe is a divine pilgrimage. Hence, bearing in mind the idea of American exceptionalism, the sanctity of Billy Pilgrim and his actions is suggested by the persona with which the author imbued him.[24] Additionally, Vonnegut compares Pilgrim to Jesus, explaining that, in principle, "Jesus really *was* a nobody, and a pain in the neck to a lot of people,"[25] which is also highly reminiscent of Billy's character. What happened after the crucifixion is described as follows: "The voice of God came crashing down. He told the people that he was adopting the bum as his son, giving him the full powers and privileges of The Son of the Creator of the Universe throughout all eternity."[26] If we accept Billy Pilgrim as a proxy for Jesus, it then seems appropriate to assume that the words of God were addressed to Pilgrim as well. An American soldier thus becomes the son of God, responsible for conveying the word of God to others; Billy's participation in the Second World War is sanctioned because he is fighting on the side of God.

Beyond these allusions to American exceptionalism, *Slaughterhouse-Five* touches heavily upon the issue of interventionism and criticizes it quite explicitly. This "anti-war" novel,[27] as it is described by the author on the first pages, focuses on the firebombing of Dresden that was carried out by British and American soldiers on 13 February 1945. Vonnegut's intent is to speak out about the attack and publicly decry it as a cruel and cruelly forgotten operation: "It wasn't a famous air raid back then in America. Not many Americans knew how much worse it had been than Hiroshima, for instance. I didn't know that, either. There hadn't been much publicity."[28] The author's attempt to restore the devastating act to memory is vigorous, if not to say revolutionary, as he turns the bombing of Dresden into a national tragedy for the United States too. When Pilgrim and the other US soldiers are taken as prisoners of war and dislocated in Dresden, they become eyewitnesses – and in some cases victims – of, to borrow from Vonnegut, "the greatest massacre in European history."[29] For Vonnegut, the horrors of war and military policy consist not only of those numerous inexplicable and unimaginably brutal offensives but also in how fast and easily they can be forgotten, how readily and obediently people agree that collateral damage is an inevitable part of every war.[30]

In connection with that, it is worth examining how Vonnegut depicts the enemy in his novel. The notion of evil *Otherness* can be found easily in the pages of *Slaughterhouse-Five*: "The [German] reserves were violent, windburned, bristly men. They had teeth like piano keys."[31] The description continues: "They were festooned with machine-gun belts, smoked cigars and guzzled boose [*sic*]. They took wolfish bites from sausages, patted their horny palms with potato-masher grenades."[32] However, Vonnegut does not seek to portray the Americans as purely good soldiers either. For instance, he introduces Howard W. Campbell, Jr, who has taken the Nazi side and now agitates the prisoners of war to join him in the fight against Russians (and communism in particular). He describes American POWs as follows: "Expect no brotherly love, even between brothers. There will be no cohesion between the individuals. Each will be a sulky child who often wishes he were dead."[33] By attributing certain shortcomings and defects of character to both sides, Vonnegut demonstrates how, as Richard Giannone puts it, they are "lost in the struggle for human decency."[34] Indeed, the chaos of war makes it absolutely impossible to strictly divide the participants into good ones and bad ones. Moreover, Vonnegut suggests that it would be wrong to group US soldiers exclusively among the good. The author's appreciation for ambiguity and the individual capacity for both good and evil is revealed in the confession that he taught his sons that "the news of massacres of enemies is not to fill them with satisfaction or glee."[35] The most prominent message of the novel is, hence, that people must not reconcile themselves to savagery and genocide.

This admonition applies to both sides. While the mass eradications of the Jewish, Polish, Russian, and Roma people were firmly allocated their rightful place in general history, the atrocities carried out by the self-styled liberators were kept secret. *Slaughterhouse-Five* reveals that US citizens only found out about the bombing twenty-three years after the massacre. The reason for this was the "fear that a lot of bleeding hearts … might not think it was such a wonderful thing to do."[36] In his narrative of trauma, Vonnegut clarifies the horror of bombing by pointing to a fictional book written by Billy's favourite author Kilgore Trout, *The Gutless Wonder*, which tells a story of robots who were dropping burning jellied gasoline on people from planes: "They [robots] had no conscience, and no circuits which would allow them to imagine what was happening to the people on the ground."[37] In this way Vonnegut makes the inhumanity of the intervention apparent. Towards the end of the novel, as he finds himself on a plane flying over Germany, he "imagine[s] dropping bombs on those lights, those villages and cities and towns" that he sees from above.[38] It is shocking to realize that the everyday existence of those who do not expect (and obviously do not deserve) so brutal a punishment might be cut short by complete strangers in the pursuit of peace.

A MIDNIGHT CLEAR: FRIENDS AND FOES

Like Vonnegut's *Slaughterhouse-Five*, William Wharton's *A Midnight Clear* is noteworthy for its reluctance to take a particular stance in favour of one army or the other. *A Midnight Clear* looks at the Second World War from a different perspective than the one offered by *Slaughterhouse-Five*: it demonstrates how both sides became victims of war. Wharton does not judge those who took part but, rather, shows how utterly impossible it was for anyone to survive if they chose to ignore the coldly binary rule of war: fight or die. This is supported by my observation that Germans – the enemy – are portrayed positively in his novel. According to Sergeant Will Knott, although the Germans are still labelled as adversaries, "we [the soldiers in the squad] almost never call the Germans KRAUT or JERRY or HUN or NAZIS, any of the usual army names. At the most, they're 'the enemy.'"[39] The author's concern is to erase the moral boundary that separates German and American soldiers. Essentially, Wharton lifts the curtain and steps onto the very slippery ice of wartime ethics. He is not interested in war as a political event viewed from several removes; rather, he descends to the level of the battlefield and poses the question: What if there is no difference between the combatants, regardless of whether they are Americans or Germans? What if the soldiers on both sides think they are doing the right thing? Therefore, instead of describing US soldiers as the ones who believe themselves to be fighting on the side of God, the author points out that it is Germans who are wearing belts with the inscription "*Gott mit uns*."[40] Wharton's character, Knott, continues: "German priests're telling them [German soldiers] *they're* fighting a holy war against us. We're busy making martyrs of each other, fighting Godlessness. Same religions sending us all to the same heaven."[41]

Wharton's impulse to turn these foes into equals attains crystalline clarity in the scene when US soldiers find the two frozen corpses of an American and a German in the forest, intentionally arranged in such a way that they seem to be dancing with each other. Therefore, the concept of brotherhood, which is the novel's leitmotif,[42] is carefully applied to the characterization of the opposing sides. Wharton's strategy is to draw the reader's attention away from the complexities of politics and propaganda and instead focus it on soldiers as the ones who are simply and straightforwardly responsible for carrying out the orders they receive from their superiors. The novel therefore seeks to appeal to a mass audience and show that most soldiers do not want to fight. Over the course of the entire novel, the Germans, knowing that the war will end soon, look for an opportunity to surrender because they do not want to die, whereas the Americans find themselves in a state of endless fear. To be sure, *A Midnight Clear*'s preamble, a poem titled "Fear," suggests from the outset that the most common and most intense feeling experienced in war is exactly that. Sergeant Knott routinely shares with the reader how afraid he is: "I'm scared all the time and can't sleep, not even on a long guard";[43] "Wilkins looks scared, but we're all looking scared most of the time";[44] "Father's not careful enough; he isn't as afraid of dying as the rest of us";[45] "We are out of breath, more from fear than anything."[46]

Because both Germans and Americans were told they were enemies, they subconsciously perceived each other as such. Therefore, the epithets directed towards Germans, such as "filthy, Nazi, Kraut-headed, super-Aryan, motherfucking bastards,"[47] that we find in the novel do not come as a surprise. However, the reader notices that propaganda fades on the battlefield and soldiers start questioning "if there is an enemy" at all.[48] In a telling scene, Will Knott observes: "I don't think we've been seeing ourselves the way we look in these mirrors; it's hard to accept. *We* look like the enemy."[49] Gradually, the author demonstrates that, just as the US soldiers were previously unable to guess the age of one of the Germans ("He could be any age from twenty-five to forty"),[50] they are now unsure if Germans are actually the enemy. This is touched upon in the conversation between two American soldiers: "Maybe they're just nice guys, Stan. There could be some good Germans."[51] The main hardship of this "nonsensical" war for all soldiers is,[52] therefore, to assess the situation adequately, which is almost impossible given that Wharton likens the war to a "safari,"[53] in which humans thoughtlessly and callously "hunt"[54] other humans and "nothing matters much except not getting killed."[55] The chaos and insanity of the war consume everyone involved; no one can truly challenge the war as it creates and dictates its own set of self-serving rules. And the most important of those rules is that there is no brotherhood among enemies, as Knott alludes to when he tries to describe the improving "relations" with the enemy: "Fraternizing always seemed the wrong word; it did not have much to do with 'brothering.' I've always felt consorting was more what was going on. We were sure consorting with the enemy that night."[56] Any remote hope of being brothers with the enemies fully fades later on: the German surrender ends tragically for the American as well as the German troops when their comrades are killed. Moreover, in this episode, Wharton spectacularly recalls the metaphor of dancing soldiers he used in the beginning. To transport the two soldiers who were wounded during the surrender, Americans do the following: "We tear the satin into strips and tie Shutzer [an American] tight to the German so they support each other … I wrap all the extra satin tatters around Shutzer and the German's hands. They're both already ice cold."[57] Thus, the two soldiers come to resemble the frozen dancing statue of two other soldiers found in the forest earlier in the novel. Ultimately, both soldiers die, signifying Wharton's failure to contravene the dictates of war, which call for one to draw a merciless line between groups of individuals, thereby creating two distinct and bitterly opposed sides.

THE NAKED AND THE DEAD: VULNERABILITY AND DEPRAVITY

A radically different concept of intervention is found in Norman Mailer's *The Naked and the Dead*, mostly because the novel deals with US involvement in the Pacific. The author portrays war as a classic struggle between humans and nature. *The Naked and the Dead* provides a triumphal meditation on war as an environment artificially constructed by people in their quest for power and might. It functions as a wartime exposé in which Mailer reveals all the brutal aspects of the fighting that necessarily accompanies the intervention.

Understanding *The Naked and the Dead* as a novel about the struggle between humanity and nature provides a solid basis for interpreting US intervention in the Second World War. For Mailer, war starts with soldiers themselves, which implies that a brotherly relationship between soldiers of opposing sides is in fact unattainable. The main obstacle to this relationship is the corruption and brutality inherent in human nature. In Mailer's novel, this obstacle enables the emergence of morally void, unsympathetic characters (even among the Americans) who represent "an active personification of evil,"[58] which in turn allows readers like Jeffrey Walsh to argue that "not a single actively good man emerges throughout *The Naked and the Dead*."[59] Although there are glimmers of hope for male friendship and redemption – for example, when two soldiers continue carrying their wounded comrade out of a sense of moral obligation despite their own incessant pain and overwhelming fatigue – the novel depicts a great number of events that demonstrate the complete absence of solidarity and cohesion between soldiers, thereby contending that, even amid the familial bonds of the military, it's every man for himself. Evidence for this is seen in the hatred of some soldiers towards their Jewish comrades Roth and Joey Goldstein; in William Brown and Stanley, who refuse to go on carrying dying Wilson and leave him for "less exhausted" Goldstein and Ridges;[60] in the imperturbable Sam Croft, for whom war becomes the most thrilling experience one could ever hope to have and who eagerly participates in operations like a callous and single-minded killing machine; and in General Cummings, whose presence in the Pacific is explained by his thirst for power and his desire to be perceived as God.

There is pretence to mask the corruption – metaphorically, for example, by keeping the floor in the general's cabin spotless, presupposing that the territory the general occupies is pure – but the ridiculousness of "the clean floor" is underlined by the author: "If you looked at it clearly without the aura of military mumbo-jumbo, it became absurd, perverted, a revolting idea."[61] Moral perversion and depravity become an integral part, or rather, the foundation of the army, which General Cummings confidently and shamelessly declares in his own novel: "Corruption is the cement that keeps the Army from breaking apart."[62] Mailer's soldiers consider the intervention to be a welcome opportunity to obtain a higher rank and recognition through medals rather than a patriotic service to the country. Moral vices find their vehicle in aggression and a struggle for more status, more glory. This struggle then becomes a form of addiction, which is evoked by the author's juxtaposition of war and alcohol: "Whisky is the kind of thing a man oughtn't to do without. That's the trouble with the goddamn war."[63] Yet the addiction manifests itself in different ways. It is not necessarily a longing for combat but also pursuit of an easy profit when, for instance, Martinez, having found the corpse of a Japanese soldier, is "filled suddenly with a lust for the golden teeth";[64] or when Minetta pretends to be insane so as not to return to duty but nevertheless imagines how, once back in the United States, his family, friends, and neighbours will admire him for "all he went through" while still remaining so "modest."[65]

In this way, the aggression that characterizes human beings in their constant battle for material well-being and social status and the resulting chaos are now transferred to the Pacific arena where the soldiers are fighting. Their enemy is, however, not Japanese soldiers – indeed, the Japanese play a secondary role in the novel – but nature itself. What at first sight seemed to be "the Garden of Eden,"[66] and "a Biblical land of ruby wines and golden sands and indigo trees,"[67] turns into a terrifying environment where "the black dead ocean looked like a mirror of the night; it was cold; implicit with dread and death."[68] The intervention in the Pacific thus still has its roots in the idea of American exceptionalism but in a sinister, fallen sense in which the soldiers aim to conquer nature and thereby put themselves on a par with God. For General Cummings, this philosophy of war consists of "achiev[ing] God,"[69] though he already believes his role to be godlike, whereas the ambition of the soldiers is to tame and overcome their natural environment. Mailer frequently goes into the particulars of the conditions that the American soldiers must endure on the island: the fauna, the terrain, the climate, and the impenetrable jungle exhaust but do not break the majority of them – at least in the beginning. The climax of the novel comes when Croft decides to climb over the highest mountain on the island together with a small unit of soldiers. Here the superiority of nature is in full force as the mountain, which is "too immense, too powerful," simultaneously provokes "awe" and "fear,"[70] thereby anticipating the humans' failure. The soldiers incur losses, grow faint from the intense physical activity; they become scared of falling and dying. They ultimately have to surrender and turn back after a hornet attack. The irony of Mailer's choice of deterrent here is brilliant: in attempting to conquer the largest natural object on the island, the soldiers prove to be unable to cope with mere insects – what might seem to be the weakest and most insignificant natural menace. It is worth mentioning that the only "nature" in the novel that is subject to the power of men is a caterpillar and a small injured bird. It is skilful contrasts like these that Mailer uses to illustrate the fragile smallness of human beings and the powerful majesty of nature. In another passage comparing the soldiers to "a line of ants dragging their burden back to their hole,"[71] the author portrays them as something insignificant and imperceptible both in the context of the larger war and in the grand scope of nature. To further intensify this portrayal, Mailer frequently refers to the soldiers as children, as if to indicate their helplessness, immaturity, and impotency.

However, the soldiers' inefficacy and their failure in their endeavour can be explained in large part by the difficulty and disorientation of being in a foreign place. The theme of alienation is therefore central to *The Naked and the Dead* and is useful in coming to understand how its author chose to portray the nature of American intervention in the Second World War. Mailer intentionally added sections called "The Time Machine" throughout the novel. On one level their aim is to provide the reader with more information on the soldiers, such as who they were and what they had done before they joined the army. But another reason for this constant addition of background information is to remind the reader that the soldiers do not belong to the Pacific, that they are indeed aliens who are meddling in foreign territory. At the end of the novel, the soldiers are forced to surrender and leave the island, which can be viewed as a military failure and a natural victory. Such themes are evident in the very title of the novel in which "naked" stands for the soldiers' vulnerability and weakness in the face of nature, and their eventual defeat is expressed in the word "dead."

CONCLUSION

As *The Naked and the Dead* and the three other novels under discussion show, the subset of American literary fiction that deals with the Second World War attempts to provide answers to multiple issues that emerged during the war: from psychological hardships and ethical quandaries to personal and collective struggles. Yet this literature also serves as a reliable tool to examine such a complex issue as American intervention in the Second World War (and since) from cultural perspectives. As the examples demonstrate, American intervention was undoubtedly necessary in the larger fight against the German Nazis and their allies, and the opposition and ultimate defeat of the Axis powers was clearly what made the Second World

War a "good war" for the American people. But the pure, high-minded sense of patriotism and the reassuring belief in fighting for the greater good are not much reflected in the literary fiction that deals with this subject. Instead, this literature focuses on issues pertinent to the day-to-day operations of the military, revealing soldiers' doubts, friendships, or betrayals; the venality and bureaucratic problems within the army; and the soldiers' difficulty in comprehending the reasons for intervention. By the same token, it depicts the destructive nature of war and all the sorrow and pain, chaos and death it causes. And it is perhaps this aspect more than any other that allows me to conclude that Norman Mailer's *The Naked and the Dead*, Joseph Heller's *Catch-22*, Kurt Vonnegut's *Slaughterhouse-Five* and William Wharton's *A Midnight Clear* are critical of US involvement in the Second World War because of the way in which it was carried out and because of the sufferings young American men were inevitably forced to face, despite the fact that the cause was arguably just.

NOTES

1 Kenneth D. Rose, *Myth and the Greatest Generation: A Social History of Americans in World War II* (New York: Routledge, 2008), 61.
2 Ibid., 69.
3 Donald Pease defines American exceptionalism as "a complex assemblage of theological and secular assumptions out of which Americans have developed the lasting belief in America as the fulfillment of the national ideal to which other nations aspire." See Donald E. Pease, *The New American Exceptionalism* (Minneapolis: University of Minnesota Press, 2009), 7.
4 Rose, *Myth and the Greatest Generation*, 194.
5 Joseph Heller, *Catch-22* (London: Vintage Books, 2004), 10.
6 Ibid., 21.
7 Ibid., 18.
8 Ibid., 19.
9 Vance Ramsey, "From Here to Absurdity: Heller's *Catch-22*," in *Seven Contemporary Authors: Essays on Cozzens, Miller, West, Golding, Heller, Albee, and Powers*, ed. Thomas B. Whitbread (Austin: University of Texas Press, 1966), 103.
10 Heller, *Catch-22*, 19.
11 Ibid., 141.

12 Ibid., 33.
13 Ibid., 101.
14 Ibid., 335–6.
15 Ibid., 36.
16 Ibid., 127.
17 Ibid., 67.
18 Ibid., 374.
19 Ibid., 375.
20 Ibid., 279.
21 Ibid., 280.
22 Ibid., 328.
23 Ibid., 206.
24 George Roy Hill's film adaptation of *Slaughterhouse-Five* (1972) elaborates on the idea even more. In the scene where Billy finds himself ambushed by American soldiers and tells them his name, they jokingly tell him to "start praying." Billy submissively kneels down and begins his prayer, while in the background we hear the voices of Americans describing how awful their situation is. See *Slaughterhouse-Five* (dir. George Roy Hill, perf. Michael Sacks, Ron Leibman, Eugene Roche, and Sharon Gans. Universal, 1972, DVD).
25 Kurt Vonnegut, Jr, *Slaughterhouse-Five or The Children's Çrusade: A Duty-Dance with Death* (New York: Dell Publishing, 1968), 109, emphasis added.
26 Ibid., 109–10.
27 Ibid., 3.
28 Ibid., 10.
29 Ibid., 101.
30 Hill's film deals at length with this issue as it showcases the pre-bombed Dresden almost too grotesquely: the streets are filled with people but the attention of the audience is drawn to a group of children running through the streets, playing and laughing. These otherwise wholesome images become perversely disturbing and cause the audience members to avert their gaze as they realize that most of the children will soon be dead. It is also worth noting that the film emphasizes the rich architecture of the city that was ultimately all but destroyed too.
31 Vonnegut, *Slaughterhouse-Five*, 64.
32 Ibid.
33 Ibid. 130, emphasis in original.
34 Richard Giannone, "Slaughterhouse-Five," in *Vonnegut: A Preface to His Novels* (New York: Kennikat Press, 1977), 87.

35 Vonnegut, *Slaughterhouse-Five*, 19.
36 Ibid., 191.
37 Ibid., 168.
38 Ibid., 211.
39 William Wharton, *A Midnight Clear* (New York: Alfred A. Knopf, 1982), 9.
40 Ibid., 87, emphasis in original.
41 Ibid., emphasis added.
42 The relationship between the solders in the squad is indeed much tighter than that of brothers. The squad is depicted quite literally as a family, as the eldest members are nicknamed Father and Mother. Moreover, Mother is frequently referred to as Mother Hen Wilkins since, as Knott puts it, "he's always hounding *us* for being sloppy, bugging us about leaving things around or not cleaning out mess kits and canteen cups … Wilkins would sneak up on everyone after breakfast and give the sniff test to see if we'd brushed our teeth" (7, emphasis in original). Father is so named on account of his taking the vows before the war. This is also symbolic because the soldiers are therefore guided by a follower of God, which means that the idea of American exceptionalism can be smoothly applied to the actions of the squad or the orders they receive. The soldiers' boundless devotion to their "parents" is made manifest when they decide to take German POWs and make Mother a hero of the operation: "He'll become a legend. Maybe we can work this up into a Congressional Medal of Honor. Wilkins's name will be inscribed on the dollar bill" (142). The risks these solders are prepared to take for their leaders is apparently substantial.
43 Wharton, *Midnight Clear*, 13.
44 Ibid., 25.
45 Ibid., 62.
46 Ibid., 78.
47 Ibid., 32.
48 Ibid., 49.
49 Ibid., 55, emphasis in original.
50 Ibid., 150.
51 Ibid., 138.
52 Ibid., 183.
53 Ibid., 102.
54 Ibid.
55 Ibid., 238.
56 Ibid., 168.

57 Ibid., 207.
58 Joseph J. Waldmeier, *American Novels of The Second World War* (The Hague: Mouton, 1969), 76.
59 Jeffrey Walsh, *American War Literature 1914 to Vietnam* (London: Macmillan, 1982), 112.
60 Norman Mailer, *The Naked and the Dead*, 50th anniversary ed. (New York: Henry Holt and Company, 1998), 631.
61 Ibid., 314.
62 Ibid., 244.
63 Ibid., 199.
64 Ibid., 214.
65 Ibid., 360.
66 Ibid., 97.
67 Ibid., 453.
68 Ibid., 454.
69 Ibid., 323.
70 Ibid., 497.
71 Ibid., 133.

14

Fighting Fascism? The Second World War in British Far-Right Memory

Paul Stocker

Lady Butler was born three decades after Waterloo; Nick Griffin, leader of the British National Party (BNP), came into the world almost fifteen years after the end of the Second World War. On 11 June 2009 in Manchester Town Hall, Nick Griffin held his arms aloft making two palm-facing "victory" gestures with his fingers. Griffin had just guided the BNP to the greatest electoral result for a far-right and neo-fascist political party in Britain in history, obtaining just under 1 million votes nationally and winning two seats in the European Parliament. Griffin's finger gesture, symbol of the international Nazi-resistance movement formed in 1941, and most prominently adopted by Prime Minister Winston Churchill, could be easily dismissed as an audacious attempt to draw comparisons between himself and a national icon. However, this chapter argues that this seemingly benign event represents something more substantial: the BNP's attempt to transition itself from fringe, neo-Nazi pariah to a politically and culturally mainstream party. It ultimately seeks to demonstrate how the BNP used the memory of the Second World War during what Nigel Copsey calls its "quest for legitimacy."[1] By analyzing the far right over the past forty years, this chapter demonstrates how this fringe but by no means insignificant political movement's memory of the Second World War has transformed from portraying Churchill as the evil defeater of the Nazis and enabler of the British Empire's destruction to a British hero and archetypal modern British nationalist.

This chapter investigates "public memory," which refers to interpretations of the past shaped by elite and "official" institutions, such as the state, but which also includes political parties.[2] This is not to be confused with private memory, which reflects individual and "micro" memories that cumulate to form to what is usually referred to as "collective memory." Often, public "official" memory and private memories contradict each other, especially when it comes to contentious visions of history. Nevertheless, "individual personal memories of an event or period are shaped by public shaped representations of the same, while public representations, equally, draw on individual memories for recognition."[3] This study analyzes the far right as a political movement and institution. It assesses its official memory of the Second World War and Churchill by looking at official pronouncements, publications, and broadcasts. Due to limitations of space, the private individual memories of far-right voters, activists, and leaders – while important – are not investigated in this study.

Official far-right memory of Churchill and the Second World War has transitioned from one constructed almost entirely in isolation towards one greatly shaped by a wider collective memory. That is to say, the far right's position has moved from directly rejecting and criticizing the narrative of the Second World War as Britain's "finest hour" towards embracing it (albeit selectively). This shift did not develop organically, nor was it evolutionary: it reflected a conscious decision by power brokers. In an attempt to be electorally successful, Nick Griffin, upon becoming leader of the BNP (1999–2013), explicitly sought to make the party more in tune with public opinion. His predecessor, John Tyndall (1982–99), was less concerned with public opinion given that he was sceptical of the democratic route to power. For Tyndall, ideological purity was more important than winning over the public. The BNP, like any other political party, is not a passive organization absorbing public sentiment for the sake of it but, rather, is pro-actively seeking to interact with members of the public and shape their outlook on political, economic, and cultural matters. Memory is thus not something that is constructed from scratch, nor is it merely imported without criticism – it is a calculated negotiation that melds the party's core ideology with public opinion.

The awkward but nevertheless apt term "fascistization" becomes important here in terms of conceptualizing how the far right co-opted the mainstream memory of the Second World War. Fascistization, as demonstrated by Aristotle Kallis, refers to the adoption of fascist ideological traits and styles by non-fascist actors as well as the co-option of fascist political movements by non-fascist regimes and governments in an attempt to mimic the appeal of fascism without subscribing to its totalitarian goals.[4] One example of this is Britain's first self-proclaimed fascist organization, the British Fascisti, which, while not genuinely revolutionary and thus not fascist, aped much of the ideology and style (as well as the name) of Benito Mussolini's fascist state in Italy in an attempt to appear modern and dynamic.[5] This chapter uses the term to reflect a related but different form of fascistization – the *re-adaptation* by (neo)fascists of non-fascist ideas and style. This is done to appear less culturally and politically alien while, at the same time, transferring their own ideology into an acceptable frame. For example, the BNP sought to "fascistize" Winston Churchill's memory by selectively quoting his views on Muslims and, in so doing, explicitly arguing that he would support the neo-fascist BNP were he alive in the twenty-first century.

BRITAIN'S "FINEST HOUR"

It could be quite easily argued that no historical event has received more attention in British public life than the Second World War.[6] As British author and comedian John O'Farrell once wrote in the *Guardian*: "As far as the British people are concerned the history of planet earth goes like this. 1) The earth cools. 2) Primitive life forms emerge. 3) Britain wins World War II. Apart from that, nothing much of any importance has happened, with the possible exceptions of England winning the World Cup and the Beatles going on the Ed Sullivan show."[7]

The narrative that underlies Britain's memory of the war is certainly an epic one. It holds that Britain stood alone in 1940 following the fall of France and faced certain defeat by the Nazi war machine. Yet, against all odds, the RAF defeated the Luftwaffe during the Battle of Britain. Bombs soon began to pummel towns and cities, but the British spirit was undiminished and the public joined together to "never surrender." By 1945, Britain, with a little help from its allies, stood victorious and had brought about peace on the continent. It is not the purpose of this chapter to outline the historical facts but, rather, to state that the popular myth of Britain's role in the Second World War is just that: a "myth," encompassing elements of both truth and conjecture. Historians have picked holes in the popular narratives of Britain's role in the Second World War for decades. Books such as Angus Calder's *The People's War* (1969) and *The Myth of the Blitz*, as well as Clive Ponting's *1940: Myth and Reality*, have all critiqued elements of popular myth – yet sober academic analysis has not depleted the public's appetite for championing Britain's "finest hour."[8]

The central figure in Britain's modern Second World War story is its wartime prime minister, Sir Winston Churchill. The continued popularity of Winston Churchill as a national hero and icon surely cannot be disputed, and his invocation has continued to be ever present since the end of the war. He is invoked in the form of Churchill the archetypal "British bulldog"; Churchill the gallant war leader; Churchill the hero who, almost single-handedly, leads Britain to victory over the Nazis; and Churchill the orator, who, with stirring speeches, lifts the spirits of the beleaguered members of a war generation, leading them to an unlikely triumph. Perhaps the best example of the continued belief in Churchill's place in British history occurred when he was voted "Greatest Briton" in 2002. The TV poll was conducted nationwide and attracted over 1 million votes, with Churchill receiving just short of 450,000 of them. Popular Labour minister Mo Mowlam, in a documentary on Churchill in the run-up to the Greatest Britons vote (which is believed to have led to a late surge in support for Churchill), argued: "If Britain – its eccentricity, its big heartedness, its strength of character – has to be summed up in one person, it has to be Winston Churchill."[9]

Churchill's memory remains salient enough that he continues to be an important political player over half a century since his death. Most recently, Churchill was invoked during Britain's so-called "Brexit" referendum. The fact that both Remain and Leave campaigns sought to claim Churchill as their own speaks volumes about the importance of possessing the "Churchill vote." Sir Nicholas Soames, Churchill's grandson, member of Parliament, and campaigner to keep Britain in the EU argued:

> The last thing on earth Churchill would have been would have been [sic] an isolationist – to want to stand apart from Europe right now at a difficult time … There is something awfully un-British, in my view, about wanting to leave. I think we stay. I think "Non" he would not think it is a good thing to leave. "Oui," I think he would have wanted to stay.[10]

This is in contrast to prominent Leave campaigner and present prime minister Boris Johnson, who said: "Britain needs to be supportive of its friends and allies – but on the lines originally proposed by Winston Churchill: interested, associated, but not absorbed; with Europe – but not comprised."[11] On another occasion, Johnson argued: "the evidence is absolutely conclusive that Churchill did not want us to join the EU or its predecessor ... he would be horrified at the loss of sovereignty this country has suffered and would definitely campaign for us to leave."[12] In addition to the palpable nostalgia for previous British triumphs, this demonstrates the continued salience of Churchill and his frequent invocation during contemporary British political discourse as well as the adaptability of his myth and memory relating to a range of issues. It is thus no surprise that Churchill continues to loom large in the history of British nationalist politics, although not always as a gallant war leader.

THE BRITISH FAR RIGHT AND MEMORY

The memory of a glorious past has always been an important strand in far-right propaganda and thought.[13] Fascist organizations of the interwar period often looked back towards Tudor England as a model for a modern fascist state. As Thomas Linehan notes, "the Tudor State's hostility to party factions and self-interested sectional interests, and its objective of national integration through authoritarian centralized government, were collectively held up as a prototype of fascist government and the modern fascist rational state."[14] Fascists of a biologically racist disposition, such as the Imperial Fascist League, looked back even further to a mythical Nordic era of racial purity. When memory was used as a function of propaganda, more recent British achievements were championed, primarily in the imperial arena. Britain's largest interwar fascist party, the British Union of Fascists (BUF), nostalgically harked back to swashbuckling imperial heroes who, it was claimed, would no doubt have been fascists had they been alive in the 1930s. In the BUF press, hagiographies of former imperial "heroes" such as Clive of India, Warren Hastings, and T.E. Lawrence were presented to demonstrate the values that had turned Britain into an imperial power. These were the values that would be observed under a fascist state – self-reliance, adventurism, and indefatigability when faced with a foreign enemy.[15] Linehan notes that these "reverential eulogies," while little more than "mythic inventions," formed part of fascism's attempt to sacralize politics.[16] These were not just heroes of British history, they were martyrs reimagined through a fascist lens.

The BNP's embrace of another British "hero" in the twenty-first century, Winston Churchill, at first seems completely unremarkable. It would seem palpably obvious that a political party that professes to be the true voice of the British nation would use its greatest hero as a symbol. However, Churchill has not always been popular among the far right. In fact, for decades before Nick Griffin's ascension to the leadership in 1999, Churchill was seen as the arch-enemy of the British extreme right. The far right was so marginal during much of the postwar period that it had a small, fringe countercultural identity that rarely concerned itself with public opinion or sensibilities. In many instances this meant a hard-line devotion to Nazism and the ideals of Adolf Hitler. It sought to demonstrate that British history, as taught and understood by the "rotten establishment," had been a lie. As this suggests, conspiracy theory pervaded every single aspect of the far right's ideology and understanding of the past. Its memory was ultimately one borne out of a wider ideology, completely isolated from both facts and conventional narratives of Britain's role in the Second World War.

The most important figure on the postwar far right prior to Nick Griffin was John Tyndall. Born in 1934, Tyndall was active within several far-right and neo-Nazi organizations (and was sent to jail for inciting racial hatred on a number of occasions) before becoming chairman of the National Front in 1972 and again in 1976. The National Front is a far-right political party, traditionally composed of hard-line conservatives as well as revolutionary fascists, which enjoyed a fleeting but limited period of success in the 1970s. Following the collapse of the National Front, Tyndall founded the BNP in 1982. Tyndall's political worldview more than adequately fits what is now a generally accepted, but minimal, definition of fascism as "palingenetic," or revolutionary, ultranationalism.[17] Throughout his long career on the far right he was supremely nationalist, racist, anti-liberal, and anti-communist – all hallmarks of both interwar and postwar fascist movements. One of the most salient features of Tyndall's political ideology was his attachment to conspiracy theory – an important feature of his criticism of Winston Churchill. Demonstrating this, in an interview with biographer David Baker, reflecting on his political tutelage by A.K. Chesterton, the interwar fascist and leader of the far-right movement the League of Empire Loyalists, Tyndall stated:

> The biggest contribution Chesterton made to my thinking was in the development of an understanding of the conspiracy theory of history. Before reading the writings of Chesterton, or any writings on a similar subject, I had a vague kind of suspicion that all kinds of plotting was going on in the world and events were occurring that were not explained in terms of normal reporting … There seemed to be a hidden factor in world politics that was bringing about, among other things, the destruction of this country.[18]

Tyndall followed a long line of fascists and Nazi-enthusiasts in Britain by identifying the "hidden factor in world politics" as Jewish in origin.

Tyndall was the most important far-right figure to denounce the Churchill myth frequently and passionately. In 1965 (shortly after Churchill's death), Tyndall was a devoted neo-Nazi who utterly admired Adolf Hitler. As the nation mourned, Tyndall pondered what the memory of Churchill would be when the immediate memory of his speeches and the emotion attached to the war died out. He then argued that Britain's attachment to Churchill was overly sentimental and that the country "must begin to consider the bare facts of the Churchill career."[19] Tyndall argued that Churchill was the puppet of World Jewry and was no British hero. His opposition to the Aliens Act, 1905, which restricted predominantly Jewish immigration, and subsequent modification of it when he was home secretary "laid open to a flood of the worst refugees of the European ghettoes, who soon afterwards began to form the backbone of the Communist and other subversive movements in our country. As is known, these were the beginnings of a lifelong collaboration between Winston Churchill and World Jewry in all spheres."[20]

According to Tyndall, Churchill's opposition to Nazism was due to his being in the pay of "international Jewry," and he contended that Churchill was ultimately fighting a Jewish war that cost Britain its Empire. On the twentieth anniversary of the Second World War's end in May 1965, Tyndall's periodical *Spearhead* would have a vastly different interpretation of events than did the rest of the public. The essential argument was that the Second World War was a Bolshevik-Jewish war that was not fought in Britain's interests. Hitler was Britain's friend, and communism and Jewry were the real enemies. It is clear that many of Tyndall's followers agreed. One letter written to ***Spearhead***, entitled "Churchill – The Jews Bricklayer," mirrored Tyndall's stance: "Britain, warped by Alien influence, driven out by short sightedness hypocrisy and folly, fought on the wrong side and in victory secured for herself decades of defeat, financial and moral bankruptcy."[21]

As leader of the British National Party, Tyndall continued his anti-Churchill rhetoric into the 1990s. Tyndall was impressed by historian John Charmley's revisionist biography *Churchill: The End of Glory*, which was published in 1993 and which criticized Churchill's conduct during the war, arguing that Britain did not need to fight Nazi Germany. Upon its release, Tyndall claimed that "a minor earthquake [had] hit the historical establishment" and that it "debunked the entire legend of Winston Churchill as a saviour of Britain in his role of war leader against Hitler."[22] Nevertheless, Tyndall did not foresee the erosion of the Churchill myth anytime soon. He recognized its continued significance within British society nearly thirty years after Churchill's death: "Do not expect Churchill's champions to be persuaded by the force of rational argument – for their commitment to him and everything he stood and fought for is essentially an emotional one."[23] He accused the British people of taking "refuge behind images of 'our finest hour,' that fading memory of the generation, now old and bewildered as it contemplates the nightmare of the modern world, which half a century ago was spurred by Winnie's speeches to march to calamity and catastrophe."[24]

Ultimately, Tyndall's criticism of the myth of Winston Churchill was based on his sincerely held belief that he was a fraud – a puppet of World Jewry. This provides a good example of the far right's reliance on conspiracy theory to explain historical events. It also reflects its detachment from mainstream popular culture. While the anger directed at someone widely believed to be a hero by the British public cannot be seen as the sole reason for the far right's failure to achieve any political success during these years, it is symptomatic of John Tyndall's approach to politics. He refused to compromise with what he perceived to be the liberal-indoctrination of British society and, thus, rarely attempted to make inroads into the mainstream. What could Tyndall possibly hope to achieve by criticizing Churchill shortly after his death as the nation mourned? Right up until his ousting as leader in 1999 and his death six years later, he remained unable and perhaps unwilling to understand contemporary Britain. As prominent BNP activist Eddy Butler puts it: "He knew how to talk to a small room of nationalists, but he didn't know how to talk to a thousand Yorkshire young geezers. He hadn't got a clue about normal people or normal politics."[25]

THE BNP, CHURCHILL, AND THE SECOND WORLD WAR

Upon his election as BNP leader in September 1999, after he had ousted the aging and increasingly ineffectual Tyndall, Nick Griffin sought to modernize the party, transforming it from a fringe neo-fascist organization into a more respectable, mainstream outfit. Griffin ruthlessly exploited Tyndall's extreme past as a devoted National Socialist (images of Tyndall dressed in Nazi-style paramilitary attire and posing in front of Nazi regalia made this accusation effortless) as the embodiment of the BNP's image problem. The new, professionalized BNP would embark on a period of limited but notable electoral success between 2003 and 2009. It won several local council seats in 2003 and narrowly missed out on a European Parliament seat in 2004. It made over thirty gains in local council elections in 2006, and, by 2008, the BNP – a party that had until recently called for the compulsory repatriation of non-whites in Britain – had over one hundred elected representatives across the country, including a London Assembly member. Its greatest success came in 2009, when it received just shy of 1 million votes and gained two seats in the European Parliament. This was an unprecedented achievement for the far right, which had been an electoral irrelevance for its entire history prior to Griffin's BNP.

While Griffin's modernization program can be seen as partially responsible for the electoral success of the BNP, rather than representing a fundamental break with the party's ideology it was, in fact, a thinly veiled make-over that sought to remodel the party's image and style. The revolutionary creed of fascism and "racial nationalism," along with accompanying esoteric conspiracy theories and anti-Semitism, would remain within the party but would be discussed behind closed doors among a small inner circle.[26] Griffin had stated early on in his attempt to rehabilitate the party's image that any facelift would be purely cosmetic. He indicated this in the party magazine *Patriot*, in which he argued that, while the party must speak a new language to the electorate: "of course, we must teach the truth to the hardcore."[27] Critical assessments of Churchill would now occur in secret, and one can only speculate as to their content. The revamped BNP would turn the far right's traditional memory of Churchill on its head.

A glimpse into the BNP's new style can be found in a guide handed out to activists that explained how to discuss the party's ideology and to rebuff accusations of extremism. One example of this is the insistence to activists that: "The BNP is *not* a 'fascist' or 'fascistic,' let alone a 'Nazi' or 'neo-Nazi' or 'national socialist' party. It should never be referred to as such by BNP activists, and anyone else who does so must be politely but firmly corrected."[28] Another tool used to counter the idea of the BNP as fascist or as Nazi was its invocation of Churchill and the spirit of the Second World War. In one telling section in the ominously sounding *Language and Concepts Discipline Manual* we find: "successful revolutions from the right have always presented themselves as restoring older traditions. Therefore, we should couch our agenda in restorationist terms whenever possible. This should not be misunderstood as meaning we have to sound like (old-style) Tories when their ideas do not coincide with ours. Ours is a populist traditionalism, not an elitist one."[29] Winston Churchill would function as the symbol of the new BNP and the subject of the party's new "populist traditionalism." Indeed, Griffin sought to draw comparisons between himself and Churchill, claiming to be the defender of "freedom" against the tyrannical and "fascist" New Labour establishment.

The BNP thus no longer sought to present sympathetic views of fascism or Nazism. It now defined itself against both and claimed to be fighting fascism, suggesting that its "struggle" was like that of Churchill's. Fascism, as presented in party literature, came in many forms and bore little resemblance to any generally accepted definition of that term. For example, one article in *Identity*, entitled "New Labour and the Return of Fascism," argued that Tony Blair's "'New Labour' project was demonstrably fascist. It was argued that fascism is 'a form of intensely idealistic – almost religious – commitment to an ideology of national rebirth, coupled with a Keynesian attitude to government spending and an extremely close relationship between the ruling political elite and certain favoured interests.'"[30] Chancellor Gordon Brown's spending proposals as well as Labour's "cosiness" with big business were seen as proof of this. In addition, Labour's social reforms were viewed, quite absurdly, as part of a process of "national rebirth" that reflected an attempt at "smashing centuries of constitutional settlements, a gigantic programme of genetic, cultural and social engineering through mass immigration and the extension of bureaucracy, and [was] openly flaunted as the wholesale remaking of every aspect of British society."[31] Further proof of the BNP's new found Churchillian anti-fascism came in its opposition to purported "Islamofascism."

The BNP frequently alleged that the party's hostility to Muslims was similar to Churchill's views on Islam and was often cited as evidence that members of the BNP were the legitimate defenders of his legacy. It argued in one article that: "Islam is a retrograde cult that relies on the people remaining stupid and brainwashed in order for it to survive" and that "the Koran is the word of a paedophiliac, narcissistic, psychopath who far from adding to the sum of human goodness or knowledge has destroyed it wherever its credo has taken root."[32] The notion of an Islamic "invasion" was specifically designed to invoke wartime nostalgia. On several occasions, Churchill's disparaging comments on the Islamic faith, written in his 1899 book *The River War*, were cited as proof that the party was staying true to Britain's most iconic leader, who "expressed thoughts that would now get him prosecuted by our totalitarian Government for inciting racial hatred."[33] This view was corroborated in a letter from a BNP supporter to *Identity*: "If Britain's most famous son was passing his judgement today, no doubt his concerns would focus on the current weakness of Christianity as a unifying force against the arrival of what is an alien religion in Britain. I expect his comments would more easily find a home with the British National Party than any other of our mainstream parties."[34]

The most prominent invocation of Churchill's memory by the BNP would come in public broadcasts and electoral material. In the run-up to the 2009 European Elections, the BNP's party political broadcast created a storm of controversy over its usage of Churchill. The voice-over on the broadcast, which comes with ominous music reminiscent of Britain under invasion in 1940 as well as of Churchill's "Blood, Sweat, Tears and Toil" speech (made to his cabinet on 13 May 1940), states:

> Our war heroes defied European dictators, riding out the storms of war to preserve our traditional British Christian values ... In World War II the sacrifice of our heroes saved not just Britain from invasion, but the whole of Europe from the evil Nazi plan

> to create a totalitarian European super-state. Our war heroes fought like lions to stop them being swamped by foreign invaders. What would the mates they left behind think if they could see Britain today? What's it coming to when you are made to feel like an unwelcome foreigner in your own country?[35]

Griffin, seeking to channel Churchill, stated on the broadcast: "your family's rights to enjoy the benefits that come from being part of this great nation were won by the blood, sweat, toil and tears of our past generations."[36] Churchill's grandson and MP Nicholas Soames said that the BNP had "behaved in a disgusting manner. They should not take my grandfather's name in vain. He would have been appalled by their views and the way they claim to represent the wartime generation."[37] The broadcast reflected the fascistization of the mainstream memory of the Second World War. Britain was presented as under "foreign invasion" not by armies but by immigrants, and it was the Churchillian spirit that would once again lead Britain to victory.

The BNP's use of Churchill became the subject of debate on Griffin's now infamous and highly controversial appearance on the BBC debate program *Question Time* in October 2009. One audience member asked: "Is it fair that the BNP have hijacked Churchill's image?" Griffin responded:

> I say that Churchill would belong in the British National Party because no other party would have him for what he said in the early days of mass immigration to this country. For the fact that they are only coming for our benefits system. And for the fact that, certainly in his younger days, he was extremely critical of the dangers of fundamentalist Islam in a way which would now be described as Islamophobic. I believe that all the effort in the Second World War and the First was designed to preserve British sovereignty, British freedom.[38]

Fellow panellists were quick to slap down Griffin – Jack Straw pointed to the millions of African and Asian soldiers who fought for Britain in the two world wars who would not be allowed to join the BNP due to its strict membership rules, which disqualified non-white members. Playwright Bonnie Greer also doubted that Churchill would even be allowed as a member, given that his mother was American with allegedly Native American ancestry.

It would be convenient to dismiss the BNP's usage of Churchill and the Second World War as an audacious piece of propaganda designed to mask the true fascist nature of the BNP. While this may be true, in terms of far-right history its significance lies in that it reflects part of the British far right's image change, from a practically open neo-fascistic to a more continental "right-wing populist" style. This trend has occurred throughout Europe over the past three decades, with many previously neo-fascist parties, such as the Front National in France, seeking political legitimacy by softening their style and ideology.[39] Adapting their memory of historical events not just to a more palatable narrative but to one that enables them to drape themselves in populist myths that are already present within the country is part of this process of transition, and it varies depending on the national context. What the increased success of far-right parties that do this shows is not only the power of national myths but also their ability to be moulded to suit a range of political causes.

CONCLUSION

The BNP's embrace of Churchill and the memory of the Second World War differs from that of other political parties. It was part of a process designed to make the party more respectable and, above all else, *legitimate* – a strategy that clearly succeeded for a short while as the party won dozens of council seats, a London Assembly seat, and two seats in the European Parliament. The BNP's invocation of the Churchill myth differs from its invocation by other politicians as it is designed to normalize the BNP in the eyes of the electorate, portraying it as a lawful and constitutional political party in keeping with British traditions. The far right, ever since its inception, has always faced accusations of being "foreign" both in terms of its political ideology and its loyalties to other extremist regimes. The BNP, clearly, was seen as legitimate by hundreds of thousands during the period when the Churchill myth was employed, indicating that the party's "modernization" program was partially successful in "detoxifying" the far-right brand.

After its effective collapse following the 2010 general election, the BNP has been a failed political project; however, its collapse, which began in 2011, had little to do with a lack of support or a revulsion towards its ideas. Rather, it had to do with personal infighting and poor management of finances. Britain First, a new group seeking to replace the BNP as the dominant far-right party, continues to use Churchill's image and speeches on social media posts and its website. Thus, although the BNP and its use of Churchill may have ceased, it is clear that far-right groups will continue to use the memory of Winston Churchill to attempt to legitimize their message in the eyes of the British public, whose love for the wartime leader shows little sign of abating.

NOTES

1 Nigel Copsey, *Contemporary British Fascism: The British National Party and the Quest for Legitimacy*, 2nd ed. (Basingstoke: Palgrave Macmillan, 2008).
2 This chapter has been particularly influenced by postwar histories that have drawn heavily on debates surrounding public memory. See Tony Judt, *Postwar: A History of Europe since 1945* (New York: Vintage, 2010); Dan Stone, *Goodbye to All That? The Story of Europe since 1945* (Oxford: Oxford University Press, 2014).
3 Lucy Noakes and Juliette Pattinson, eds, *British Cultural Memory and the Second World War* (London: Bloosmbury Academic, 2013), 5.
4 Aristotle A. Kallis, "'Fascism,' 'Para-Fascism' and 'Fascistization': On the Similarities of Three Conceptual Categories," *European History Quarterly* 33 no. 2 (2003): 219–49, 10.1177/0265691403033 2004.
5 Paul Stocker, "Importing Fascism: Reappraising the British Fascisti, 1923–1926," *Contemporary British History* 30, no. 3 (2016): 326–48.
6 Malcolm Smith, *Britain and 1940: History, Myth and Popular Memory* (London: Routledge, 2000); Mark Connelly, *We Can Take It! Britain and the Memory of the Second World War* (Harlow: Pearson Longman, 2004); John Ramsden, *Man of the Century: Winston Churchill and His Legend since 1945* (New York: Columbia University Press, 2002); Darren Kelsey, *Media, Myth and Terrorism: A Discourse-Mythological Analysis of the "Blitz Spirit" in British Newspaper Responses to the July 7th Bombings* (Basingstoke: Houndmills, 2015).
7 John O'Farrell, "Heard the News? The War Is Over," *Guardian*, 7 June 2000.
8 Angus Calder, *The People's War: Britain, 1939–1945* (New York: Pantheon Books, 1992); Angus Calder, *The Myth of the Blitz* (London: Jonathan Cape, 1991); C. Ponting, *1940: Myth and Reality* (Chicago: I.R. Dee, 1991).
9 "Churchill Voted Greatest Briton," BBC News, 24 November 2002.

10 "EU Referendum: Churchill would Back Remain, Soames Says," BBC News, 10 May 2016.
11 Boris Johnson, "There Is Only One Way to Get the Change We Want – Vote to Leave the EU," *Daily Telegraph*, 16 March 2016.
12 David Maddox, "'We Are with Them, But Not of Them' Even Sir Winston Churchill Opposed Membership of the EU," *Daily Express*, 2 June 2016.
13 Richard Thurlow, *Fascism in Britain: A History, 1918–1985* (Oxford: Blackwell 1987); Thomas Linehan, *British Fascism, 1918–1939: Parties, Ideology and Culture* (Manchester: Manchester University Press, 2000); Nigel Copsey, *Contemporary British Fascism: The British National Party and the Quest for Legitimacy*, 2nd ed. (Basingstoke: Palgrave Macmillan, 2008).
14 Linehan, *British Fascism*, 14.
15 E.D. Hart, "Men Who Built the British Empire," *Action*, 15 May, 1937, 8–9.
16 T. Linehan, "The British Union of Fascists as a Totalitarian Movement and Political Religion," *Totalitarian Movements and Political Religions* 5, no. 3 (2004): 409.
17 R. Griffin, *International Fascism: Theories, Causes and the New Consensus* (London: Arnold 1998); See also R. Griffin, "The Primacy of Culture: The Current Growth (or Manufacture) of Consensus within Fascist Studies," *Journal of Contemporary History* 37, no. 1 (2002): 21–43.
18 Transcript of interview with John Tyndall conducted by David Baker, 4 April 1978, 4, A.K. Chesterton Collection, A.11, Bath University Archive.
19 J. Tyndall, "Laid Bare: The Churchill Record," *Spearhead*, February 1965, 4.
20 Ibid.
21 Anon., "Churchill – The Jews' Bricklayer," *Spearhead*, March 1965, 7.
22 J. Tyndall, "A Legend Falling Apart," *Spearhead*, February 1993, 4.
23 Ibid., 8.
24 Ibid.
25 D. Trilling, *Bloody Nasty People: The Rise of Britain's Far Right* (London: Verso 2012), 63.
26 Copsey, *Contemporary British Fascism*, 102.
27 N. Griffin, *Patriot*, April 1999, 4.
28 BNP Policy Research, *Language and Concepts Discipline Manual*, 2005, http://wla.1-s.es/bnp-language-discipline-2005.pdf, emphasis in original.
29 Ibid.
30 "New Labour and the Return of Fascism," *Identity*, April 2006, 5. See also J. Bean: "New Labour's Playing Hitler's Game," *Identity*, October 2004, 3.
31 Bean, "New Labour's Playing Hitler's Game."

32 "Islamic Terrorism and the Invasion of the West," *British National Party Website*, 3 October 2012.
33 D. Hamilton, "The Magpie Society," *Identity*, 8–10; "Churchill on Islam," *Voice of Freedom*, March 2006, 4.
34 "Letters," *Voice of Freedom*, January 2005, 12.
35 British National Party: Party Political Broadcast, *YouTube*, 2009, https://www.youtube.com/watch?v=ohaahorst6g.
36 Ibid.
37 "Churchill Family Denounces BNP," *Manchester Evening News*, 26 May 2009, http://www.manchestereveningnews.co.uk/news/greater-manchester-news/churchill-family-denounces-bnp-920517.
38 BBC Question Time, *BBC iPlayer*, 22 October 2009, http://www.bbc.co.uk/programmes/boonft24.
39 Ruth Wodak, Majid Khosravinik and Brigitte Mral, eds, *Right-Wing Populism in Europe* (London: Bloomsbury Academic, 2013).

15

The National World War II Museum, New Orleans: An Architectural Interpretation of War

Victoria Young

How do we present, protect, and preserve the history and memories of war? We write books, collect oral histories, make movies, art, and memorials. We also create national museums that educate, commemorate, and inform, and, as Jay Winter notes, "museum directors and designers fulfill a critical social task" as they deal with the depiction of war.[1] The built form of war museums is critical in fostering new emotional connections and visitor insights as architecture takes on symbolic and practical functions. Site planning, spatial sequencing, and shifting levels of artificial and natural light can all heighten and connect stories. The stimulation of all the senses is important too – with touch, smell, and sound all playing important roles in shaping visitor experience and memory.

The architectural language of war museums is varied. Sometimes important extant structures are reused and adapted. For example, the Army Museum in Paris was established in 1905 in the Hôtel National des Invalides, a building project initiated by Louis XIV in 1670 as a home and hospital for aged and infirmed soldiers. The Imperial War Museum (IWM) in London has renovated existing buildings for its home since its foundation in 1917 during the First World War. The museum opened in June of 1920 in the rebuilt Crystal Palace in Sydenham before moving to the western galleries of the Imperial Institute in South Kensington. In 1936, the IWM moved into the former Bethlem Royal Hospital building on Lambeth Road, with the most recent renovations completed in 2014 by Foster and Partners.

In the United States, the National Museum of the Pacific War is located in Fredericksburg, Texas, the boyhood town of Fleet Admiral Chester W. Nimitz, commander in chief of the Pacific Fleet during the Second World War. In a creative reuse of a historic building, the museum opened in 1967 in an 1852 hotel built by German immigrants. In 2009, Richter Architects designed an addition that helped attract new visitors through the display of additional artefacts.

Although the practicality of using old buildings has been important in the creation of war museums, new buildings also shape the experience with powerful forms that represent dislocation and reunion in a contemporary language. Daniel Libeskind's 2002 completion of the Imperial War Museum of the North in Manchester, England, took elements of a shattered globe and unified them in the building's forms. The Canadian War Museum, designed by Moriyama and Toshima and completed in 2005, tells the story of Canada's involvement in military conflicts through concrete and copper architectural shapes that house galleries and research institutes.

This chapter examines the process behind the foundation, design, and completion of the United States's National World War II Museum in New Orleans, Louisiana.[2] The story centres itself on a client with a vision and an architect making manifest abstract elements of the vision and the story of the Second World War, from devastating loss to liberation, hope, and peace. Like other national museums, the National World War II Museum must not only commemorate and celebrate but also educate future generations. How can architecture do this? It often starts off with a strong client who has a dream.

THE FOUNDING OF THE MUSEUM: FROM D-DAY TO THE SECOND WORLD WAR

In the late 1980s, historians Stephen Ambrose and Gordon "Nick" Mueller devised a plan to create a D-Day museum.[3] Ambrose had been actively researching the Second World War since the mid-century. In 1964, he took a position as the associate editor of the Dwight D. Eisenhower Papers at Johns Hopkins University and was commissioned to write a biography of the former president, five-star army general, and supreme commander of the Allied forces in Europe during the Second World War.[4] Ambrose and Mueller taught history together at the University of New Orleans (UNO). Over thirty-three years at the institution, Mueller, a professor of European diplomatic history, eventually rose to the status of dean and vice-chancellor. In the early 1990s, the university asked Mueller to create a research and technology park on a thirty-acre site across from the campus on Lake Pontchartrain. Mueller recalls that Ambrose said to him at the time: "You're doing this research park, and you've got land. I've done all the research on the D-Day book I'm writing, and I've got artifacts ... so let's create a small museum."[5]

But why start a war museum in New Orleans? According to Ambrose, during his first meeting with Eisenhower in 1964, after finding out that Ambrose was from New Orleans, the former president asked him if he knew Andrew Higgins. Ambrose said that Higgins had died seven years before he arrived in town in 1959, and Eisenhower said, "That's too bad. You know that he is the man who won the war for us."[6] Without his various landing craft vehicles, particularly the Landing Craft Vehicle Personnel, a boat that could go in over an open beach during amphibious landings, Eisenhower believed the entire war strategy would have been different. Hitler called Higgins the "American Noah."[7] So with Higgins, Ambrose, and Mueller all tied to New Orleans, the museum landed in the American South and not in Washington, DC, the nation's capital.

Ambrose and Mueller dreamt of a museum that "reflected their deep regard for the nation's citizen soldiers, the workers on the Home Front, and the sacrifices and hardships they all endured to achieve victory."[8] Their vision became a reality on 6 June 2000, with the opening of the National D-Day Museum in the Warehouse District of New Orleans. Upon recommendations from the museum's Board of Directors, they moved the site from the proposed location on Lake Pontchartrain near the University of New Orleans closer to the French Quarter and into the emerging Arts District, where repurposed warehouses and other structures housed the Ogden Museum of Southern Art, the Louisiana Civil War Museum, the Contemporary Art Center, and the Children's Museum. The New Orleans architectural firm of Alfred Lyons and William Hudson reused, renovated, and added on to an abandoned Weckerling Company brewery building at 945 Magazine Street from 1996 to 1998. Exhibits based on Ambrose's collection were developed by the New York-based design firm of Ivan Chermayeff and Tom Geismar.

Ambrose and Mueller soon realized that just telling the D-Day story was not enough to convey the immense scope of the war. Now president and CEO of the museum, Mueller recalls that support for expansion came from two congressional war heroes: Senator Ted Stevens from Alaska, recipient of a Distinguished Flying Cross, and Senator Daniel Inouye of Hawaii, who had won the Medal of Honor. Both said: "Washington is never going to get around to a World War II museum. We'd rather help you enlarge this museum to tell the whole story of the war."[9] And on 30 September 2003, the United States Congress awarded D-Day the designation of "America's National World War II Museum."[10]

SELECTION OF THE ARCHITECT

The museum did not wait for its designation as the nation's official World War II museum, however, to begin its physical expansion. In October 2001, it acquired more than five acres of land between Magazine and Camp Streets, across Andrew Higgins Drive from the original D-Day building.[11] The museum hired Lord Cultural Resources Group, a cultural planning professional practice based in Toronto, and in the late spring of 2002, it hosted a planning charrette, with experts in the fields of history, museology, design, and marketing, to consider exhibits, buildings, and the proper rate of expansion. In February 2003, Lord and Associates and the museum shared an informational dossier with selected architectural firms, including Frank Gehry, Cesar Pelli, and Robert A.M. Stern, and invited them to submit portfolios in the first stage of the competition.[12]

From this group emerged a short list of five firms, Eskew, Dumez and Ripple of New Orleans; Smith Group of Detroit; and from New York City, Polshek Partnership, Rafael Viñoly Architects, and Voorsanger Architects. Halfway through the competition Eskew, Dumez and Ripple (EDR) pulled out as an individual competitor and joined with Polshek. A final stage of interviews consisted of Polshek and EDR, Smith Group, and Voorsanger. In May 2003, the jury awarded Voorsanger Architects, led by principal and founder Bartholomew Voorsanger, the commission.[13] Voorsanger had taken the proposed single-building concept and broken it down into multiple structures placed underneath a canopy. This provided an opportunity for the museum to phase in the project and complete pavilions as it could afford them. "We knew how difficult it was for Steven Ambrose to raise money for the initial museum structure to get built – it was eight years of agonizing pain," says Voorsanger. "Rather than a singular monolithic building built at once, our plan was to build pavilions over a period of six to eight years, allowing for independent fundraising and naming opportunities for each building, and for the story to be told chronologically. The story of World War II is still very political and in flux, so independent buildings also allow for more flexibility as thinking evolves about how to present [the] events of the war."[14] In addition to the architectural ingenuity of its design, the firm respected the project's budget. Voorsanger created a true campus that would support the desired mission of the museum. But who was this relatively unknown architect poised to create a new architectural identity for the museum, the city of New Orleans, and the nation?

BARTHOLOMEW VOORSANGER, ARCHITECT

Raised in San Francisco, Bartholomew Voorsanger received a bachelor's degree in architecture with honors from Princeton University in 1960 and then left for France to study at the École des Beaux-Arts Fontainebleau in 1961. He returned after a year and went on to earn a master's degree in architecture from the Graduate School of Design at Harvard University in 1964. He worked for three years with Vincent Ponte, an urban planner in Montreal, Quebec, before joining I.M. Pei's New York office in 1968, where he was in charge of many projects during his decade-long tenure, including the fifty-seven-story Canadian Imperial Bank of Commerce in Toronto (1972) and most of the firm's work in Iran. While working for Pei, Voorsanger accepted a commission for a house on Martha's Vineyard. It was his first solo project and he hand-drew all the renderings in his time away from Pei's office. Voorsanger's design won the American Institute of Architects' award for the National Better Homes and Living competition in 1976, and he came to realize that he wanted to be the design principal of his own firm rather than a lead associate for Pei. In 1978, he joined with fellow Pei associate Ed Mills to form Voorsanger and Mills. Immediately they gained attention in the architectural world for their New York City designs of New York University's Midtown Center (1980), Le Cygne Restaurant (1982), and the Brooklyn Museum addition competition (1986).[15] Eventually, Voorsanger and Mills split, the firm was reorganized as Voorsanger and Associates Architects PC in 1990, and they still maintain their practice in New York City.

Bart Voorsanger had practised architecture for nearly forty years when he won the World War II competition. During this time he worked on a number of cultural projects, including the 1997 design for a renovation of and addition to the Asia Society and Museum in New York City. He lost the important Brooklyn Museum addition competition in 1986, designed an unbuilt project for the Bayly (now Fralin) Art Museum at the University of Virginia in 2000, and completed the iron and glass garden court at the Pierpont Morgan Library and Museum in 1991. These projects, along with several important house designs in California, Colorado, Montana, Arizona, and New York, reveal his design sensibility: "I always look at three aspects – the practical, the tectonic, and the art. The idea is never to overemphasize one over the other but to keep them in balance."[16] His buildings are specific to their environment and its landscape, and they respond to nature as a force that guides design. He is also keenly aware of the spatial experience one has on a site or in a building. Modernist stylistic sensibilities, with their angularity, love of materiality over applied ornament, and functionality, are central to his aesthetic.

DESIGNING THE MUSEUM

Voorsanger was the only member of his design team that had served in the military, a brief six-month stint in the early 1960s as a second lieutenant in the army. He felt it essential that his team speak with a veteran, so he brought navy lieutenant-commander Joseph Vaghi to New York to talk with the group.[17] Voorsanger also drew upon his own travels through important war sites, being particularly moved by his visit to Normandy and the American Cemetery, where the rows of white crosses and Stars of David create a powerful landscape of commemoration.

Voorsanger approached the World War II project with a philosophical mindset focused on celebrating liberation and seeking peace rather than revelling in the dramas of war itself, and he sought in its architecture a complex of structures that verged on intimidation.[18] Voorsanger believes that a museum design's most important element is the visitors' experience of entry and moving through space. Thus, for the 2003 competition entry, Voorsanger and his team created a continuous route called the "Path to Victory." This intentional path organizes the spatial experience as visitors traverse the central green space, or Parade Ground, and enter into the exhibit pavilions, a globe-shaped theatre, and a dining/entertainment area.[19] Covering the entire site, a canopy helped to unify the individual elements with a "large scale, recognizable image signifying a national museum."[20] In 2004, Voorsanger joined with local architect Ed Mathes of Mathes Brierre Architects to create the limited liability company of Voorsanger Mathes for the project, with Voorsanger in charge of design and Mathes the architect of record and construction leader.[21]

With this first plan in place things seemed bright for the start of construction. But in late August of 2005, the museum entered into battle with a fierce foe, Hurricane Katrina. The storm devastated the area. The Board of Trustees asked President Mueller for five different scenarios regarding the master plan, ranging from dropping it altogether to building everything as planned. Mueller proposed a middle ground that would "continue the master plan but eliminate some of the cost – reduce the size of some buildings, cut one out." Mueller told the board: "This is tough, but it's not like landing on Omaha Beach. There's nobody shooting at us. If it takes us a bit longer, so be it."[22] And they went back to work with a redesigned version of the complex, one that eliminated some of the square footage by the removal of a pavilion; however, most important, this version would start construction with buildings that would generate revenue immediately, with a theater, gift shop, and dining venues at the top of the list rather than a historical research institute as originally planned.

BUILDING THE MUSEUM

The Solomon R. Victory Theater was the first building completed in the redesign and it opened with veterans and other dignitaries in attendance on 6 November 2009. The exterior of precast concrete panels, irregular in their placement, is jagged, edgy, and modern. Inside runs the 4-D movie experience, *Beyond All Boundaries*, created by the museum and the Hettema Group, led by Phil Hettema, with executive producer Tom Hanks.[23] Venues like the American Sector restaurant, run by famed chef John Besh, and the Stage Door Canteen provided dining and entertainment options and income. Office space rounded out the functional elements of this pavilion.

The next building completed was the US Freedom Pavilion: The Boeing Center, dedicated on 18 January 2013. Architecturally, its position at the southwest corner of the site pushes it up to Interstate 10, and this forces the building's design to negotiate the industrial feel and rapid movement of the highway with the green space and serenity of the parade ground on the east. Voorsanger's team responded by having the building's footprint turn the corner in consort with the off-ramp from I-10 and through a series of randomly arranged, anodized, powder-coated aluminum panels. With a shallower depth of corrugation than regularly found, the building is active and interesting for users on the ground and on the elevated roadway. The impenetrable exteriority of the pavilions of the site's perimeter, whether fabricated in corrugated metal or concrete, gives way on the parade ground side to an architecture of glass and steel that is transparent and connects the building's interior with the exterior. In addition to traditional displays of artefacts, the building features macro exhibits, some of which are hung from the ceiling, including a Boeing B-17 Bomber and a B-25 Mitchell. Visitors gain access to the planes on a series of walkways that cut across the building at different floor heights.

This kit of architectural parts, metal, concrete, and glass is the principal language of the site, but at times it is supplemented with colour in interior elements, as seen in the Campaigns of Courage Pavilion, opened in December 2014, on the northwest corner of the site. The two-story entry contains a wall of dark orange-coloured Kinon panels, a stairway, and a German Messerschmitt Bf 109 fighter.[24] Black box architectural space beyond houses the *Road to Berlin* on the first floor and, on the second floor, the *Road to Tokyo* exhibits, designed by the team at Patrick Gallagher and Associates in Bethesda, Maryland.[25] In the *Road to Berlin* you will find among the displays an "Air War Gallery" that features an interactive table, planes flying overhead in the ceiling cut-out, displays of photographs, artefacts, and a nostalgic cluster of pin ups to complete the picture of daily life for America's bomber pilots. The *Road to Tokyo* retraces the "grueling trail from Pearl Harbor to Tokyo Bay," and it opened one year later.

In the space between the Campaigns and Freedom Pavilions will be a building dedicated to liberation and the closing months and aftermath of the war.[26] Included with exhibitions will be a chapel with thirty-four-foot (10.3 metres) tall movable walls that will allow the parade ground and chapel to embrace each other when the panels slide open, creating an exterior extension of the sacred space, a type of memorializing element to the site. Voorsanger points out that: "The chapel has to ultimately become an experience that veterans, their families, and the young can participate in and be transported to the privacy of their emotions, their thoughts and their remembrances. All need to be celebrated with special dignity and commitment."[27]

The building originally deleted from the master plan after Katrina will likely be the next completed. The Hall of Democracy (October 2019) will house special exhibition space and the research centre that Nick Mueller had always wanted. It will also be the centre point of all digital research initiatives of the museum. Its scale is consistent with that of the other pavilions as well as with the historic structures preserved on the Magazine Street elevation to which it is adjacent. On the interior an expanse of glass allows this building to engage directly with the parade ground and the architectural element that will most strongly define this place once completed – the canopy.

THE ICONIC ELEMENT: THE CANOPY OF PEACE

A canopy is a covering, usually a fabric supported by poles, suspended above a bed, important person, or a sacred object. It is also a crown or a cover, like that provided by the leafy upper branches of trees in a forest. The part of a parachute made of nylon or silk that opens and fills with air is also a canopy, as is the transparent cover of the cockpit of an airplane. These are all appropriate symbolic connections for this impressive architectural icon-to-be at the World War II Museum. Voorsanger understands the powerful way this built structure will be able to signify meaning and symbolism. "The canopy emerged as an architecture of national scale, a strategic element, unifying, protective, metaphorically shielding our troops. Its evolved design is emblematic of strength (steel), of passive cover (white sails), appropriately now named the Canopy of Peace. We wanted this metaphor to be an abstraction, not to be literal, or it would lose its power."[28]

When completed in 2021, this structure became, arguably, the most prominent element in the cityscape of New Orleans, replacing the Superdome and St Louis Cathedral. Its power comes from its scale, 485 feet (147.8 metres) long, 127 feet (8.7 metres) wide, and 146 feet (44.5 metres) tall, fabricated from galvanized steel infilled with tensile fabric, Teflon-coated fiberglass panels. The structure is illuminated in various ways, adding another layering of visual power onto this iconic architectural element. The canopy was provided shelter and shading not only for the structures of the complex but also for the people who use it. Architect Frank Lloyd Wright once said that the presence of a building is in its roof – and what a roof the canopy provides for the museum and cityscape of New Orleans.

Voorsanger Architects has explored the use of the canopy as a design inspiration in subsequent museum designs, including the 2010 competition for a museum of the Second World War in Gdańsk, Poland, where the overhead canopy at World War II became a steel veil that wrapped around and defined the building. This is the "Canopy of Hope," according to Voorsanger.[29] And with its 2008 winning competition entry to design a national military museum for the United Arab Emirates in Abu Dhabi, Voorsanger's "Canopy of Peace" at World War II and "Canopy of Hope" at Poland became the entire building itself and perhaps a canopy of power.[30]

CONCLUSION: AN ARCHITECTURAL INTERPRETATION OF WARS – AND PEACE

As Nick Mueller stated on the unveiling of the $10 million gift from Boysie and Joy Bollinger to fund the creation of the iconic canopy, "the Canopy of Peace, reaching skyward for all to see, symbolizes the hope and promise unleashed by the end of the World War II hostilities ... [T]his structure testifies to the rich blessing of liberty and democracy. It also serves to unify the diverse campus of America's National World War II museum in the enduring spirit of the wartime slogan, We're all in this together!"[31] Now more than a decade in the works and at a cost of $350 million, the National World War II Museum campus is a testament to the power of memory and the history of our experience in the war, and the persistence with which they are made manifest in built form. And in the search for an architectural interpretation of the complexities of war, the design mandate, according to Voorsanger, was also emotionally complicated, "The history of World War II, bravery of our troops, the aftermath of victory, how frightening close at times we could have lost, remains frontal through the entire Master Plan evolution." It was part of a process of design that sought "to make sacred seeking the peace, seeking liberation, not the celebration of war."[32] And the modern forms of Voorsanger's pavilions, protected by the canopy, make this possible.

NOTES

1 Jay Winter, "Museums and the Representation of War," *Museum and Society* 10, no. 3 (2012): 150.

2 This chapter outlines a larger book project in process that will tell the architectural story of the National World War II Museum.

3 Stephen Ambrose, "The National D-Day Museum," in *To America: Personal Reflections of an Historian* (New York: Simon and Schuster, 2001), 203.

4 *The Supreme Commander: The War Years of General Dwight D. Eisenhower* was published by Doubleday in 1970. Ambrose went on to pen additional works on the war, including *Band of Brothers, E Company, 506th Regiment, 101st Airborne: From Normandy to Hitler's Eagle's Nest* (1992); *D-Day, June 6, 1944: The Climactic Battle of World War II* (1994); and *Citizen Soldiers: The US Army from the Normandy Beaches to the Bulge to the Surrender of Germany, June 7, 1944* (1997). Ambrose served as executive producer of the HBO miniseries *Band of Brothers* (2001) and consulted on the feature film *Saving Private Ryan* (1998).

5 "Interview with World War II Museum President Nick Mueller," *HistoryNet*, http://www.historynet.com/interview-with-wwii-museum-president-nick-mueller.htm.

6 Ambrose, "National D-Day Museum," 201.

7 Jerry E. Strahan, *Andrew Jackson Higgins and the Boats That Won World War II* (Baton Rouge: Louisiana State University Press, 1994), 3 and 166.

8 "About the Founder: Stephen E. Ambrose, PhD," http://www.nationalww2museum.org/about-the-museum/about-the-founder.html.

9 "Interview with World War II Museum President Nick Mueller," *HistoryNet*.

10 108th Congress, House Resolution 2658 Department of Defense Appropriations Act, section 8134. Ambrose did not live to see D-Day become the Second World War, having died on 13 October 2002 from lung cancer.

11 See city tax and land purchase records for 4 October 2001 by the National D-Day Museum Foundation, Inc.

12 Elizabeth Muellener, "Battlefield Promotion. You've Got to Hand It to the Founders of the D-Day Museum. Their Bold New Plan for Tripling the Size of the Museum to Create an Instant Architectural Landmark and One of the World's Premiere War Museums Certainly Has the Town Buzzing." *Times-Picayune* (New Orleans, LA), 14 March 2004, Living Section E 1.

13 *Architect's Newspaper* states that Voorsanger also beat out Antoine Predock, Davis Brody Bond Architects and Planners, Michael Maltzan Architecture, and Wendy Evans Joseph Architects. See: "World War II Remembered: Voorsanger Named Architect of National D-Day Museum," *Architect's Newspaper*, 9 March 2004, 6. The selection committee, chaired by noted local architect Arthur Davis, included: Nick Mueller; Don Gatzke, former dean at Tulane University's School of Architecture; Jim Sefcik, former director of the Louisiana State Museum; and members from the Board of Trustees. See "Design Museum Team Selected for National D-Day Museum in New Orleans," *Lord Cultural Resources Newsletter* (Spring 2004), http://www.lord.ca/Media/2004Spring-Lord_Newsletter.pdf.
14 Mike Singer, "Flexible Pavilion Plan Allows New Orleans World War II Museum Expansion Interrupted by Katrina to Get Back on Track," *Practicing Architecture*, The American Institute of Architects, 2011 Convention edition.
15 Andrea Truppin, "Voorsanger and Mills: The New Integration," *Interiors* (June 1985): 109–25.
16 Vladimir Belogolovsky, "One-on-One: Architecture of Emotion and Place – Interview with Bartholomew Voorsanger, FAIA, MAIBC," *ArchNewsNow*, http://www.archnewsnow.com/features/Feature367.htm.
17 At age twenty-three, Vaghi was the youngest beach master on Omaha Beach during the D-Day invasion. Beach masters were the first ones off the landing craft. They headed ashore to direct the arrival of new troops and the evacuation of the wounded and dead. They were often known as "traffic cops in hell." See Emily Langer, "Joseph Vaghi Dies: Navy Beach Master Who Helped Lead the Invasion of Omaha Beach on D-Day Was 92," *Washington Post*, http://www.washingtonpost.com/local/obituaries/joe-vaghi-dies-navy-beachmaster-who-helped-lead-the-invasion-of-omaha-beach-on-d-day-was-92/2012/09/08/b2e7222c-f5e3-11e1-91cb-58c92a8a140e_story.html.
18 Bartholomew Voorsanger, in discussion with author, 13 December 2014.
19 See http://www.lord.ca/Media/2004Spring-Lord_Newsletter.pdf for an image of the 2003 design.
20 "Voorsanger Architects PC, National World War II Museum Masterplan," http://www.voorsanger.com/recent_projects/3.
21 The team for this project is as follows: Voorsanger Architects: Design Principal Bartholomew Voorsanger FAIA; Lead Designers Martin Stigsgaard and Masayuki Sono; Design Team Peter Miller, James Macdonald, Radoslaw Krysztofiak, Andrea Wiedemann, Mark Wagner,

Reema Pathak, Van Hsin-Hung Tsao, Issei Suma, Won Jun Jung, and Anastasiya Konopitskay. For Mathes Brierre Architects: Principal Edward C. Mathes; Team Peter Priola, Scott Evans, Tony Alfortish, John Gray, Johannah Fernandez, Vivien Yu, Joyce Bergman, Marc Robert, and Stephanie Kerr.

22 "Interview with World War II Museum President Nick Mueller," *HistoryNet*. Mueller went on to say: "And that's what we agreed to do. We had some good fortune after a year or two-commitment from major private donors, a big federal gift and the benefit of the new market tax credits."

23 http://www.thehettemagroup.com/beyond-all-boundaries, for a link to a "making of" video on "Beyond all Boundaries."

24 http://www.nww2m.com/2014/09/countdownrtb-bf109/.
Kinon is a hand cast, resin panel composite that is applied to a base material that, along with the addition of decorative materials, creates unique surfaces for floors, walls, furniture, and more.

25 Patrick Gallagher and his team won the exhibition design competition at the same time that Voorsanger won the architectural competition. See http://gallagherdesign.com/projects/the-national-world-war-ii-museum/

26 See http://www.nationalww2museum.org/expansion/liberation-pavilion.html.

27 Bartholomew Voorsanger, e-mail to author, 8 June 2016.

28 Ibid.

29 Voorsanger Architects PC, "Polish Museum of the Second World War," http://www.voorsanger.com/recent_projects/6.

30 See http://www.lord.ca/Media/2004Spring-Lord_Newsletter.pdf for an image of the 2003 design.

31 Nick Mueller, speech at the event to receive the Bollinger donation, 24 March 2015.

32 Bartholomew Voorsanger, e-mail to author, 8 June 2016.

4

Remembering and Forgetting War

16

War on Memorialization: Constructive and Destructive Holocaust Remembrance on American Sitcoms, 1990–2000s

Jeffrey Demsky

After Victoria Young's consideration of the elaboration of war and peace discourses through the concept of museography, I have chosen to tackle another angle of Second World War remembrance, focusing on the implications of the sometimes casual, sometimes consensual, and sometimes controversial evocation of the Holocaust remembrance in American situation comedies in the relatively recent period from 1990 to the 2000s.

In fact, just twenty-five years before the history's centennial, American Holocaust memorialization faces an unlikely commemorative contest. The crux of the tension is the fraying relevance of the so-called Nuremberg liberator narrative.[1] More fashioned than actual (and silent regarding the fact that the US government provided Europe's Jews slight aid prior to 1945), this trope celebrates American rectitude for defeating Nazism and ending the Holocaust.[2] Although bowdlerized, throughout the twentieth century's second half, so-called rescuer lore proved popular, affirming the "tales of pluralism, tolerance, democracy, and human rights that postwar America told about itself."[3] However, increasing numbers of twenty-first-century Americans display ennui towards the legacy. All histories struggle to remain relevant, even seemingly eternal pasts like the Holocaust. Holocaust-themed jokes on television sitcoms, in line with the larger "Ha Ha Holocaust" cultural trend,[4] demonstrate this distancing. If unlikely, the aphorism, that history presents first as tragedy and next as farce,[5] appears to predict this flippancy towards a once sacrosanct past.

Whimsy does not inherently belittle the Holocaust's seriousness, although it can, and instead might remind audiences that laughter is the most humane part of that suffering.[6] Holocaust joking is also a unique post-memory form, remembrance that empowers cultural producers to witness the history "through imagination and adoption."[7] While not comedic, poet Sylvia Plath's "Daddy," written shortly before her suicide, models this technique. "A Jew to Dachau, Auschwitz, Belsen. I began to talk like a Jew. I think I may well be a Jew."[8] That the contemporary producers effect comedic ruptures, rather than poetic confessionals, does not negate their post-memory sway. A joke's wit cannot sting its target, just as Plath could not fully reveal her pain, without first validating the truth of the subject it hopes to demystify. Such tweaks, however, can also destabilize facts by fabricating alternative scenarios that supplant historical realities, an especially insidious threat among youthful viewers. If unanticipated, sitcom skits laughing about Nazism and the Holocaust may represent a new sign of the times. As the once dominant Nuremberg liberator narrative wanes, the brash comedy airing on American television might prove salutary, recalling this history into contemporary discourses, helping to ensure that people remember it at all.[9]

Shows like *Seinfeld*, *Curb Your Enthusiasm*, *Robot Chicken*, and *The Simpsons* broadcast what could be called "constructive joking." This device finds writers using jokes to spur, not spurn, faithful Holocaust remembrance. However, a more destructive form of joking also exists. Shows like *Family Guy*, *South Park*, and *American Dad* produce skits that deride both the memorialization's status in American society and the actual Jewish destruction. This sort of joking does not offer satire, parody, or cultural critique. It mocks. My method does not position one comedic impulse – constructive or destructive – over the other; rather, their co-existence is my focus, as they appear to constitute opposing ends of the same memorial continuum. Before closing, I present an example of constructive and destructive Holocaust humour complementing each other, if unwittingly, to achieve a hybrid remembrance. Analyzing comedic Anne Frank representations on three American sitcoms shows how brazen humour draws her past into the present, invigorating her story, while also altering its factuality.

Jeffrey Shandler explains how American television networks, as early as the 1940s and 1950s, grew Holocaust awareness.[10] CBS Television's *Who Killed Anne Frank* (1964), as well as its contemporary

Nazi-related episodes of *The Twilight Zone*, *Star Trek*, and the wider *Hogan's Heroes* series (1965–71),[11] stoked audiences' imaginative intersections with this not-too-distant past. The *Holocaust* (1978) miniseries, which Elie Wiesel dismissed as a "soap opera,"[12] again connected hundreds of millions of people to an Americanized tale in which the US and its democratic allies vanquished the Nazi's program of murder. In terms of scholarly literature, cheeky displays of sitcom post-memorialization index Norman Finkelstein's arguments about "exploited" Holocaust mourning.[13] They intersect with Oren Baruch Stier's thinking about Holocaust icons.[14] The subverting reconstructions also build out Jeffrey Shandler's study of politicized Anne Frank representations,[15] Tim Cole's "selling and packaging" insights,[16] and Gavriel Rosenfeld's findings about the "normalization" of Hitler and the Holocaust in contemporary pop culture.[17]

As with all art, Holocaust humour is dicey. Tragicomedies like *Life Is Beautiful* (1998) and *Train of Life* (1999) did not offend audiences because its creators dealt compassionately with the history of the event and the strain its details place on those who inherit it.[18] However, more recent sitcom interventions, like the 2019 *Historical Roast* episode featuring Anne Frank, elicited harsh public scolding.[19] This potential to sow seditiousness is precisely why the lowbrow objet d'art is worthy of greater retrospective analysis. Sitcom skits lampooning the Holocaust are not ephemera; rather, they are a harbinger, revealing the twenty-first-century challenges and opportunities this memorialization faces. Humour matters. Its words, images, and impacts cannot be undone.[20] Sitcom skits have the ability to help ordinary people learn about serious topics. Alternatively, teasing Nazi/Holocaust memorialization may also have the exact opposite impact, subverting its importance, as onlookers conclude that laughing about it is all you can do. This is the combustible foundation upon which American Holocaust sitcom humour rests. It is a risky business, but a flourishing one nonetheless.[21]

CONSTRUCTIVE HOLOCAUST HUMOUR ON AMERICAN SITCOMS

Humour and the Holocaust have always been intertwined. Prisoners relied on "gallows" humour as a coping mechanism for their everyday ordeals.[22] After the Second World War, both Jewish and non-Jewish survivors found in levity a means of sharing their anguish and achieving healing.[23] However, former captives' uses of wit to express their suffering, for example Tadeusz Borowski's *This Way for the Gas, Ladies and Gentlemen*,[24] is significantly different from contemporary American television writers laughing at these experiences. Television's duelling status as both a guardian of traditionalism and a presenter of subversive content is conducive to Holocaust humour.[25] Situation comedy, with a heavy reliance on facile stereotypes about everyday experiences, especially supports this method. Take, for example, *Seinfeld's* (1989–98) "Soup Nazi" (1997). The script's banal framework, in this case some friends going to a restaurant, enables complex Holocaust joke work that conveys real glimpses of historical suffering. In counterposing Nazi-era line-ups with current-era line-ups, only now set before an authoritarian chef, the show's writers confront viewers with this history's factuality. Observing the dread that Jerry and George display in their interactions with "dictatorship" is constructive to Holocaust memorialization because it offers post-witnesses a new way to approach and remember this past.

The *Curb Your Enthusiasm* (2000–present) "Survivor" (2004) script reaches a similar end. Like "Soup Nazi," the sitcom format, again, offers a banal setting – a family luncheon – against which to unloose sly Holocaust joking. The table banter between the rival "survivors," one who participated on the identically named reality television show and the other a former Holocaust prisoner, reveals a new complication – namely, twenty-first-century misremembering. The younger "survivor" regales his dining partners with harrowing stories of poisonous jungle spiders. The Holocaust survivor looks on with dismay. "That's a very interesting story," he eventually explodes. "But I was in a concentration camp! You never even suffered one minute in your life compared to what I went through!"[26] Notably, the younger man refuses to drop his claims to suffering, displaying "disagreement humor,"[27] in line with Mel Brooks's insight that "tragedy is when I cut my finger and comedy is when you fall into an open sewer."[28] The episode, however, ultimately delivers a sombre truth. Other twenty-something-year-olds, the current generation that this "survivor" represents, may well lack the cognitive commitment to uphold the older man's legacy.

Robot Chicken (2005–present) is a popular Claymation comedy series that directly targets this youthful demographic. Its approach to Holocaust humour involves simultaneously acknowledging the factual history of the Jewish genocide but joking about it nonetheless. "Major League of Extraordinary Gentlemen" (2011) finds Ben Stiller's watchman character from *Night at the Museum* (2006) strolling down a city street searching for his new assignment's building. When the directions bring him to the front of the US Holocaust Memorial Museum – keeping in mind the film's comedic premise is that the historical collections awaken at night – the Stiller figure screams, "Oh, hell no!"[29] "Punctured Jugular" (2012) models similar light-heartedness, conflating Holocaust remembrance with a spoof of *The Smurfs*. The skit depicts evil wizard Gargamel, the Smurfs' sworn enemy, standing trial for perpetrating a Smurf genocide. Upon receiving his guilty verdict, his character exclaims, "Guess it is me and Hitler now!"[30] Equating remembrance of the factual Jewish destruction to an imagined Smurf annihilation vulgarizes the history. Nonetheless, the skit still confirms, and might even introduce, this long bygone era to youthful viewers, recognizing some positive commemorative end.

Reflecting its haute pedigree,[31] the writers of *The Simpsons* (1989–present) tread lightly near this past. Several episodes spoof Adolf Hitler, not his genocide, stoking time-honoured American traditions tracing back to Charlie Chaplin's *The Great Dictator* (1940) and Ernst Lubitsch's *To Be or Not to Be* (1942).[32] "Bart vs. Australia" (1995) follows this path. Hoping to verify the Coriolis Effect, he randomly dials a South American phone number. A man drawn to resemble an older-looking Hitler, standing next to a Mercedes Benz with an "Adolf 1" licence plate, prepares to answer the call. Before he can do so, however, another figure enters the frame. It is a similarly aged male, a swastika button pinned to his vest, delivering a Nazi salute and declaring, "Buenas noches, mein Führer!" Distracted by the phone's ringing and his interlocutor, *The Simpsons*' beleaguered Hitler figure hastily responds, "ja, ja."[33]

Such a ruse manages to "get a laugh" on Hitler. The joking mostly serves to confirm that fifty-plus years after Hitler's death, snickering at him remains part of the American national identity.[34] However, owing to *The Simpsons*' longevity as a series, the show's writers have gradually sharpened their Holocaust satires to reflect contemporary trends. An example appears in "The Spy Who Learned Me" (2012). Its script parodies Morgan Spurlock's *Super-Size Me* (2004) documentary, chronicling the negative health effects associated with daily fast food consumption.[35] The show's Declan Desmond character, a muckraking videographer, trains his camera on exposing

Springfield's sordid fast food industry. The unexpected opening to Holocaust humour occurs when he completes his film's Academy Award submission form. The only two possible genre boxes listed on the application are "Holocaust" and "non-Holocaust."[36] This subtle jab suggests weariness with the subject's perceived cultural oversaturation as hundreds of Holocaust films have appeared globally during the last fifty years.[37] This joke also recovers the well-worn "Jews control Hollywood" trope, implying the furtive interest in – and financing behind – the abundance of Holocaust-themed stories.[38]

The British sitcom *Extras* (2005–07) also critiques this commerce. Take, for example, a 2005 episode starring Kate Winslet, portraying a Second World War-era nun who saved endangered Jews.[39] During a scripted break in the filming, still wearing her habit costume, she speaks off-stage with fellow cast members. One assures her, "I think you doing this is so commendable, using your profile to keep the message alive about the Holocaust."[40] Unexpectedly, Winslet replies, "Thank God, I am *not* doing it for that. I don't think we really need another Holocaust film. How many have there been? We get it. It was grim. Move on."[41] While her colleagues stare at her, their mouths agape, Winslet shares her true motives. "No. I am doing it because I have noticed that if you do a film about the Holocaust you are guaranteed an Oscar ... That's why I am doing it. Schindler's-bloody-List. *The Pianist*. Oscars, coming out their ass."[42] This impulse exemplifies a less constructive form of remembrance, although it is important to note that the writers' target is the memorialization's perceived politicization, not its historical veracity or victims' suffering.

Sitcoms like *South Park* (1997–present) operate on both sides of this line. The series depicts the lives of four grade-schoolers, and their neighbours, living in a quiet Colorado town. One of the boys, Kyle Broflovski, is Jewish, likely an eponymous character for the show's co-creator, Matt Stone.[43] Kyle struggles to negotiate his Jewish identity, a process aggravated by the cascade of harassment his bigoted classmate, Eric Cartman, directs towards him. Although just a fourth-grader, Cartman is an anti-Semite. Throughout the show's run, writers have used his character to provoke taboo conversations,[44] specifically ones about Nazism and the Holocaust. "Pinkeye" (1997) features Cartman wearing an Adolf Hitler Halloween costume to school.[45] After seeing his get-up, teachers and administrators predictably recoil. The principal orders Cartman to her office where he watches an anti-Nazi documentary. Taking this step models a constructive and traditional response to American Holocaust education.[46] Cartman's flippant response, however, destabilizes the attempted correction. Throughout the documentary, still outfitted as Adolf Hitler, he expresses glee at the marching Nazi columns. At one point, the show's animators depict him drifting into a dream sequence where he becomes the Nazi Führer, standing atop a rostrum and screaming German-sounding gibberish.[47]

If puerile, this mockery does not lampoon the Nazis' genocide. This is the boundary, I suggest, separating constructive and destructive comedic memorialization. However, *South Park*'s "Passion of the Jew" (2004) models a damaging approach.[48] The show's writers prepared the script in the midst of contentious public discourse about the film *The Passion of the Christ* (2004), conversations centred on director Mel Gibson's decision to foreground the role of Jews in the deicide.[49] Remembering Jesus's suffering is fundamental to Christian collective identity. It is this familiarity that writers exploit, as a sitcom device, for unpacking Holocaust blaspheming.[50] Initially, the script does not betray this intention. The action instead features dozens of townsfolk, all perfect strangers, assembled at a backyard party. Their common wish is "figuring out how to use *The Passion* to enrich everyone's lives ... organize the Christian community."[51] The door to anti-Semitism and Holocaust humour opens when it turns out that Eric Cartman had arranged this impromptu community gathering. *The Passion of the Christ* had similarly sparked his interest in "organizing" Christians to "enrich life," although his vision involved reviving Hitler's Jewish genocide.

Viewers join Cartman inside his bedroom, dressed in a Nazi-styled costume, rehearsing his pitch before a mirror. "Töten Sie die Juden!" (Kill the Jews), he screams in fluent German, cracking a black leather riding crop. "Wir können nicht stillstehen, bis sie alle tot sind!" (We can't stop until they're all dead).[52] Emboldened, he leaves his bedroom and joins the get-together. "Now, we all know why we're here," he opens his remarks, "and I believe that we all know what needs to be done." "We sure do," a woman replies approvingly, although her enthusiasm supports Christian charity rather than mass murder.[53] Oblivious to the distinction, Cartman continues, "I think it is best that we do not talk out loud about it until we have most of them on the trains heading to the camps."[54] While some of his guests murmur confusion, the force of his delivery, and urgency of his pleas for Christian solidarity, steers the group's energy towards sinister ends.

By the episode's denouement, a crowd assembled initially for charity instead marches into town, behind Nazi Cartman, reciting German-language chants of "Es ist Zeit für Rache!" (It is time for revenge).[55]

Across cultures, a familiar idiom teaches people about the differences between "laughing with" and "laughing at" people. This adage is particularly true with regard to Holocaust sitcom humour. "Passion of the Jew" (2004) reflects the latter impulse, unpacking a subtext that implies that Jews' role in the crucifixion elicited their destruction. This predicate sanctions ever more flippant deviations from the "reverence and love" memorializing tradition.[56] In addition to complicating American commemorative discourses, this joking also affects European encounters. Many nations other than the United States air American sitcom Holocaust humour: the episodes stream unfettered on the World Wide Web. This is the unlikely point at which twenty-first-century digital communications can license a profane merger between Jewish American sitcom writers and European xenophobes.[57] No American sitcom encapsulates this potential more than *Family Guy* (1999–present). A need exists to detail its many Nazi/Holocaust interventions as both a destructive memorialization form and as being in line with Pablo Picasso's insight that "destruction is the first act of creation."[58] While jarring, the series' impertinent renditions may provide guideposts for sustaining interest in this memorialization's future.

DESTRUCTIVE HOLOCAUST HUMOUR ON AMERICAN SITCOMS

In 2013, Rob Eshman, editor-in-chief of the *Jewish Journal of Greater Los Angeles*, published "Seth MacFarlane: Not an Anti-Semite."[59] Defending the *Family Guy* sitcom creator, he attempted to explain MacFarlane's frequent comedic use of anti-Semitic tropes. The joking is not disdainful but, rather, intended to "diffuse prejudice … through mockery, exaggeration, and sometimes by just bringing prejudice to light."[60] At times, *Family Guy* scripts achieve these Bakhtinian disruptions,[61] especially in their Hitler lampooning. "Death Is a Bitch" (2000) depicts the dictator as a talk show host, angling to see his guest's posterior. "Untitled Griffin Family History" (2006) finds "Addy" in a hen-pecked family life, unable to manage his Reich. "Road to Germany" (2008) features the Stewie Griffin baby character costumed as Hitler, time-travelling to the Nazi era. He eventually encounters his muse and the two engage in Chaplinesque miming. If seemingly harmless, rendering the Nazi dictator a relatable contemporary figure is commemoratively dangerous. It stokes the impression that Hitler was a caricature rather than a life-threatening person.[62] This analysis also relates to YouTube "Hitler Rants," redubbed movie clips that portray the dictator enraged at such frivolities as late pizza deliveries and *Harry Potter* scripts.[63]

When *Family Guy* writers shift from lampooning Hitler to peddling anti-Semitism and barbing the Holocaust instead, their humour is more openly damaging. "Hot Pocket Dial" (2015) conflates Auschwitz with Disney World.[64] The show's oafish protagonist, Peter, loudly voices his dismay when arriving at the death camp's iconic gate, clearly spelled out in block capital letters for unknowing viewers. Wearing colourful beach/vacation attire, he reminds his wife, "You said that we were going to a place that Walt Disney built." "No," she replies, "I said a place that he supported."[65] In a nod to genocide humour, she further reminds her dim spouse, "Don't go on the train rides."[66] *Family Guy* writers recall Walt Disney's alleged biases in additional episodes.[67] "Stewie Griffin: The Untold Story" (2005) features Disney/Jew humour. "Nanny Goats" (2017) finds Mickey Mouse murdering the "Jew mouse" Fievel Mousekewitz,[68] intended to parody the film *The American Tail* (1986). Writers' insistence on plying this humour reminds current audiences that some per centage of Christian Americans, both during Disney's era[69] and in current times,[70] struggle to view Jews outside of anti-Semitic constructs.

"Stewie, Chris, and Brian's Excellent Adventure" (2015) reveals additional brashness.[71] Its script finds Stewie and his talking dog Brian in pre-First World War Europe. They spot a street performer wearing a toothbrush mustache and bowler hat. Surmising it is Charlie Chaplin, Stewie hands him money. "Sir," he implores, "Please take this money and use it to pursue your dreams." "Danke," the vamp replies. "I think that was a young Hitler," Brain exclaims. Unfazed, Stewie replies, "Yeah, but that was before all the crazy stuff."[72] Again, this joking both certifies and deflates Holocaust memorialization. Affirming the veracity of the "crazy stuff" is both the joke's predicate and its puncture. This is a common *Family Guy* technique. "Long John Peter" (2008) has a Kristallnacht joke. The show's Jewish character, Mort Goldman, shrieks the term after a pirate gang smashes through his storefront's window. "McStroke" (2008) features a Dachau quip. "Family Goy" (2009) spoofs

Schindler's List. Reimagining Peter as Amon Göth, he is drawn shirtless, a cigarette dangling from his mouth, holding a rifle and fixing Mort Goldman in its sights.

Indeed, while onlookers can debate whether or not Seth MacFarlane is anti-Semitic, there is no question that many *Family Guy* episodes peddle anti-Semitism. This tendency bolsters the impression that the show's related Holocaust comedy is mean-spirited rather than *avant-garde*.[73] "Believe It or Not, Joe's Walking on Air" (2007) uses the name Anne Frank as a descriptive adjective for identifying a cheap and socially awkward Jewish woman. In "Love, Blactually" (2008) writers infuse the bigotry into Cyrano de Bergerac's clandestine discourses with Christian de Neuvillette. Instead of pitching woo, Cyrano whispers harsh condemnations of Jews, saying that they had "disproportionate control of the world's wealth … starting every major war since the dawn of time."[74] As the two men argue over the dialogue's tone, Roxanne asks what is going on. *Family Guy* writers exit this skit with the concealed and frustrated Cyrano screaming, "Mind your own business, you stupid Jew."[75] Since none of this activity matches the episode's wider story, or has anything to do with Edmond Rostand's text, the appropriation is as gratuitously absurd as it is needlessly anti-Semitic.

A dialogue from "Trump Guy" (2019) indicates that the series' writers contemplate their flippancy. Arguing with Peter about whose reckless public speech damages civic society more, the cartoon President Trump figure assails, "Many children have learned their favorite Jewish, black, and gay jokes by watching your show."[76] Peter's response is telling, pleading: "In fairness, we've been trying to phase out the gay stuff."[77] Such a rejoinder implies that the show's producers have (similarly) reviewed their Holocaust and Jewish joking and judged them to be acceptable. Indeed, Seth MacFarlane has created additional sitcoms promoting this formula. His *American Dad* (2005–present) resembles *Family Guy*, including its nuclear family ensemble, overbearing father, and speaking animal characters. The series also plies Holocaust humour via the goldfish character named Klaus, who communicates in crisp German-accented English. In "Not Particularly Desperate Housewife" (2005), he makes an Anne Frank joke. Again, in "Bully for Steve" (2010), Klaus advocates "A racially pure Europe." In "Kung Pao Turkey" (2013), responding to the show's patriarch's frustrations with his Asian in-laws, Klaus proposes an "an organized and systematic" solution.[78]

American Dad also has a Jewish character, a teenaged boy nicknamed Snot. Show writers direct him towards both irreverent and grieving Holocaust remembrance. "I Ain't No Holodeck Boy" (2014) finds Snot playing a video game entitled "Nazi Natal Nightmare."[79] The action permits him to time-travel to Austria during the late 1880s, pursuing a mission to kill Klara Hitler. "You bastard!" Snot yells as he tries to bludgeon the baby Hitler avatar. "You killed my grandparents!"[80] After Snot fails at his mission – and Hitler lives – Snot's tone becomes notably less strident. "I know he is going to be a monster," he remarks, "but I do love babies."[81] Such dialogue has the effect of negating his earlier outrage, implying to viewers that this horrific past could stand to accommodate a little contemporary humour. In the same way, displaying its *Family Guy* extraction, *American Dad* also broadcasts unashamed anti-Semitic joking. "Paranoid Frandroid" (2018) depicts the show's matriarch, Francine, worrying about "George Soros controlling the Federal Reserve from an underground bunker."[82] Her husband Stan, however, assuages such fears with bigotry: "He is Jewish, Francine. That is what he is supposed to be doing. He is helping us."[83]

Both *American Dad* and *Family Guy* promote countless additional examples of this levity. The comedy openly assails the "never forget" duty to remember that formed during the twentieth century's second half, instead modelling a twenty-first-century "stop reminding" plea. The proliferation and popularity of sitcom skits spoofing Nazi/Holocaust history indicates that audiences now seek to *enjoy* their Holocaust encounters. These unlikely interventions are less a matter of fault than fact. Contemporary post-witnesses do not want to see black-and-white images depicting gas chambers, crematoria, and corpse piles of long gone strangers. Destructive creative memorializing, however, can possibly draw these people to study the history. Alternative representations, especially those packaged with humour, might entice potential learners by presenting the trauma with enough emotional distance to renew contemplation.[84]

REPRESENTING ANNE FRANK VIA HUMOUR ON AMERICAN SITCOMS

Despite her global celebrity, Anne Frank is a protean figure. Of course, various evidence sources certify her life and times. Nonetheless, as Ilana Abramovitch,[85] Alvin Rosenfeld,[86] and Oren Baruch

Stier remind us,[87] her eternal "young girl" persona has supplanted memory of her life as an actual person. "If Anne Frank could return from among the murdered," Lawrence Langer once judged, "she would be appalled at the misuse to which her journal entries had been put."[88] Indeed, disparate actors conflate her hopeful image into many competing narratives. During the 1960s, West Germans took (implied) solace from her forgiving nature.[89] South Africans later recognized in Anne's writings a narrative for discussing Apartheid.[90] North Korean educators find in her life a call to arms against the US.[91] In the US, the novel/film *Anne B. Real* (2003) transposes Anne into a teenaged Afro-Latina protagonist named Cynthia Gimenez.[92] Such disparate memorializing does not supplant Anne Frank's factual existence; instead, the repurposing brings her story into understandings of the times in which she is read rather than the times during which she wrote.

Other interventions, however, are less positive. Palestinian Arabs and their allies regularly mock Anne. In 2006, responding to the *Jyllands-Posten* Muhammad cartoons controversy,[93] the Arab European League website posted "Hitler Goes Dutroux," a cartoon referencing the notorious Belgian child molester.[94] In this instance, the drawing depicted Adolf Hitler in bed with Anne Frank. His dialogue instructs, "Write this one in your diary!"[95] In 2008, a street artist in Amsterdam painted Anne's smiling girlish face wearing a *keffiyeh* on several of the city's buildings.[96] Turning the most famous Jewish Holocaust victim into a symbol of Palestinian nationalism, redeployed against perceived "Israeli-Nazism,"[97] encapsulates the "array of paradigmatic roles into which she has been cast."[98] Such ambiguity provides a useful context for examining her representation on American sitcoms. Not surprisingly, various series have parodied Anne. In line with the genre's superficiality, their joking, as Edward Portnoy explains, primarily messages present-day obliviousness towards this past, not opprobrium.[99]

Family Guy's "If I'm Dyin', I'm Lyin" (2000) was the first sitcom to create this joke work. It features a black-and-white "cut away" scene featuring mindless Peter Griffin seated in an attic. He is set to the side, looking at a huddled group of men, women, and teens, drawn as observant Jews. In the centre sits a girl with braided hair. She is clutching a book marked "Diary."[100] The visual punch line arrives via three interlocking images. First, viewers witness the Nazis storming into the house. Next, the camera pans upward to show those hiding in the attic. It is at this most dramatic point that writers unveil the final frame, Peter munching on potato chips. His crunches are very loud. The soon-to-be victims stare in horror, but he continues to snack.[101] The inanity of this scene renders the clip so clearly absurd that writers might well have concluded it was unlikely any viewer would mistake the rendition for fact. However, they may also have been offering additional commentary. Twenty-first-century Americans might consent to remember this past, indeed, representing it on popular sitcoms, but they will do so on their own terms, freed from normative expectations for piety.[102]

Robot Chicken also worked up an Anne Frank skit. YouTube hosts the clip, outfitted with French subtitles, which has attracted almost thirty thousand viewers.[103] The episode is "Toy Meets Girl" (2005).[104] In this repurposing, Anne is happy to live in an annex behind her father's storefront. She confides to her dairy that it is "awesome" because "everything I want is right here."[105] Following her jubilance, the camera pans across a tastefully decorated room to reveal her companion, Peter Van Dann, reading a scroll that appears to be the Torah. When the Nazis arrive, Anne is ferocious. Equipped with a full complement of *Home Alone* (1990) traps – paint can on rope and slippery ice puddles – she single-handedly stymies a Storm Trooper outfit.[106] *Robot Chicken* writers close their fantasy with Anne and Peter, clad in modern couture, embracing atop an Amsterdam river crossing. "You're really something, Anne Frank," Peter declares, over-emphasizing her name for both comedic and historical effect. "I'm just me," she replies, tossing back her hair and flirtatiously curving her leg in the air.[107]

What renders this imagining potentially baneful is that the show's likely youthful and non-Jewish viewers might form the mistaken impression that during the Second World War Anne Frank lived in a studio apartment with her boyfriend. While Nazism was posed to her (and by extension other Jews) as an inconvenience, it was not a mortal threat, and, indeed, was one that she handily dispatched. This damaging portrayal, however, could also clear a pathway towards an innovative classroom teaching tool. Instructors could counterpose the sitcom representation against a second clip of the actual annex, its many inhabitants and crowded spaces, incorporating a digital media that is readily available on the Anne Frank House's website.[108] Learning about this past via farcical and factual lore is perhaps the best potential outcome for engaging twenty-first-century learners. Indeed, *South Park* writers also build on this trend, and this is the final Anne Frank sitcom example that I will analyze. The show is "Major Boobage" (2008), with a script that ostensibly centres on stopping the town's children from huffing cat urine.[109] Indeed, the authorities decide to round up and deport the town's cats.

Nothing obvious suggests Holocaust joking. However, when authorities arrive at Eric Cartman's home, searching for cats, Anne Frank comedy results. Indeed, the anti-Semitic youngster channels a "Miep Gies" impulse, frustrating the officers' search. After it is safe, as plangent, minor-mode background music crescendos, he pulls down the latch to his attic's ladder. As he climbs the steps, the camera pans upward to reveal that he is hiding a cat. "Kitty, you have to live in the attic for now," he explains, "Write a diary."[110] By relating Anne's tale to that of a stray cat, *South Park* writers might be satirizing any number things. Of course, the real-life Anne began many diary entries by writing "Dear Kitty." Employing such homage demonstrates writers' familiarity with the canon.[111] However, while displaying their academic literacy, the sitcom's creative team also points out that contemporary readers are not receiving this text as dutifully as their parents did. Alvin Rosenfeld terms this commemorative spacing the "Anne Frank we remember versus the Anne Frank that we forget."[112] I add to this idea the possibility that the duelling reconstructions – such as irreverent sitcom skits – can bolster what appears more and more likely a vulnerable twenty-first-century legacy.

Seven decades past the Second World War, living in a global "fake news" age,[113] ours is an obvious toxic atmosphere for apt Holocaust memorialization. Such contests, however, provide stewards of this history opportunities to rejuvenate the lore's packaging in order to ensure its continued relevance. Professors should discuss with their students how spoofing figures like Anne Frank and Adolf Hitler, if possibly entertaining, also requires the producers of this material to affirm and demonstrate the history's truthfulness.[114] Even upholding the core fact of genocide may not be enough to sustain constructive future memorialization. The toughest part of predicting future paths for American Holocaust remembrance is determining whether twenty-first-century domestic discourses will continue to abide by their current post-truth trajectory.[115] Indeed, fictiveness and iconoclasm define the lingering Age of Trump. It is the truest danger to Holocaust remembrance – and to all inherited narratives. Absent sturdy intellectual and institutional moorings that nurture academic investigation and stoke public memorializing, substituting bastardized accounts that insinuate facts and invite fictions may ultimately conclude the "war" on American Holocaust memorialization.

NOTES

1 Peter Maguire, *Law and War: International Law and American History* (New York: Columbia University Press, 2010), viii.
2 Daniel Plesch argues that Allied leaders displayed meaningful wartime concern about avenging Jewish losses. He notes that in March 1945, a month before his suicide, the UN War Crimes Commission prepared to deliver at "least seven separate war crimes indictments against Adolf Hitler." See Daniel Plesch, *Human Rights after Hitler: The Lost History of Prosecuting Axis War Crimes* (Washington, DC: Georgetown University Press, 2017), 162.
3 Michael Berenbaum, *After Tragedy and Triumph: Essays in Modern Jewish Thought and the American Experience* (New York: Cambridge University Press, 1990), 41–2.
4 Aviva Atlani, "The Ha-Ha Holocaust: Exploring Levity amidst the Ruins and Beyond in Testimony, Literature and Film" (PhD diss., University of Western Ontario, 2014).
5 Slavoj Žižek, *First as Tragedy, Then as Farce* (New York: Verso, 2009), 1.
6 Maurizio Viano, "Life Is Beautiful: Reception, Allegory, and Holocaust Laughter," *Film Quarterly* 53, no. 1 (1999): 34, https://doi.org/10.2307/3697210.
7 Lillian Kremer, *Witness through the Imagination: Jewish American Holocaust Literature* (Detroit: Wayne State University Press, 1989).
8 James Young terms Plath a "Holocaust Jew," someone who appropriates this history to represent her or his own suffering. See James Young, *Writing and Rewriting the Holocaust: Narrative and the Consequences of Interpretation* (Bloomington: Indiana University Press, 1988), 99.
9 Sarah K. Pinnock, "Atrocity and Ambiguity: Recent Developments in Christian Holocaust Responses," *Journal of the American Academy of Religion* 75, no. 3 (2007): 516, https://doi.org/10.1093/jaarel/lfm038.
10 Jeffrey Shandler, *While America Watches: Televising the Holocaust* (New York: Oxford University Press, 1999), 259.
11 For specific discussion, see Robert Clary, *From the Holocaust to Hogan's Heroes: The Autobiography of Robert Clary* (Lanham, MD: Taylor Trade Publications, 2008), 172–3.
12 As quoted in Shandler, *While America Watches*, 168.

13 Norman Finkelstein, *The Holocaust Industry: Reflections on the Exploitation of Jewish Suffering* (New York: Verso, 2nd ed., 2003), 4.
14 Oren Baruch Stier, *Holocaust Icons: Symbolizing the Shoah in History and Memory* (New Brunswick: Rutgers University Press, 2015), 8, 11.
15 "Introduction," *Anne Frank, Unbound: Media, Imagination, Memory*, ed. Barbara Kirshenblatt-Gimblett and Jeffrey Shandler (Indianapolis: Indiana University Press, 2012), 7.
16 Tim Cole, *Selling the Holocaust: From Auschwitz to Schindler – How History Is Bought, Packaged, and Sold* (New York: Routledge, 1999), 81.
17 Gavriel Rosenfeld, *Hi Hitler! How the Nazi Past Is Being Normalized in Contemporary Culture* (Cambridge: Cambridge University Press, 2015), 28.
18 Walter Metz, "'Show Me the Shoah!': Generic Experience and Spectatorship in Popular Representations of the Holocaust," *Shofar* 27, no. 1 (2008): 10, http://www.jstor.org/stable/42944683.
19 Itay Stern, "'Sick and Unacceptable': Netflix's Satirical 'Roast' of Anne Frank Draws Fire," *Haaretz*, 29 May 2019, https://www.haaretz.com/life/television/premium-sick-and-unacceptable-netflix-s-satirical-roast-of-anne-frank-draws-fire-1.7303599.
20 Jeffrey Demsky and Liat Steir-Livny, "Holocaust Jokes on American and Israeli Situational Comedies: Signaling Positions of Memory Intimacy and Distance," in *The Languages of Humor: Verbal, Visual and Physical Humor*, ed. Arie Sover (London: Bloomsbury Academic, 2018), 71.
21 Jeffrey Demsky, "Searching for Humor in Dehumanization: American Situational Comedies, the Internet, and the Globalization of Holocaust Parodies," in *Analysing Language and Humor in Online Discourse*, ed. Rotimi Taiwo (Hershey, PA: IGI Publishing, 2016), 12.
22 Russell Heddendorf, *From Faith to Fun: The Secularisation of Humour* (Cambridge, UK: Lutterworth Press, 2009), xv.
23 Ruth Wisse, *No Joke: Making Jewish Humor* (Princeton, NJ: Princeton University Press, 2013), 179.
24 Tadeusz Borowski, *This Way for the Gas, Ladies and Gentlemen* (New York: Penguin Books, 1959).
25 Liat Steir-Livny, "Holocaust Humor, Satire, and Parody on Israeli Television," *Jewish Film and New Media* 3, no. 2 (2015): 212.
26 *Curb Your Enthusiasm*, season 4, episode 9, "The Survivor," Larry David, 7 March 2004, HBO.
27 Eric Romero and Kevin Cruthirds, "The Use of Humor in the Workplace," *Academy of Management Perspectives* 20, no. 2 (2006): 60.
28 As quoted in Andy Egan, "There's Something Funny about Comedy: A Case Study in Faultless Disagreement," *Erkenntnis* 79, no. 1 (2014): 73.

29 *Robot Chicken*, season 5, episode 6, "Major League of Extraordinary Gentlemen," Brendan Hay, 13 February 2011, Adult Swim.
30 *Robot Chicken*, season 6, episode 3, "Punctured Jugular," Zeb Wells, 30 September 2012, Adult Swim.
31 Edward Fink, "Writing *The Simpsons*: A Case Study of Comic Theory," *Journal of Film and Video* 65, nos 1–2 (2013): 43–55, https://doi.org/10.5406/jfilmvideo.65.1-2.0043.
32 Adrian Daub, "Hannah, Can You Hear Me? Chaplin's *Great Dictator*, 'Schtonk,' and the Vicissitudes of Voice," *Criticism* 51, no. 3 (2009): 451–482, https://www.jstor.org/stable/23131524.
33 *The Simpsons*, season 6, episode 16, "Bart vs. Australia," Matt Groening, 19 February 1995, Fox.
34 Demsky and Steir-Livny, "Holocaust," 75.
35 Lisa Young and Marion Nestle, "Portion Sizes and Obesity: Responses of Fast-Food Companies," *Journal of Public Health Policy* 28, no. 2 (2007): 238–48, https://doi.org/10.1057/palgrave.jphp.3200127.
36 *The Simpsons*, season 23, episode 20, "The Spy Who Learned Me," Matt Groening, 6 May 2012, Fox.
37 Lawrence Baron, "Incorporating Film into the Study of the Holocaust," in *Essentials of Holocaust Education: Fundamental Issues and Approaches*, ed. Samuel Totten and Stephen Feinberg (New York: Routledge, 2016), 169–88.
38 Harold Brackman, "The Attack on 'Jewish Hollywood': A Chapter in the History of Modern American Anti-Semitism," *Modern Judaism* 20, no. 1 (2000): 1–19, https://www.jstor.org/stable/1396627.
39 *Extras*, season 1, episode 3, "Kate Winslet," Ricky Gervais and Stephen Merchant, 9 October 2015, BBC Two Television.
40 Ibid.
41 Ibid.
42 Ibid.
43 Robert Samuels. "Freud Goes to South Park: Teaching against Postmodern Prejudices and Equal Opportunity Hatred," in *Taking South Park Seriously*, ed. Jeffrey Weinstock (Albany: State University of New York Press, 2008), 103.
44 Ibid.
45 *South Park*, season 1, episode 7, "Pinkeye," Trey Parker and Matt Stone, 29 October 1997, Comedy Central.
46 Lawrence Baron, "The First Wave of American "Holocaust" Films, 1945–1959," *American Historical Review* 115, no. 1 (2010): 90–114, http://www.jstor.org/stable/23302762.

47 *South Park*, "Pinkeye."
48 *South Park*, season 8, episode 3, "The Passion of the Jew," Trey Parker, 31 March 2004, Comedy Central.
49 Pamela Grace, "Sacred Savagery: *The Passion of the Christ*," *Cinéaste* 29, no. 3 (2004): 13–16.
50 Michael DeLashmutt and Brannon Hancock, "Prophetic Profanity: *South Park* on Religion or Thinking Theologically with Eric Cartman," in *Taking South Park Seriously*, ed. Jeffrey Weinstock (Albany: State University of New York Press, 2008), 178–9.
51 *South Park*, "Passion."
52 Ibid.
53 Ibid.
54 Ibid.
55 Ibid.
56 Hasia Diner, *We Remember with Reverence and Love: American Jews and the Myth of Silence after the Holocaust, 1945–1962* (New York: New York University Press, 2009).
57 Demsky, "Searching," 11.
58 As quoted in Mark Runco, *Creativity: Theories and Themes: Research, Development, and Practice* (London: Academic Press, 2014), 142.
59 Rob Eshman, "Seth MacFarlane: Not an anti-Semite," 25 February 2013, *Jewish Journal*, https://jewishjournal.com/opinion/rob_eshman/113318/.
60 Ibid.
61 Lillian Boxman-Shabtai and Limor Shifman, "Evasive Targets: Deciphering Polysemy in Mediated Humor," *Journal of Communication* 64, no. 5 (2014): 977, https://doi.org/10.1111/jcom.12116.
62 Rosenfeld, *Hi Hitler*, 310.
63 Liat Steir-Livny, "Is It OK to Laugh about It Yet? Hitler Rants YouTube Parodies in Hebrew," *European Journal of Humour Research* 4, no. 4 (2017): 107, https://doi.org/10.7592/EJHR2016.4.4.steir.
64 *Family Guy*, season 14, episode 7, "Hot Pocket Dial," Seth MacFarlane, 22 November 2015, Fox.
65 Ibid.
66 Ibid.
67 Scholars disagree about whether or not Disney was anti-Semitic. Marc Eliot argues that he held the period's dominant Protestant attitudes, mindsets that cast Jews unfavourably. See Marc Eliot, *Walt Disney: Hollywood's Dark Prince* (London: Andre Deutsch 1994), 173. On the other hand, Karl Cohen argues that no evidence confirms Disney held anti-Semitic views, stressing that his former Jewish employees never

charged that he did. See Karl Cohen, *Forbidden Animation: Censored Cartoons and Blacklisted Animators in America* (Jefferson, NC: McFarland, 2004), 73.

68 *Family Guy*, season 16, episode 3, "Nanny Goats," Seth MacFarlane, 15 October 2017, Fox.

69 Lynne Olson, *Those Angry Days: Roosevelt, Lindbergh, and America's Fight over World War II, 1939–1941* (New York: Random House, 2013), 378.

70 Richard Rubenstein, "Antisemitism and the Contemporary Jewish Condition," in *Not Your Father's Antisemitism: Hatred of the Jews in the Twenty-First Century*, ed. Michael Berenbaum (St Paul, MN: Paragon House, 2008), 31–4.

71 *Family Guy*, season 13, episode 7, "Stewie, Chris & Brian's Excellent Adventure," Seth MacFarlane, 4 January 4 2015, Fox.

72 Ibid.

73 *The Forward* and Mark Pinsky, "Is That Funny Animated Show *Family Guy* Secretly Antisemitic," *Haaretz*, 21 March 2014, https://www.haaretz.com/jewish/is-family-guy-anti-semitic-1.5337047.

74 *Family Guy*, season 7, episode 1, "Love Blactually," Seth MacFarlane, 28 September 2008, Fox.

75 Ibid.

76 *Family Guy*, season 17, episode 11, "Trump Guy," Seth MacFarlane, 13 January 2019, Fox.

77 Ibid.

78 *American Dad*, season 9, episode 5, "Kung Pao Turkey," Seth MacFarlane, 24 November 2013, Fox.

79 *American Dad*, season 9 episode 13, "I Ain't No Holodeck Boy," Seth MacFarlane, 23 March 2014, Fox.

80 Ibid.

81 Ibid.

82 *American Dad*, season 13, episode 2, "Paranoid Frandroid," Seth MacFarlane, 12 February 2018, Fox.

83 Ibid.

84 For related analysis, see Rachel Brenner, "Holocaust Culture in Perspective: Evading the Holocaust Story and Its Legacy of Responsibility," *Dapim* 26, no. 1 (2012): 125–50, https://doi.org/10.1080/23256249.2012.10744418.

85 Ilana Abramovitch "Teaching Anne Frank in the US," in *Anne Frank, Unbound: Media, Imagination, Memory*, ed. Barbara Kirshenblatt-Gimblett and Jeffrey Shandler (Indianapolis: Indiana University Press, 2012), 176.

86 Alvin Rosenfeld, *The End of the Holocaust* (Bloomington: Indiana University Press, 2011), 146–9.

87 Stier, *Holocaust Icons*, 121.

88 As quoted in Lawrence Langer, "The Uses and Misuses of a Young Girl's Diary," in *Anne Frank: Reflections on Her Life and Legacy*, ed. Hyman Aaron Enzer and Sandra Solotaroff-Enzer (Urbana: University of Illinois Press, 2000), 203–5.

89 Robert Sackett, "Memory by Way of Anne Frank: Enlightenment and Denial among West Germans, Circa 1960," *Holocaust and Genocide Studies* 16, no. 2 (2002): 366–93, 243–6, https://doi.org/10.1093/hgs/16.2.243.

90 Shirli Gilbert, "Anne Frank in South Africa: Remembering the Holocaust during and after Apartheid," *Holocaust and Genocide Studies* 26, no. 3 (2012): 366, https://doi.org/10.1093/hgs/dcs058.

91 Abramovitch, "Teaching," 176–7.

92 For analysis, see Leshu Torchin, "Anne Frank's Moving Images," in *Anne Frank, Unbound: Media, Imagination, Memory*, ed. Barbara Kirshenblatt-Gimblett and Jeffrey Shandler (Indianapolis: Indiana University Press, 2012), 120–1.

93 Jytte Klausen, *The Cartoons That Shook the World* (New Haven: Yale University Press 2009), 18.

94 For discussion, see Evelien Gans, "Hamas, Hamas, All Jews to the Gas: The History and Significance of an Antisemitic Slogan in the Netherlands, 1945–2010," in *Perceptions of the Holocaust in Europe and Muslim Communities*, ed. Günther Jikeli and Joëlle Allouche-Benayoun (New York: Dordrecht Springer, 2013), 100.

95 For a copy of the image, see Edward Portnoy, "Anne Frank on Crank: Comic Anxieties," in *Anne Frank, Unbound: Media, Imagination, Memory*, ed. Barbara Kirshenblatt-Gimblett and Jeffrey Shandler (Indianapolis: Indiana University Press, 2012), 321.

96 Rosenfeld, *End of the Holocaust*, 152.

97 For analysis of the construct, see Sara Roy, "A Response to Elie Wiesel," *Journal of Palestine Studies* 44, no. 1 (2014): 133–4, 133.

98 "Introduction," *Anne Frank*, 12.

99 Portnoy, "Anne Frank," 311.

100 *Family Guy*, season 2, episode 9, "If I'm Dyin' I'm Lyin'," Seth MacFarlane, 4 April 2000, Fox.

101 Portnoy, "Anne Frank," 317.

102 Joost Krijnen, *Holocaust Impiety in Jewish American Literature: Memory, Identity (post-) Postmodernism* (Boston: Brill, 2016); Matthew Boswell,

Holocaust Impiety: In Literature, Popular Music and Film (London: Palgrave Macmillan, 2012).

103 "Hilary Duff in the Diary of Anne Frank." *YouTube*, uploaded by Cameron Rodriguez, 22 May 2016, https://www.youtube.com/watch?v=4xqS1_TXGik.

104 *Robot Chicken*, season 1, episode 11, "Toy Meets Girl," Matthew Senreich, 1 May 2015, Adult Swim.

105 Ibid.

106 Portnoy, "Anne Frank," 316.

107 Ibid.

108 "The Secret Annex." *AnneFrank.org*, https://www.annefrank.org/en/anne-frank/secret-annex/.

109 *South Park*, season 12, episode 3, "Major Boobage," Trey Parker, 26 March 2008, Comedy Central.

110 Ibid.

111 Demsky, "Searching," 6.

112 Rosenfeld, *End of the Holocaust*, 162.

113 Matthew D'Ancona, *Post Truth: The New War on Truth and how to Fight Back* (London: Ebury Press, 2017).

114 Jeffrey Demsky, "We Are a Long Ways Past *Maus*: Responsible and Irresponsible Holocaust Representations in Graphic Comics and Sitcoms Cartoons," *The Palgrave Handbook of Holocaust Literature and Culture*, eds. Victoria Aarons and Phyllis Lassner (Cham, Switzerland: Palgrave Macmillan, 2019), 537.

115 Nancy Rosenblum and Russell Muirhead, *A Lot of People Are Saying: The New Conspiracism and the Assault on Democracy* (Princeton: Princeton University Press, 2019), 27.

17

Of Wars, Scars, and Celluloid Memory: Representations of War in Sri Lankan Cinema (2000–10)

Vilasnee Tampoe-Hautin

The Sri Lankan film industry, which many describe as modest and ailing, has mirrored the country's past and recent ethno-politics, founded on the Sinhala/Other distinction. Innumerable conflicts have indeed marked Sri Lanka's colonial and post-independence periods, including the 1915 Pogroms, the communal riots of 1956, and, more recently, the protracted war against terrorism that ended in 2010. It is in light of this troubled inheritance that this chapter aims to discuss representations of the Sri Lankan ethnic conflict (1983–2010) through the medium of film and to probe the question of art forms in relation to one of the first wars waged and won against one of the first terrorist organizations in modern times, the Liberation Tigers of Tamil Eelam (LTTE).

Yet, for lack of material and hindsight, to discuss Sri Lankan cinema in relation to a war that came to an abrupt and brutal end only recently is not an easy task. Scholars in Sri Lanka have nevertheless produced illuminating and ground-breaking studies on this conflict, braving war-induced restrictions, including censorship of their research findings and publications. Therefore, this chapter is also written in recognition of the invaluable contribution to the academic study of the Sri Lankan war by Neloufer de Mel, Robert Crusz, Sunila Abeyesekara, Priyantha Fonseka, and others.[1] There still remain uncharted waters and much territory to be mapped for those experts and non-specialists who are outside Sri Lanka. The economy of Sri Lankan cinema was likewise hard hit by what is today seen as Asia's longest-running conflict in modern times, with mass killings, bomb scares, and wanton destruction of cinema halls and other public places.[2]

Yet this theatre of assassinations and violence also gave birth to a Sinhala cinema with a distinct character and flavour, and this is the focus of this chapter: at the turn of the new millennium, several Sri Lankan directors used the war setting both as fiction and reality to reconstitute it through a narrative of film images, producing a number of internationally acclaimed films. Prasanna Vithanage, Asoka Handagama, and Vimukthi Jayasundara rose to prominence during the late 1990s and drew the attention of international film juries not only to their work but also to the horrendous war that had been ravaging the island since 1983. The cinema of Vithanage, Handagama, and Jayasundara came as a major breakthrough in what had become a barren cultural landscape, deeply scarred by more than a decade of political instability, economic duress, and restricted mobility.

Recruiting actors from among war victims, defying the multiple constraints and prohibitions of the day, including restricted movement, lack of subsidies, and state censorship, these directors located their films, in both content and actual shooting, in an unsafe militarized environment. Making the absurdity of war its major theme, their cinema explored the psychological and physical dilemmas of people caught up in the throes of the conflict. This new Sinhala cinema exposed the profound changes the civil war had brought about both in the Sri Lankan social, political, and cultural landscape and in the national and collective consciousness. Moreover, the stringent state-imposed emergency regulations curtailing the media reporting of the war affected the Sri Lankan film industry's production, distribution, and exhibition sectors. Films were either censored or banned, or their producers forced into intellectual exile, raising the question of freedom of expression for Sri Lankan film directors, which I also examine here.

THE SINHALA-TAMIL DIVIDE

It is not possible, within the scope of this chapter, to examine at any length the subject of Sri Lankan ethnicities, about which there is an abundance of scholarly works, covering topics ranging from the genesis of the Sinhala-Tamil enmity to its evolution during the colonial and post-independence periods. Nevertheless, it is worth briefly recalling here some aspects of Sri Lanka's ethnic problem, which hinges on Sinhala-Tamil antagonism. Sri Lankan ethnicities can be explained not only by colonial practices but also by the island's geographical and demographic distribution, characterized by two cultural communities occupying distinct spaces, resulting in an almost exclusively Tamil and Muslim North and North–East and a predominantly Sinhala South and South West. This population distribution had already caught the eye of early colonial administrators. The first European reference to this demographic specificity dates back to 1799 when Hugh Cleghorn,[3] upon arriving in Ceylon, wrote in his report to the British Colonial Office: "Native Inhabitants – Two different nations, from a very ancient period have divided the island. First the Sinhalese, in the southern and Western parts, from the river Walawe to that of Chilaw; and secondly, the Malabars in the Northern Eastern Districts. These two nations differ in their religion, language and manners."[4] The Cleghorn Minute (whose original has disappeared) remains to this day pertinent and controversial in Sri Lanka. The author's vision of two distinct nations, and his attribution of the northern part of Sri Lanka to the Tamils, has served to buttress and legitimize separatist demands by some components of its Tamil population.

Furthermore, from the mid-nineteenth century onwards, the British colonial regime proceeded to classify the island's inhabitants according to pre-defined "races." The typologies, aimed at facilitating the census of population, were largely inspired by polygenic and linguistic theories that were in broad circulation at the time. Historian Nira Wickramasinghe notes that earlier attempts at understanding and recording the configuration of the island's population were carried out in 1814 and 1824. These surveys had already made reference to the criteria of caste and religion. In fact, until 1824, the Sinhalese and Tamils were classified as belonging to castes and not to ethnic groups.[5] The term "race" appeared for the first time during the census operations of 1871, and later in 1881, which mention twenty-four races and seventy-eight "nationalities," with interconnections between these categories. It was only in 1911 that the term "nationality" ceased to be used in favour of the term "race."[6]

As recalled by Eric Meyer, the colonial practice of applying predetermined racio-linguistic criteria to the inhabitants led to the emergence of two monolithic groups, the Sinhalese and the Tamils.[7] Their relative numerical strengths and weaknesses in turn gave rise to the problematic majority-minority dichotomy, drawing the comment from Darini Rajasingham-Senanayake that, "despite sharing common cultural traits and a common history, the Sinhalese and the Tamils are entrenched in two distinctive groups, based on ethno-racial and linguistic criteria."[8]

ETHNICITY AND EARLY SINHALA CINEMA

Notwithstanding, until the 1960s, Sri Lanka's film industry had managed to rise above the antagonisms that had gradually divided the island's population. Cinema in colonial and postcolonial Ceylon was a stomping and trading ground for enterprising individuals from all horizons, with interesting collaborations between various cultural identities. For example, in 1947, *Broken Promise*, shot in South India and produced by S. Nayagam, an Indian residing in Sri Lanka, lifted the curtain on Sri Lankan cinema, in the Sinhala language. The same year, the first Sri Lankan Tamil sound film, *Deivaneethi*, was directed and produced by W.M.S. Tampoe and released in South India.[9] Most of the subsequent Sinhala films were directed and produced by Sri Lankan Tamils and Muslims as well as South Indians: A. Gardiner, K. Gunaratnam, J.B. Cader, L.S. Ramachandran, A. Bunker Raj, W.M.S. Tampoe, and R.C. Tampoe.

These pioneers of Sri Lankan cinema, hailing from non-Sinhala minority communities, also fuelled Sinhala cultural patriotism during the late 1950s, leading to demands for a film industry with more participation by ethnic Sinhalese.[10] Nationalist claims were based on the grievance of a quasi-total exclusion of indigenous Sinhalese from the local film industry, even if such a foreign monopoly was inevitable given that the cinema was invented in the West and was introduced into the country primarily by itinerant film operators from Europe, Great Britain, and the United States. Subsequently, until the 1960s, film exploitation remained in the hands of the colonial elite, the Anglicized middle classes, and non-Sinhala communities engaged in trade and commerce. It took nearly sixty years for the tide to turn as Sinhala Buddhist ethno-nationalism spilled over to the cultural domain, with the call to "sinhalize" the film sector and make it the standard-bearer of the newly independent nation, Ceylon.[11]

Released from what was perceived as economic and cultural imperialism, in particular by South India and the US, the Sri Lankan film industry has come a long way. It consists in the production mostly of commercial films in the Sinhala language almost exclusively for a local market, though some have reached overseas markets and received plaudits at international festivals. Despite the ubiquity of the entertainer, Sri Lankan audiences do have a penchant for social realism, and Sri Lanka is one of the few countries in South and Southeast Asia where people throng with equal enthusiasm to see realist movies as they would a three-hour song-and-dance spectacle.[12]

SINHALA CINEMA OF THE WAR YEARS

This next part of this chapter focuses on two films *This Is My Moon/Me Magay Sanday* (2001) and *Death on a Full Moon Day/ Pura Handa Kaluwara* (1998) by Asoka Handagama and Prasanna Vithanage, respectively, as examples of representations and memorialization of the ethnic war. Both films continue the tradition of "realist" Sinhala cinema that grew out of the documentary and neo-realist genres, developed from the 1960s onwards by Sri Lankan directors like Lester James Peries, D.B. Nihalsinghe, Pragnasoma Hettiarachchi, Dharmasiri Pathiraja, and Vasantha Obeyesekara. *This Is My Moon* was Asoka Handagama's response to the desolate land that Sri Lanka had become in the 1990s, disfigured by the armed conflict. The film is about a Sinhalese soldier deserter from a rural background who returns to the army, as entreated by his parents, so that his salary can support the family again. A woman he rapes, identified in the film as Tamil, follows him to his village, a predominantly Sinhala community, where his fiancée also happens to live. A second story unfolds in parallel whereby two village youths vie with each other to get into the army in order to impress the girl with whom they are both in love. As recalled by Neloufer de Mel, "Handgama masterfully foregrounds the relationship of militarism and male sexual prowess. That the lad who does join the army gets the girl underscores the investment within militarization of military vocations and masculinities as economically viable and virile."[13] Vithanage's *Death on a Full Moon Day* is likewise set in a Sinhala village where Vannihami, a blind farmer, lives, and whose son is presumed to have been killed in action. A sealed coffin draped with the Sri Lankan flag is returned to a village in mourning. But as the burial takes place, the father is convinced that the body in the coffin is not his son. Army orders had previously been issued that the coffin should remain sealed and the body should not be revealed to the family. Then a letter from Bandara is delivered to the family in which the young soldier promises to come home on leave, in time for his sister's wedding, leaving his father all the more perplexed.

Through Vannihami's dilemma, the film contemplates the issue of the probity of the state and the army and the impossibility, during the war years, for most villagers to retrieve the bodies of their children who had died in combat. As pointed out by human rights campaigner Sunila Abeyesekara, funeral rites for soldiers killed in action were often performed for blocks of wood, which replaced corpses in the sealed coffins.[14]

In both films, the directors engage with memory and commemoration as they place the ethnic conflict at the heart of their work. The process of memorializing implies the recording on celluloid of the atrocities and futilities of this war. If memory, as Henry Rousso suggests, is a phenomenon that is related to the present, a living active trace carried by subjects, then Handagama and Vithanage, by focusing on a war whose end is not yet in sight, contribute to its memory through "a reconstructed or reconstituted present which is organized around a complex network of images (and of image memories), of sensations and emotions, of forgetting, of denials, of repressions and of their return."[15] By bringing the historic conflict to the screen while it is still unfolding, the directors join the past with the present, mingling fact and fiction. Shot in the North and North-East of Sri Lanka in the periphery of raging battlegrounds, the films have documentary value with their offer of near authentic situations, even if Vithanage could not entirely fulfill his wish to brave any hazard to shoot his film as he was asked to keep out of the war zones and refused permission to employ army personnel for his film.[16]

Neloufer De Mel points out that, although the two films vary in aesthetic form and pace, they are custodians of popular memory, an important function of Third Cinema.[17] For film historian Wimal Dissanayake, Vithanage's works are a composite of three strands of memory, identified as personal memory, social memory, and cultural memory, generating guilt and anxiety. This idea is corroborated by Vithanage himself, who insists that the ideas of the coping with memory, feelings of guilt, and the tormenting power of the mind are central to his films.[18]

The two films expose the way a paradise island becomes an inferno, devastated by slow but steady militarization with war-induced psychological and physical trauma holding centre stage. The war has a profound and shattering impact on the mentalities and mores of the inhabitants. New lifestyles, lax sexual mores, and economic dynamics penetrate the villages, notably in the Vanni,[19] which become war-stricken regions. Heroes, and anti-heroes, often of Sinhala origin, are trapped between two worlds, one peaceful and the other militarized, and between two cultural environments, their own and that of their Tamil fellow citizens.

The films, as examples of Third Cinema, are equally notable for their ground-breaking techniques. Priyankara Fonseka, in his insightful research on silence and cinema, highlights the "complex strategy of mutism that underpins these films."[20] *Death on a Full Moon Day* and *This Is My Moon* are singular in the way their characters, atmosphere, and environment remain distant and glacial in the face of a raging war. Background music and sound are also kept to a minimum. Fonseka, in his analysis, argues that the silence observed by the Tamil girl in *This Is My Moon* stands for loss of identity, while Vannihami, the father, is both blind and taciturn. Voluntary mutism then is at once a refuge and a weapon that empowers the weak and the oppressed, while trauma-induced mutism is a way to communicate pain. The directors also evince a fondness for village and jungle settings, where silence usually reigns, an echo of Leonard Woolf's *Village in the Jungle* (1911), one of the rare English-language novels on colonial Ceylon written by an Englishman.[21] To this characteristic stillness and aloofness may be added Handagama's and Vithanage's penchant for empty static frames. With reference to *This Is My Moon*, Handagama points out to film critic Robert Crusz that "there are no dissolves no fade ins no fade out. This film applies only the most primitive technique of shot joining, the Cut. No panning no zooming no tracking. This film uses the most fundamental way of framing, the Static frame. No magic. No effects. It employs the simplest of techniques a rhythmic flow of static images to maintain the tempo."[22] Handagama likewise insists that "there are no plot points, no suspense, standing against the rules of Syd Field, the guru of scriptwriting, this anti-Hollywood style film will, I hope, be powerful enough to keep the audience tagged to the film right throughout."[23] The film was not without its critics: indeed, Thomas Sotinel laments the director's tendency to avoid close-ups, a technique that recurs systematically in the film,[24] while Panini Wijesiriwardana considers that Handagama is too preoccupied with "visual forms at the expense of thoughtful character and plot development," and with a number of gimmicks.[25]

Despite these reservations, the directors picked up the gauntlet of bringing a horrific thunderous war to the screen through a quasi-silent visual text, portraying and memorializing a new social reality in the face of the failure of prevailing traditional forms and narratives to carry out the task.[26] Neloufer De Mel makes the pertinent remark that Handagama is nevertheless "tapping into an ancient indigenous art form and keeping faith with third cinema's goal of using narrative frames and images that are indigenous."[27] The late Lester James Peries,[28] the doyen of Sri Lanka cinema, qualified Handagama's *This Is My Moon* as a "third revolution in Sri Lankan cinema." Finally, not only do these directors challenge the Hollywood model but they also wage another war, of words and images, against the Sri Lankan state.

BANNED AT HOME, ACCLAIMED OVERSEAS

In their interviews with this author, the two directors laid bare the difficult conditions in which their films were created, produced, and distributed as they struggled through financial straits and faced censorship constraints.[29] Despite the artistry of these films, domestic reception too was lukewarm, with a public that now stayed away from cinema halls and a state opposed to the propagation of war films, both within and beyond the island's shores, for fear of tarnishing the island's image. It had warned several Sri Lankan directors that treating the subject of war would oblige them to choose between self-censorship and the risk of being taken to court.

In fact, Vithanage's *Pura Handa/Death on a Full Moon Day* was to become another battleground, this time between the Supreme Court, the Sri Lankan state, and the associations for the defence of the freedom of expression. The director was informed by the Competent Authority of the Government Information Department that several scenes would have to be removed under emergency regulations as the film's portrayal of soldiers was likely to "affect the morale of the security forces, especially the exhumation of the coffin and its follow up scenes."[30] This was further justified by the fact that "the motion picture is able to stir up emotions more deeply than any other form of art and its impact on a nation that has experienced, and is still experiencing a war situation of this dimension, which is raging, could be counterproductive to the efforts of the government."[31]

Consequently, in 1999, the public release of Vithanage's film was deferred, whereas it had already suffered a three-year wait on the notorious National Film Corporation queue system. Attempts by government to impose a permanent ban on the film failed, as the courts invalidated the laws on censorship that the state had promulgated. The director finally obtained the necessary clearance for the release of his film in October 2001. Financed by the NHK Japan national TV, *Pura Handa/Death on a Full Moon Day* won the Grand Prix at the Festival of Amiens and was consecrated as the most successful film in the history of national Sri Lankan cinema, breaking a number of records in the country. The triumph of the film led the author to collaborate with the Italian director Umberto Pasolini to make *Machan* (2008), presented at the 65th Venice Film Festival and crowned with success. For his part, Asoka Handagama was the victim of indirect censorship as only part of the loan he had applied for from the People's Bank was granted, leaving him with an unfinished film. As noted by Robert Crusz, while the rule of releasing only part of the loan was mainly a safeguard against misuse of funds by the director, at the preview of the film, bank authorities and officials found its contents unacceptable. Several scenes were deemed to be injurious to a Sinhala Buddhist nation and to Sri Lanka's image. Bank authorities had thus become censors, moralizers as well as politicians.[32]

As Neloufer De Mel observes, "The film was not publicly banned or even denounced in a widespread fashion," but the withdrawal of financial support and other delaying strategies drove the director to self-censor, "sending out a signal to other artists that they should think twice before emulating Handagama, or mocking the Sinhala Buddhist nation."[33] Although Handagama received support from intellectuals and the bank was finally pressured into releasing the funds, Vithanage and Handagama were hindered in their attempt to bring the war to the screen in a way that went against official discourses. However, state censorship to prevent the moral and political subversion allegedly transmitted by these films was denounced by the Sri Lankan Free Media Movement, which declared that a conservative government was hindering the development of avant-garde cinema.

Additionally, the new Sinhala cinema's penchant for themes considered as "taboo" in Sri Lanka, as well as challenging conventional male-female relationships, did nothing to ease the tension between the state and artists. Vimukthi Jayasundara's *The Forsaken Land* (2005) focuses on isolation and solitude, leading the main female character, the frustrated Punchi Kirrilli (Little Bird), to hang herself at the end of the film. In Handagama's *Letter of Fire* (2005), an adolescent discovers female nudity by watching his mother undressing. Lester James Peries, however, described the film as a serious piece of work even if it ruffled sensitivities, pointing out the state's lack of democracy and judgment in its attitude towards Sri Lankan creativity.[34] Other films, like Sanjeewa Pushpakumara's *Flying Fish/Ingilena Maluwo* (2011),[35] were censured and censored by the Sri Lankan authorities because of scenes judged as being highly objectionable in a Buddhist society for containing suggested incest, homosexuality, or voyeurism.

Despite the multiple impediments, both *Death on a Full Moon Day* and *This Is My Moon* won international acclaim and plaudits, as did *Flying Fish* at a later date, for bringing the atrocities of the Sri Lankan armed conflict to the outside world, an achievement all the more impressive as the directors use their intellectual and aesthetic powers of expression to put across their messages, and avoid political or ideological sword-crossing.

CONCLUSION

In the wake of a twenty-six-year conflict between extreme factions of ethnic Sinhalese and Tamils, it may be pertinent to ponder the future of Sri Lankan cinema as the country still struggles to regain its political, economic, and social balance. What are the chances of Sri Lankan cinema being resurrected and of thriving as a fully fledged art and industry? Can it aptly mirror the Sri Lankan nation in its newfound unity? In light of the ongoing "healing" process, with ethnic reconciliation, social peace, political stability, and economic prosperity on the list of goals to be achieved, how far will Sri Lankan cinema, both in Sinhala and Tamil, become emblematic of "unity in diversity" and not in rivalry?

In all evidence, the young directors evince a strong desire to combat xenophobia and racism by reconstructing a country torn in two. One notes a move away from the nationalistic "pro-Sinhala" stance that characterized the early phase of the industry's existence during the 1950s and 1960s. The case of Handagama's *Ini Avan* (2012) is illustrative of the aspiration for inter-ethnic amity and harmony: the director, who is an ethnic Sinhalese, shot the film in the bastion of Tamil culture, the Jaffna peninsula. *Ini Avan*, with its dialogues in Tamil subtitled in Sinhala, may be seen as an attempt to bridge cultural and linguistic gaps between the two communities as well as a brave and worthy contribution to the post-conflict healing process with which the country is engaged. It also straddles two geo-cultural spaces – the Sinhala South and the Tamil North. In the same way, Vithanage's *August Sun/Ira Mediyama* explores the suffering of the Muslim community of Sri Lanka, forced out of its traditional place of habitation by the LTTE. In an interview, Vithanage explains to Robert Crusz that the focus on Muslims in a Sinhala film, directed by an ethnic Sinhalese, is "a rare event."[36]

But the film also comes in the wake of previous, albeit isolated, attempts at achieving inter-ethnic harmony on and off the screen, such as the celebrated *Ponmani* (1978), a film on Sri Lankan Tamils shot in the Jaffna Peninsula by an ethnic Sinhalese, Dharmasena Pathiraja. Sumitra Peries's *Yahaluwo/Friends* (2003) mobilizes a multicultural cast, with a screenplay by a Sri Lankan Muslim, Vahima Sahabdeen, and dialogues and songs in Sinhala and Tamil. It remains to be seen whether more filmmakers of talent and commitment to inter-ethnic peace will join the ranks of these pioneers and their contemporaries.

Finally, what should be made of the decline in war films now that peace has been won? It would appear that post-conflict Sinhala cinema is keeping its distance from themes that evoke the Sinhala-Tamil rift. War-inspired films have given way to grand Buddhist frescoes and historical dramas, such as *Aba* (2001), *Mahnidagamanaya* (2013), *Siri Parakum* (2013), and *Daladagamaya* (2014). These, in turn, have been joined by a prolific production of children's movies, while others imitating American gangster films have also flooded the local market. Sri Lankan cinema seems to have quickly turned the page on one of the most tragic chapters of the country's recent history. The author's discussions held in December 2019 with Asoka Handagama, Vimukthi Jayasundara, and Prasanna Vithanage seem to confirm this trend away from war-inspired cinema, despite the release of their *Her, Him, The Other* in 2018 – an anthological retrospective of the ethnic war.[37]

Admittedly, the growth of the Sri Lankan cinema industry, at the level of production and distribution but also with regard to the conservation of film heritage, continues to be seriously hindered by state indifference, cultural politics, economic privations, and (though to a lesser extent) Sinhala-Tamil antagonisms. Likewise, on the international scene, Sri Lanka continues to remain in the shadow of the gigantic Indian film industry, even though the latter owes its success, at least economically speaking, primarily to commercial entertainers such as the Bollywood movies. Finally, as concerns the production of films on the Sri Lankan ethnic war, it would seem that memorializing and commemorating the conflict through the filmic mode is (also) now a thing of the past. Duty to forget rather than the duty to remember?

NOTES

1 So far only a handful of scholars, without whose works this chapter would not have been possible, have engaged with the question of war in relation to art forms. I would here like to acknowledge the insightful research on war-related Sri Lankan films by Neloufer De Mel, Robert Crusz, and Ian Conrich; historians Michael Roberts, Nira Wickramasinghe, Kumari Jayawardena, and Jayadeva Uyangoda for their work on Sri Lankan ethnicities; more recent studies include S.L. Priyantha Fonseka's innovative PhD research on Sinhala cinema in the war context as well as Annemari de Silva's analysis of the limits of creative expression and censorship in Sri Lanka, published in 2017.

2 The Sri Lankan ethnic war left a trail of disaster at the economic level. In Jaffna (where the LTTE headquarters were located), most cinema halls were either shut or destroyed, like the Regal, which was gutted by fire and shelled in 1985. The North and North-East were particularly hard hit by war conditions, and, for twenty-six years, the public stayed away from potentially dangerous venues, which included cinema halls, for fear of bombs. See Senaka Bandaranayke (chair), Cyril Gunapala, and Gamini Weragama, chapter 4, "The Contemporary Predicament," in *Report of the Presidential Committee on the Rehabilitation and Development of the Film Industry in Sri Lanka* (Colombo: n.p., 15 January 1997), 13–16.

3 Hugh Cleghorn was sent to Sri Lanka (or Ceylon) in 1795 to expel Swiss mercenaries who were protecting the Dutch governor, and he took over the colony in the name of Great Britain.

4 Colonial Office Records, 61, 55, 17 February 1795.

5 These early statistics also included Portuguese and Malays.

6 Nira Wickramasinghe, *Sri Lanka in the Modern Age: A History of Contested Identities* (London: C. Hurst and Co., 2006), 48–9.

7 Eric Meyer, *Sri Lanka entre particularismes et mondialisation* (Paris: La Documentation Française 2001), 32.

8 Darini Rajasingham-Senanayake, "Identity on the Borderline: Modernity, New Ethnicities, and the Unmaking of Multiculturalism in Sri Lanka," in *The Hybrid Island: Culture Crossings and the Invention of Identity in Sri Lanka*, ed. Neluka Silva (Colombo: Social Scientists' Association, 2003), 42.

9 See Vilasnee Tampoe-Hautin, "Cinema in Mainstream and Minority Contexts: From *Samuthayam* (1960) to *Ponmani* (1978): The Plight of Sri Lanka's Tamil Film Industry," in *Carnet de Veille Scientifique et d'actualités sur Sri Lanka et ses diasporas*, ed. Eric Meyer, Delon Madavan and Gabrielle Dequirez, http://slkdiaspo.hypotheses.org.

10 See Vilasnee Tampoe-Hautin, "Of Ethnic Battles, Truces and a National Cinema: The 'Bolly-and-Hollywood Syndrome' as a Catalyst in the Construction of Sri Lankan National Cinema (1930–1990)," in *Asian Cinema* 19, no. 2, ed. I. Conrich and N. Gillet (Philadelphia: Asian Cinema Studies Society, Temple University 2008), 14–34.

11 Vilasnee Tampoe-Hautin, *Ethnicity, Politics and Cinema in Sri Lanka: from Global to "Gama"* (Colombo: Tower Hall Publications, Aitken Spence, 2015), 97.

12 For more on the subject, see Diongu B. Nihalsingha, *Public Enterprise in Film Development: Success and Failure in Sri Lanka*, chap. 9, "The Formula and the Non-Formula Films in Sri Lanka" (Victoria, BC: Trafford Publishing, 2006), 91–9.

13 Neloufer De Mel, *Militarizing Sri Lanka: Popular Culture, Memory and Narrative in the Armed Conflict* (Colombo: Sage, International Centre for Ethnic Studies 2007), 224.

14 See Sunila Abeyesekara, "Imaging the War in the Sinhala Cinema of the 1990s," in *Cinesith* 1, ed. Robert Crusz (Colombo: Asian Film Centre, 2001): 4–13.

15 Henry Rousso, *La Hantise du Passé* (Paris: Textuel, 1998), 16. My translation of "une présence reconstruite ou reconstituée qui s'organise autour d'un écheveau complexe d'images (et de mémoires d'images), de sensations, d'émotions, d'oublis, de dénis, de refoulements et donc leur éventuel retour."

16 Robert Crusz, "Diving Deep: Interview with Prasanna Vithanage," *Cinemaya* 45 (Colombo: Asian Film Centre, 1999), 32–3.

17 De Mel, *Militarizing Sri Lanka*, 221.

18 Wimal Dissanayake, *Profiling Sri Lankan Cinema* (Colombo: Asian Film Centre, 2000), 68.

19 The Vanni, or Wanni, is a term applied to an area in the North of Sri Lanka, roughly covering the districts of Mannar, Killinochchi, Mullaitivu, and Vavuniya.

20 S.L. Priyantha Fonseka, "Silence in Sri Lankan Cinema: From 1990 to 2010" (PhD diss., University of Sydney, 2015), 183.
21 Leonard Woolf lived and worked in Ceylon as a colonial administrator during the first decade of the twentieth century. He left the island in 1907 and later married Virginia Stephen. Woolf's novel *Village in the Jungle* was published in 1913 and was adapted to the screen by Lester James Peries. The film bears the title *Baddegama* (1972). Woolf made his final visit to Sri Lanka in 1961.
22 Robert Crusz, "The Forms of Contentious Content: A reflection on Asoka Handagama's *This is My Moon*," *Cinesith* 1 (Colombo: Asian Film Centre, 2001), 25.
23 Ibid.
24 Thomas Sotinel, "'This Is My Moon' Comedy Strange and Wild on the Devastations Intimate Civil War," *Le Monde*, 11 January 2003, quoted in De Mel, *Militarizing Sri Lanka*, 223.
25 Panini Wijesiriwardana, "No Substitute for Thoughtful Character Development," World Socialist Website, 12 June 2002, quoted in De Mel, *Militarizing Sri Lanka*, 223.
26 Chandima Nissanka, "The New Moon of a Future Cinema," *Sithijaya*, quoted in De Mel, *Militarizing Sri Lanka*, 223.
27 De Mel, *Militarizing Sri Lanka*, 222.
28 Lester James Peries died on 29 April 2018.
29 I am deeply indebted to Prasanna Vithanage and Asoka Handagama for the interviews they granted me in Colombo in August 2014.
30 Supreme Court Judgment, Case 516/2000: p. 2, quoted in De Mel, *Militarizing Sri Lanka*, 233.
31 Ibid., 8, quoted in De Mel, *Militarizing Sri Lanka*, 234.
32 See Crusz, "Contentious Content," 23–32.
33 De Mel, *Militarizing Sri Lanka*, 229.
34 Author's interview with Lester James Peries, Colombo, August 2011.
35 Pushpakumara made his début with a film that sparked controversy in Sri Lanka but received accolades overseas (winning more than fifteen international awards). The young director explores human relationships in a war-ravaged environment and engages with the multiple topics of voyeurism, passion, humiliation, and torture. It draws from his own experience and memories as a child who witnessed the horrors of war and its effects on human behaviour. As Pushpakumara explained during an interview with Inde-Réunion at the Nantes Festival des Trois Continents, shooting the film came as a therapy for and a release from

his psychological traumas, enabling him to "cure" his soul. See https://www.indereunion.net/actu/Sanjeeva/interPushpakumara.htm.

36 Robert Crusz, "Vithanage, Prasanne: 'Facing My Own Contradictions' Interview," *Cinesith* 3 (Colombo: Asian Film Centre, 2003), 49.

37 The three films, each shot separately by Jayasundara, Handagama, and Vithanage, bring together victors and victims of the war (including former militants of the LTTE). They offer multiple, intertwined stories and perspectives, straddling the past and the present. The entire cast of actors, the crew, and the directors themselves (who are also part of the screenplay) play their roles but also share their past experiences, knowledge, and memories of the horrendous Sinhala-Tamil struggle, thus conferring biographical and documentary value to *Her, Him, the Other*.

18

The Spanish-American War on Film: An International Approach

András Lénárt

After Vilasnee Tampoe-Hautin's consideration of the mapping of war histories in an internal, contemporary, Sri-Lankan context, in this chapter I examine multifocal exploitations of a single historical event of crucial, symbolic, and political importance at various times and for diverse purposes by cinemas from three countries directly concerned to: create a favourable national (self)image, reinforce national dogma and internal propaganda, and/or present a dramatic backdrop to purely fictional scenarios.

As a significant international event that involved various countries, the Spanish-American War and its aftermath have appeared in the culture of the three nations involved: the United States of America, Cuba, and Spain. The aim of this chapter is to introduce the general filmic representations of this war in the field of international movies, highlighting some notable examples filmed in/by these three countries, all of whom have made movies dedicated to this topic. Their approach is quite divergent: they deal with the war according to their subjective, often biased, interpretation of the events, in accordance with the effect that this war has had on their nation. Historiography – written usually in Spanish and English – has dealt with this war abundantly, but in the case of the movies only a few works have been published, the most important being Santiago Juan-Navarro's essay,[1] which discusses those Spanish films that focus on this war. This chapter offers an overview of the movies produced by these three nations in which the Spanish-American War appears as a main topic or a subplot, pointing out – in some cases – the political, historical, and social context of the shooting, from the beginning of the twentieth century until the present day.

In 1898, the Spanish-American War was the final stage of the emancipation process engaged in by the Spanish colonies in Latin America. It was a milestone not only from the historical perspective but also in the field of international film culture as various countries' film industries were involved. As cinema was born at the end of the nineteenth century, the Spanish-American War was the first armed conflict that appeared on the screen almost simultaneously with the conflict itself. Since then, all of humanity's wars have been the subjects of movies, but the attention paid by cinema to the Spanish-American War is much less than the attention it paid to other wars – an absence that this chapter seeks to address.

THE CONTEXT: A FEW FACTS ABOUT THE SPANISH-AMERICAN WAR

During the Spanish-American wars of independence (1808–33), the majority of the Spanish possessions gained their self-determination, with only Puerto Rico, Cuba, and the Philippines remaining under Spanish rule. At the end of the nineteenth century, the last colonies began their struggle to achieve independence.

The United States of America entered this armed conflict under the Monroe Doctrine (1823), through which it was determined to prevent European interventions in American affairs, claiming that the Caribbean area had always belonged to its sphere of interest. Indeed, throughout the nineteenth century, the question of the status of Cuba and Puerto Rico was frequently raised among powerful groups in US administrations. As a consequence of the Spanish-American War, the United States managed to extend its influence over the Caribbean and part of the Pacific. By virtue of the Treaty of Paris (1898), Spain sold the Philippines, Puerto Rico, and Guam to the United States for $20 million, and Cuba became independent. These territorial changes had some negative effects, like the Philippine-American War (1899–1902) and the US military government that was established in Cuba and, in compliance with the Platt Amendment, refused to grant the island full sovereignty and guaranteed the USA's right to intervene in Cuban issues. From the European perspective, the outcome of this war marked the end of Spanish power and influence in the region, leading to the ultimate downfall of the Spanish colonial empire. Not surprisingly, in Spain the conflict is often referred to as the "Disaster of 98," a shock that served as a breeding ground for the birth of the "Generation of '98," a group of intellectuals (philosophers, writers, and poets, such as Miguel de Unamuno, Pío Baroja, Azorín, and Ramiro de Maeztu) who, in their essays, poems, and novels, looked for a possible way out of the political and moral crisis induced by the distress of this war. They tried to identify the moral, cultural, and historical reasons that had led to this disaster. One of these authors, Ramiro de Maeztu, was to play a crucial role in the establishment of Fascist ideology in Spain during the 1920s and 1930s. In *Defence of Hispanidad* (*Defensa de la Hispanidad*, 1934),[2] he considers the Spanish race as a highly developed representative of Christian civilization, arguing that it must therefore recover from its traumatic past. Some intellectuals, like Azorín (pseudonym of José Martínez Ruiz) in his novel *The Will* (*La voluntad*, 1902),[3] identify the poverty of Spain's culture and aesthetics as the main reason for its national tragedies.

FILM AND HISTORY

Historical films reflect the mentality of a certain nation as well as its attitude towards the historical event depicted. As with conventional historical sources, it is essential for the researcher (and also for the viewer) to be able to approach the material with critical distance in order to decide whether the audio-visual document has any value. Indeed, the stories told cannot be treated as true reflections of the period, and the events represented in these films must be seen as little more than subjective interpretations. Some historical films exaggerate the importance of certain details or figures, others offer what might be called "a free interpretation of true events," while yet others simply falsify. The figure of Christopher Columbus is a perfect example of all these attitudes as the movies concerning him show that his treatment depends on the nation, the period, and the filmmaker. Ridley Scott's famous *1492: Conquest of Paradise* (1992) portrays Columbus in a way that contradicts preconceptions and tries to adhere to historical facts; the Mexican *The Greatness of America* (*Cristóbal Colón/La grandeza de América*, José Díaz Morales, 1943) modifies some details in order to offer a Latin American viewpoint; while the Spanish *The Dawn of America* (*Alba de América*, Juan de Orduña, 1951), made under the dictatorship of

Francisco Franco, simply falsifies even the best-known facts in an attempt to present Columbus as a Spanish (not Italian) hero who led the Spanish Christian soldiers to Latin America with the purpose of fulfilling God's will.

According to French film historian Marc Ferro,[4] it is impossible to "write" history genuinely by making films. The filmmaker, even if aspiring to be authentic, is forced not only to reinterpret the existing and codified visions of historians but also to add his or her personal judgment. Nevertheless, this does not mean that films cannot help us in mapping the past. The personality, the political views, the knowledge, and the implicit or explicit intentions of the filmmaker, as well as the given social and political circumstances, all have a direct influence on the representation. The director sorts, emphasizes, and conceals some fragments of history; it is therefore necessary to evaluate the films and not to accept them unconditionally. It is also a complex question as to whether representations of history should be presented in such a way that they appear to be true. In *Sociologie du Cinéma*, Pierre Sorlin argues that the screen does not show us how the world is but, rather, what people think/thought of the world in a certain age.[5] This statement is also relevant to traditional historiography. Historical films usually reflect the period when the film was shot and not the one that it represents. A film usually explains the prevailing attitude towards the national or foreign history of a particular society, and it also tells us what the production crew thought it important to show. Thus, historical films teach us about both the past and the present. The bibliography on these issues is quite rich: historians and film historians, such as Ferro, Sorlin, Robert A. Rosenstone, and José María Caparrós Lera, are pioneers in the mapping of the relations between film and history.

THE WAR IN US CINEMA

Screen images of the Spanish-American War were the first major attractions that helped boost the popularity of vaudeville theatres, leading to the false conclusion that they would be the perfect venues for such screenings – something that proved not to be the case. After the end of this over-excitement concerning the conflict, audience enthusiasm decreased and, for a short period of time, it seemed that the Spanish-American War would be the first and also the last such event to attract people's attention in this way.[6]

The first contemporary films about this war can be found on the website of the US Library of Congress, a rich archive whose homepage offers a collection of sixty-eight motion pictures, produced between 1898 and 1901 by the Edison Manufacturing Company and the American Mutoscope and Biograph Company. These "actuality films" typically show armies, war correspondents, parades, re-enactments of battles, and various related events.[7] In some of them, like *Burial of the "Maine" Victims* or *Wreck of the Battleship "Maine,"* the USS *Maine* has a central role. This battleship was destroyed by a mysterious explosion in Havana Harbour in 1898, an event that served as a casus belli and resulted in the United States entering the war. The cause of the explosion remains controversial, but the contemporary yellow press, notably the sensationalist newspapers of William Randolph Hearst (*New York Journal*) and Joseph Pulitzer (*New York World*), accused Spain, and they managed to convince American public opinion of their point of view. Hearst's activity, especially his journalists' and photographers' manipulations of and misinformation about the first events of the Spanish-Cuban conflict, are now regarded as the birth of classic propaganda. Nevertheless, taking advantage of the USS *Maine* tragedy was only the last ingredient of an orchestrated propaganda campaign that eventually led to the declaration of war.

Other short US actuality films deal with military preparations (e.g., *US Cavalry Supplies Unloading at Tampa, Florida*; *Roosevelt's Rough Riders Embarking for Santiago*); the Cuban war itself (*Pack Mules with Ammunition on the Santiago Trail, Cuba*) or its re-enactment (*Raising Old Glory over Morro Castle*); the homecoming and the subsequent parades (*Close View of the "Brooklyn" Naval Parade*); and the Philippine-American War that broke out after the signing the Treaty of Paris (*25th Infantry*). The online collection compiled by the US Library of Congress contains all sixty-eight films, sometimes in an incomplete version, accompanied by short descriptions and extended explanatory essays on the films and also on the historical context.

Some film historians, like M. Paul Holsinger, claim that the first war movie ever made was *Tearing Down the Spanish Flag* in 1898 (J. Stuart Blackton and Albert E. Smith). It was shot on a roof in downtown Manhattan a few days after the US declared war on Spain. We see hands pull down a Spanish flag from a flagpole and then run up a US flag. The symbolism of this patriotic scene was quite explicit as audiences could imagine that those hands belonged to the American nation, which had set the oppressed Caribbean people free. This simple film gained popularity, leading the two filmmakers to shoot other pictures, like *The Battle of Manila Bay* (1898), which was recreated in Blackton's bathtub with the assistance of his family and of Smith,[8] and further short films about various events that occurred during the Spanish-American War, sometimes using original footage. Their company, Vitagraph Studios, was, until the 1920s, one of the most important firms specializing in war films. Various other short films, usually documentaries, depict these events. This quasi-factual approach contrasts with the practice in the majority of American feature films of the postwar years, where the war serves as a background story or appears in fragments and flashbacks, leaving the main plot to focus on a romantic relationship between a soldier and a nurse, a combatant and his beloved on the home front, and so on. *Across the Pacific* (Roy del Ruth, 1926), a lost romantic silent film, is typical, for it centres on an American soldier who joins the US Army in the Philippines and gets involved in a love triangle between his sweetheart at home and a half-caste woman.[9]

A much more enduring appeal in the field of war films was achieved by the famous "Rough Riders" (the First United States Volunteer Cavalry), who, under the command of Theodore Roosevelt, were the subject of various films throughout the twentieth century. Their participation in the Spanish-American War, especially in the famous Battle of San Juan Hill (1 July 1898), brought them enormous popularity on the front pages of newspapers and beyond. The life of the former assistant secretary of the navy, who later became the twenty-sixth president of the United States, intrigued Hollywood. *The Fighting Roosevelts* (William Nigh, 1919), based on a screenplay authorized by the president himself, follows Roosevelt's life from his infancy until his arrival in the White House. His involvement in the Spanish-American War also appears in this dramatization.

One of the most revealing films concerning Roosevelt's military exploits is Paramount's *Rough Riders* (aka *The Trumpet Call*), directed by Victor Fleming in 1927. The script's historical advisor was Hermann Hagedorn, the secretary of the Theodore Roosevelt Association, and the US Army put twelve hundred soldiers at the disposal of the future director of *The Wizard of Oz* (1939) and co-director of *Gone with the Wind* (1939). The film was shot in San Antonio, in parallel with William Wellman's *Wings* (1927). The entire movie has not survived, but fragments may be located at the Library of Congress and at the Museum of Modern Art in New York. Besides the war preparations and the actual battles, this film also contains a small romantic plot: two of the soldiers are in love with the same woman. The action sequences reflect the director's perfect planning, including the famous charge of the Rough Riders in the Battle of San Juan Hill.[10] For both the studio and the US government, this movie was intended to represent an official account of the events and, in particular, of the role of Theodore Roosevelt, hence the special attention paid to it. After a nineteen-minute biopic, *Teddy, the Rough Rider* (Ray Enright, 1941), which won an Academy Award for Best Short Subject, the same topic reappeared in 1997 when John Milius directed a two-part miniseries under the well-known title *Rough Riders*, originally aired on the TNT cable channel. Milius and TNT's media mogul Ted Turner launched this project together. As Milius had always idolized Roosevelt and Turner was quite interested in dramatizing American military history, this plan seemed to be perfect for them. Furthermore, they found a justification in the present: according to some sources, it was a "general groundswell of right-wing frustration over President Bill Clinton's perceived misuse of the armed forces, especially after the disastrous Battle of Mogadishu on October 3, 1993, that drove the United States out of Somalia."[11] Theodore Roosevelt's memoir, *The Rough Riders* (1899), served as a primary source for the screenplay, and the crew paid homage to Roosevelt's militarism and masculinity. As Robert Niemi points out, in concert with the president's memoir (but in contrast to the truth), the film describes the US at the end of the nineteenth century as a country in which Anglos (both North and South), Native Americans, Afro-Americans, and Hispanics cooperated with each other for the sake of a common purpose. In this ideal world, no racial, social, and/or class tensions could be detected.[12] It also presents a number of historical inaccuracies as it carefully avoids mentioning such things as the passing of the Platt Amendment. These war films transformed Roosevelt and the Rough Riders into iconic figures of American history, even creating details and events that had nothing to do with historical fact. They made use of the representations that constituted the US "iconography" that evolved around American values and achievements, resulting in national symbols. The Rough Riders – mainly due to literature and cinema – became emblematic national heroes and their portrayal set the standards for the depiction of American military personnel and conflicts in future war films.

War has always proved to be a popular topic for the film industry, even when, as is often the case, it is no more than a pretext for entirely fictional heroic action. Both real-life and fictionalized characters appear in George Marshall's *A Message to García* (1936), inspired by Elbert Hubbard's essay written under the same title in 1899. According to the plot, President William McKinley commissions the American agent Andrew Summers Rowan to deliver a message to Calixto García, one of the leading figures in the Cuban fight for independence. Cuban patriots, an American marine, and a British merchant help him get to his destination, while Spanish troops and a mercenary try to prevent him from completing his mission. After lots of hardships and struggle, the Cuban insurgents come to the rescue of Rowan, so he manages to get the message to García. Although the film has a loose connection with real events (a message sent by the president to García), the circumstances are clearly fictitious, with the explicit intention being to galvanize the story. The same war serves as a background for *Santiago* (Gordon Douglas, 1956), an adventure film that relates the story of a gunrunner (played by Alan Ladd) who is hired to smuggle guns and ammunition to Cuba in order to aid the rebels on the island against the Spanish Army. Compiled with modest historical concern, the plot centres around the adrenalin-packed adventures of the main character, making use of Ladd's popularity during the 1940s and early 1950s. Among the historical inaccuracies, one is quite conspicuous: Cuba's national hero, the political theorist and poet José Martí, appears in the film, despite the fact that he died in 1895, three years before the Spanish-American War broke out. Some films deliberately insert anachronisms, like Alex Cox's acid western about the American filibuster *Walker* (1987), in order to point out historical absurdities or provocative similarities between different ages. However, in *Santiago*, Martí's appearance seems to have no explicable purpose apart from the desire to insert a widely known Cuban figure into the plot. In other US productions, like *Pursued* (Raoul Walsh, 1947) or *Stars and Stripes Forever* (Henry Koster, 1952), the Spanish-American War is also present, but usually as part of a secondary storyline.

THE WAR IN SPANISH CINEMA

Spanish cinema has paid little attention to the Spanish-American War. This conflict made its appearance in the cinematic world, especially during the dictatorship of General Francisco Franco (1939–75), but only to reinforce the relevant ideological goal of the official guiding principles. Francoist film policy – similar to that of other authoritarian regimes – exercised strict control over filmmakers and determined what topics were to be depicted. Propaganda was a core element of the Spanish films of the period, a key aspect of those that dealt with national historical events or figures.[13] This marks a fundamental difference between the film policy of a dictatorial country like Franco's Spain and that of a democratic country like the United States. In the latter case, box office expectations and results are of the utmost importance, with political and ideological considerations playing a major role only in cases of strained international relations (e.g., between and during the two world wars and the Cold War).

The most substantial and powerful result of the Spanish dictatorship's film policy was *Race* (*Raza*, José Luis Sáenz de Heredia, 1942), widely regarded as the essential Francoist movie. The story of the film presents the loyal incarnation of an ideological collage of the regime and, at the same time, the idea of *Hispanidad*,[14] the very essence of the Hispanic race. The main characters are the members of the paradigmatic Francoist family: the father is a hero who sacrifices his life for his country and his principles in the Spanish-American War; the mother holds the family together and is respected as a saint; while the young ones dedicate themselves to their military, political, and ecclesiastical vocations. Don Pedro Churruca, the head of the family,[15] loses his life in a heroic battle in Cuba during the war of 1898,[16] and this disaster is the origin of his children's fight to survive Spain's forty years of struggle. The film demonstrates that the Spanish-American War was the tragedy of the whole nation but makes it clear that those Spaniards who participated in it were are all martyrs; moreover, the Spanish Civil War (1936–39) is viewed as a re-enactment of the Cuban war – a battle between good and evil but on different soil. In the original novel,[17] written by General Franco himself under the pseudonym "Jaime de Andrade," the author roughly sketches the events, mainly emphasizing the conversations; however, with regard to its origins and plot, *Race* can be regarded as the idealized autobiography of Franco. The last name "Andrade" alludes to one of Franco's ancestors,[18] and all the primary components of the story have something to do with the general's past, his family, and his moral precepts. The main dialogues consist of explicit ideological and propagandistic messages that praise patriotism, the Spanish nation, the Hispanic race, and the supremacy of Catholicism and that constantly belittle the enemy. In short, the original novel is a dramatized political pamphlet and the film is its visual adaptation. Considering the most important propaganda films in the world, we may affirm that, given its declared purpose and unanimous messaging, *Race* is one of the best films a dictatorship has have ever made.

Three years later, the same director (José Luis Sáenz de Heredia) used the Spanish-American War as a kind of "decorative element" for his movie *Bambú*, a musical melodrama centred around the singer and actress Imperio Argentina, a distinguished and popular star of the period. According to the (fictional) story, a Spanish volunteer is posted to Cuba, where he falls in love with a local girl – Bambú – portrayed according to widespread stereotypes of the Cuban people. The film lacks explicit political references because one of the most important producers of the 1940s and 1950s, Cesáreo González, wanted to distribute it on both sides of the Atlantic Ocean. This meant that it was crucial to avoid topics that were likely to offend the sensitivities of Latin American audiences.[19] In fact, of course, this producer's decision fitted perfectly with the Franco's ideology (i.e., *Hispanidad*), according to which there was a strong and unbreakable bond between the peoples of the Hispanic race, the Spanish nation, and Latin American societies.

Another battle served as the central plot of *Last Stand in the Philippines* (*Los últimos de Filipinas*, Antonio Román, 1945), the most important Spanish film to deal with the events of the Spanish-American War. It tells the fact-based story (although with fictitious episodes and characters) of "The Last Ones of the Philippines" – the Spanish troops who fought in the Siege of Baler against the supporters of independence. The narrator of the film provides patriotic comments that emphasize the Spanish soldiers' moral and physical perseverance in this dreadful situation. This type of narration, spoken by an authoritative male voice, adds ideologically driven nationalist and patriotic verbal propaganda to the moving images and was a standard way for Francoist filmmakers to influence the public. According to the plot, the Spanish soldiers stationed in the Philippines formed a harmonious relationship with the local people – one of them even maintains a romantic affair with one of the Aboriginal women. When the revolt breaks out, sixty soldiers take shelter in a church, which, from that moment, is treated as a fortress, both by the soldiers and the insurgents. The narrator states that signing the Treaty of Paris was one of the most distressing moments in the nation's history. However, the soldiers in the church do not believe that such a treaty was signed: they cannot accept that Spain would have approved such a humiliating and shameful agreement. They hold out in their fortress for eleven months (337 days), disregarding the orders that come from Spanish headquarters, considering them to be false and misleading. Finally, after having read articles from Spanish newspapers, they resign themselves to the harrowing reality. When they return to Spain, they are praised for their staggering patriotism. Román's film perfectly fits the Francoist interpretation of national history: the Spanish politicians who signed the Treaty of Paris sold out and betrayed their nation; only the firm resistance of this small group of soldiers, ready to sacrifice their lives for the sake of their country, is worthy of admiration. According to this explanation, the Spanish armed forces are the supreme embodiment of the motherland's virtues. In nationalist and conservative circles, the struggle of the heroes of Baler was regarded as one of the most heroic and glorious episodes in the nation's history.[20] Although the Spanish-American War is the historical background to these events, Spain's enemies in this film are the native Tagalogs, and the United States is judged favourably (an American battleship even assists the Spanish soldiers). This is a classic example of how past (war) history is rewritten to serve a present purpose: after the Second World War, the Spanish government wanted to present Spain as democratic, and for this it needed the USA's support and recognition.

THE WAR IN CUBAN CINEMA

One of the negative consequences of the Spanish-American War for Cuba was the Platt Amendment, which became part of the Cuban Constitution in 1901. It laid down seven conditions for the withdrawal of US troops from the island, meaning that, while Cuba formally gained independence, its people felt betrayed, and, throughout the twentieth century, the US's frequent interference in Cuban affairs, justified by the amendment, set the seal on the deteriorating relations between the two sides. Although the majority of the

Platt Amendment's provisions were revoked in 1934 by virtue of the Cuban-American Treaty of Relations, within the scope of Franklin D. Roosevelt's Good Neighbor policy, relations worsened again during the Cold War. Since the end of 2014, however, a new period of rapprochement has opened up between the two countries.

In Francisco Franco's Spain, so in Fidel Castro's Cuba the national film industry's main task was to assist and promote the regime's ideology. The Cuban Institute of Cinematographic Art and Industry (ICAIC) oversaw the accomplishment of this task. The Ten Years' War (1868–78) and the uprising that began in 1895 received special attention from filmmakers, with the explicit purpose of exploiting these events as means of producing political propaganda. As with Franco's Spain, Castro's Cuba strove to establish direct and unambiguous connections between the experiences of the remote past (the wars portrayed in films), the recent past (the triumphant revolution in 1959), and the present (Fidel Castro's Cuba). The notion of independence and the desire to be free to manage internal affairs without foreign intrusion are constant elements of Cuban history and, consequently, of Cuban cinema.

For Fidel Castro, the Cuban War of Independence did not begin with the events of 1895 but with the insurrection of Carlos Manuel de Céspedes in 1868. The Cuban national hero José Martí is a central figure in various films.[21] In 1969, Manuel Octavio Gómez shot one of the most important Cuban films, *The First Charge of the Machete* (*La primera carga al machete*), which depicts the battle between Cuban and Spanish troops for the town of Bayamo. Like others mentioned, this movie also builds an imaginary bridge between different historical periods in order to establish generalizations and make apt allusions. The Ten Years' War (1868–78), the so-called Little War (1879–1980),[22] the Cuban War of Independence (1898), the struggles against Gerardo Machado (1925–33) and Fulgencio Batista (1940–44; 1952–59), Fidel Castro's revolution of 1959, and the permanent resistance to the United States's constant interventions all appear as part of a long Cuban fight for national liberation and sovereignty.[23] Gomez's film garnered outstanding international success not only in Latin America but also at the Venice Film Festival, and it still ranks among the most significant works in Cuban film history.

Lucía (Humberto Solás, 1968), a substantial cinematic triptych, features what were to be regarded as the most significant episodes in Cuban history (in 1895, the war against Spain; in 1932, the fight against Gerardo Machado; in the 1960s, the literacy campaign after the victory of Fidel Castro's revolution). Across all these historical periods a woman named Lucía appears, and she becomes the symbol of the Cuban nation's development, from the war of independence until the 1960s. According to Santiago Juan-Navarro, "the lack of specificity in the date of the final episode, marked by a question mark, puts the trilogy within the eschatological vision of socialism: the conception of history as a project that has its ultimate climax in the construction of an ideal society in an indeterminate future."[24] He also points out that the characters of this film follow the guidelines for protagonists proposed by the Hungarian philosopher and literary historian György Lukács in his theory of the novel and the historical drama (in *The Historical Novel*,[25] Lukács writes about anonymous beings who embody the historical consciousness of a nation during a period of change).

The passage of time, along with Spain's return to democratic rule and the gradual softening of the Cuban regime, have slowly allowed the Spanish-American War to lose the propaganda value that it once had for both countries. In turn, this has permitted filmmakers to lift the emotional weight attached to the subject and to treat what remains a major historical event with a somewhat lighter touch. The events in the Philippines are evoked in two episodes ("Time of the Brave," part 1 and part 2) of the second season of the popular Spanish television series *The Ministry of Time* (*El ministerio del tiempo*, 2015–) set during the siege of Baler, while a remake of Antonio Román's film *1898: Los últimos de Filipinas*, directed by Salvador Calvo, was released at the end of 2016 with a shift in focus from nationalism to adventurism. In 1997, two brothers, Teodoro and Santiago Ríos, directed *Mambí*, a co-production between Cuba and Spain, and here, again, a significant change in emphasis can be observed. The main character is Goyo, a young Spanish day labourer from the Canary Islands, who takes part in the Spanish-Cuban conflict but who falls in love with a Cuban woman, a mambí,[26] during the hostilities and changes sides. Spanish troops imprison him but he manages to escape when the USS *Maine* is sunk by an explosion. The film premiered in 1998, on the occasion of the hundredth anniversary of the end of the Spanish-American War, but, in a sense, this was a lost opportunity in the relations between the two countries and their respective film industries: the quite simple story and narrow budget prevented the Ríos brothers' film from becoming the pre-eminent audio-visual representation of the independence of Cuba.

CONCLUSION

The Spanish-American War, as the first war that was filmed on site by the (generally American) correspondents and filmmakers, was a milestone in the relations between film and history. The filmed sequences of this war are among the first audio-visual sources of information pertaining to it, and even historians use them in their research projects. We can affirm that history has become an inseparable part of cinema as scenes of this conflict were projected in vaudeville theatres. In the hostilities of the twentieth century, films played a crucial role in documenting, propagandizing, and even entertaining in both local and world wars. In the past 115 years, albeit from different perspectives, the Spanish-American War has appeared in various American, Spanish, and Cuban films. Thus, this war serves as a perfect example of how different national film industries handle the same historical event from completely different points of view. This provides us with indispensable information not only about the past (the specific historical event) but also about the present attitudes of any given nation. In Rosenstone's terminology,[27] these works are a revision of the past, offering a new interpretation, a new language, and a new conception of how the past is important to the present.

NOTES

1 Santiago Juan-Navarro, "Las guerras de independencia en las cinematografías de Cuba y España: una visión transatlántica," *Circunloquios: Revista de investigaciones culturales* 1 (2014): 15–31.
2 Ramiro de Maeztu, *Defensa de la Hispanidad* (Madrid, 1934), http://guardiadelahispanidad.files.wordpress.com/2009/09/defensa-de-la-hispanidad.pdf.
3 José Martínez Ruiz, *La voluntad* (Madrid: Cátedra, 1997).
4 Marc Ferro, *Cinema and History* (Detroit: Wayne University Press, 1988), 162–3.
5 Pierre Sorlin, *Sociologie de Cinéma* (Paris: Aubier Montaigne, 1977), 33.
6 Douglas Gomery, *Shared Pleasures. A History of Movie Presentation in the United States* (Madison: University of Wisconsin Press, 1992), 15–17.
7 "The Spanish-American War in Motion Pictures," Library of Congress Digital Collections, https://www.loc.gov/collections/spanish-american-war-in-motion-pictures/.

8 M. Paul Holsinger, *War and American Popular Culture: A Historical Encyclopaedia* (London: Greenwood Press, 1999), 190–1.
9 Kenneth White Munden, ed., *The American Film Institute Catalog of Motion Pictures Produced in the United States: Feature Films, 1921–1930* (Berkeley: University of California Press, 1997), 4–5.
10 Michael Sragow, *Victor Fleming. An American Movie Master* (Lexington: University Press of Kentucky, 2013), 119–22.
11 Robert Niemi, *History in the Media: Film and Television* (Santa Barbara: ABC-Clio, 2006), 39.
12 Ibid., 40–1.
13 For more details on the Francoist film policy, see András Lénárt, "Ideology and Film in General Francisco Franco's Spain," Öt kontinens 2 (2013): 323–36, https://edit.elte.hu/xmlui/handle/10831/22150?locale-attribute=en.
14 The term "Hispanidad" includes not only the Spanish nation but also the peoples of Latin America. They share the same past, traditions, culture, language, and blood. According to the Francoist regime, the Hispanic race is superior to those of other nations.
15 The name of this character – Cosme Damián Churruca – evokes the Spanish hero of the Battle of Trafalgar.
16 According to Hungarian historian and Hispanist Iván Harsányi: "Franco carved a father for himself who was to his own taste." See Mária Ormos and Iván Harsányi, *Mussolini – Franco* (Budapest: Pannonia, 2001), 272.
17 Jaime de Andrade, *Raza: Anecdotario para el guión de una película* (Madrid: Ediciones Numancia, 1942).
18 Román Gubern, *Raza: Un ensueño del general Franco* (Madrid: Ediciones 99, 1977), 12.
19 Juan-Navarro, "Las guerras," 17–18.
20 José Javier Esparza, "Los últimos de Filipinas: Los héroes de Baler," *El Manifiesto*, 24 June 2008, http://www.elmanifiesto.com/articulos.asp?idarticulo=2421.
21 But as they are not what could be called war films, they have no place in this chapter.
22 The Little War was the second armed conflict between Spain and Cuba, and was the immediate sequel to the Ten Years' War.
23 See the detailed analysis of the film in Santiago Juan-Navarro, "La primera carga al machete de Manuel Octavio Gómez: cine, mito y revolución," in *Cinéma et Révolution cubaine*, ed. Julie Amiot and Nancy Berthier (Lyon: Université Lyon 2, 2006), 105–13.

24 Juan-Navarro, "Las guerras," 22.
25 György Lukács, *The Historical Novel* (London: Merlin, 1962 [originally published in Russian in 1937]).
26 The term "mambí" refers to the Cuban guerrillas who fought against Spain between 1868 and 1898.
27 Quoted by Robert Burgoyne, "Introduction: Re-enactment and Imagination in the Historical Film," in *Leidschrift. Verleden in beeld. Geschiedenis en mythe in film* 24–3 (2009): 14.

19

Wings (William Wellman, 1927) and *Broken Lullaby* (Ernst Lubitsch, 1932): The Psychological Drama of Memory and the Modern Pacifist Narrative

Raphaëlle Costa de Beauregard

By reflecting on the complexities of remembering, forgetting, and coping with trauma in movies produced in the immediate aftermath of the Great War, this chapter examines films that offer a more contemporary perspective, focusing on the psychological drama of memory and the anti-hero narrative. Films dealing with war memory in the aftermath of the First World War are a restricted corpus, but they share a set of striking similarities, ranging from a history of nationalism in Europe to a history of the emergence of a new ideal of civilization. This "war to end all wars" provided public opinion with a new frame of mind that entailed a shift in moral values – that is, from one's duty to kill the enemies of one's nation to one's duty to remember a "lost generation," as the survivors called themselves. The following debate focuses on some dichotomies that inform memory narratives devoted to the reinvention of peace in post-First World War ideology – that is, tales of forgetting versus memorizing, of collective memory versus individual memory, as well as the trauma narrative versus the narrative of survival. Although this chapter focuses primarily on *Wings* (1927), directed by William Wellman,[1] and *Broken Lullaby*,[2] directed by Ernst Lubitsch (1932), this does not preclude references to other contemporary films on the same subject, notably King Vidor's *The Big Parade* (1925)[3] and Lewis Milestone's *All Quiet on the Western Front* (1929).[4] These four post-1918 films share a common characteristic: they are a pacifist reflection upon the tragic events of the First World War in Europe.[5]

REMEMBERING VERSUS FORGETTING

Remembering as a narrative paradigm is central to these initial reflections on *Wings*. Death is given a lurking presence early in William Wellman's film when immediate memory changes the images of the world surrounding the two American friends Jack and David. As youthful cadets training for future combat flights over the front line, they meet a more experienced pilot, Cadet White (Gary Cooper), who is killed in a loud crash almost immediately afterwards (VHS 28:14). As William Wellman Jr tells us from conversations with his father, the film director, the choice of Gary Cooper was made on the basis of a special character effect that was to be created: "Cadet White would have to be able to make an immediate and lasting impression. His only help would come from the onscreen words: 'I've got to go up and do a flock of eights before chow' and 'Luck or no luck, when your time comes, you're going to get it.'"[6] Once he has been killed, the two friends are asked to collect "his things." With the news of his crash, the objects lying around take on a new significance: they are framed in close-shots, the change in scale extracting them from the general overview and turning them into a collection of intimate mementos – a family photograph cohabiting with a pair of shoes and slippers, a scarf, and a chocolate bar, which Jack refrains from touching, remembering White had fingered it before leaving (VHS 32:09). As they are soon busy with actual encounters with the enemy, the two friends often reminisce about the dead pilot's declaration that he despised the superstitious habit of talismans, while we have been shown several close-shots of David's own teddy-bear (VHS 16:30). White's story – literally a memento mori (remember you must die) – becomes written on other things around them, such as the Red Cross van with its label "crash-tools," while the order "report to the dead-line" now has an ominous ring. Because the close-shots suggest the characters' gaze upon them, these images have become imprinted in the memory of the viewer, who shares Jack's and David's reminiscences of the past and fears for the future.

As an antithesis to such memories, *Wings* then invites us to share a moment of peace and forgetfulness during Jack's leave in Paris. The camera focuses on Jack spending a night at the Folies Bergère and playing gaily, producing bubbles from a glass of champagne (VHS 1:09:60). His escape into a world of forgetfulness materializes in these bubbles as they quickly become imaginary ones that he believes he sees all around him. However, forgetting becomes an impossible task once he kills his friend David by mistake (VHS 2:07:57). Believing David to have been shot down by a German pilot – who has dropped a deceitful message saying as much – Jack seeks revenge by attacking a solitary German plane. Unfortunately, this very plane has been stolen from the enemy by David in an attempt to return to his US camp. This impossibility of forgetting is dramatized by scenes taking place upon his return to his hometown once the war is over. As David's grieving parents watch Jack's triumphal parade along the main street from their window, we are shown the gulf that David's death has created between two visions of the reality of war: a public memorial celebration and private grieving are visualized within the same screen by the in-depth editing of the two scenes (VHS 2:12:05). A second element is added to this memorial ritual, as Jack then calls upon David's parents and gives them two mementos from their dead son: his Military Cross awarded by the grateful French and his small teddy bear. Simultaneously, the emblem of Jack's youthful infatuation with speed and adventure, the shooting star painted on his car by Mary (Clara Bow), is now bereft of its past glory (VHS 2:15:51).

In *Wings*, forgetting is thus opposed to memorizing in a series of scenes that gradually lead to Jack's understanding that war cannot, and should not, be forgotten since it has turned his image of the heroism of aerial "dog-fights" with enemies who believe in the laws of chivalry – Jack's life is spared by the German pilot Kellermann who realizes his opponent's machine gun is jammed (VHS 0:45:29) – into a nightmare of unintended murder. Jack is now compelled to reshape his romantic dream into a view of war as an error leading to fratricide.

When we turn to *Broken Lullaby* remembering and forgetting are immediately evoked. Indeed, the film opens with a military ceremony celebrating victory in Notre-Dame Cathedral in Paris and calling for forgetfulness (DVD 01:11). We then discover a young Frenchman, Paul, who is not wearing a uniform and seeks help from a priest, wishing to confess that "he has killed a man" in breach of the Fifth Commandment "thou shalt not kill" (DVD 03:50). They disagree about what a man's duty is: for the priest, it is to kill the enemies of the nation, while for Paul this cannot be. Official duty thus clashes with the conscientious objector's own sense of duty. The two approaches to the war experience introduce the theme of memory with acute intensity since Paul's mind is haunted by a picture engraved in his memory that he cannot dismiss. He has a visual hallucination of his dying foe's uncomprehending gaze upon him – an unbearable experience of eye-contact (DVD 06:49).

The scene then shifts to a parallel time sequence showing Walter Hölderlin's grieving family in a small German town (DVD 13:43). Here too people are suffering from the impossibility of forgetting. Upon Paul's arrival in the German town at the end of the war, ritual is the form in which the duty never to forget is expressed. On the one hand, grieving fathers meet at an inn and drink to their dead sons' memory, promising revenge (DVD 42:30), and, on the other hand, grieving mothers lay flowers on their dead sons' graves (DVD 27:07–27:13). Paul shares in their mourning by bringing flowers too. The film thus clearly opposes the need for forgetting and its impossibility, translating it into two opposite desires: revenge versus mercy.

Several narrative forms related to the issue of memory are then developed in the ensuing episodes. Paul has come to tell the true story of his actual meeting with the Hölderlins' son, whom he killed, as a catharsis of his guilt complex. The memory he has of this encounter is a full narrative in its own right, which the film shows through a flashback. During their tragic confrontation, Paul had read in Walter's letter to his fiancée Elsa his secret doubts about the war and his love for the French. Paul had also realized they shared a common love of music. Consequently, once he confronts the family's emotional response, he allows himself to invent a story of friendship in which he embodies a fictional character for himself, that of Walter's music student friend in Paris (DVD 33:57). The film then introduces a different but complementary memory paradigm, that of Walter's letters kept by Elsa (DVD 1:00:10). As she selects the last one that she received and begins to read it aloud, Paul cannot refrain from saying aloud its last lines, which he has memorized, thus betraying his presence at the moment of Walter's death. The letter brings the dead man back to life, so to speak, reactivating Paul's neurosis and making him at last confess the truth to Elsa. But it is his truth, "I killed a man," that he confesses, not the official confession, "war is war."[7]

A final narrative form is introduced in Lubitsch's film when Walter's father offers Paul his son's violin (DVD 1:07:56). The Frenchman then plays Robert Schumann's "Träumerei" (Dreaming) – on the violin,[8] and Elsa sits at the piano to play an accompaniment, while Walter's parents close their eyes as if in a vision that their son has returned alive for, when Paul is given Walter's violin, he is called "my son" by the old man. For Schumann, the piece meant a desire for fatherhood, but for Walter's parents it means a return to parenthood, while for Elsa – and Paul, who complies with her wish – it means their intention never to forget Walter (DVD 1:10:00) and to keep him alive in their memory by their shared secret, the truth about his death at the hands of a Frenchman – Paul.

Memory provides a variety of psychological drama narratives in both films, all focusing on individual memory. However, while the new duty of remembering introduced by these narratives originates in individual memory, it also appears as the product of a more general narrative, which is told in terms of collective memory.

COLLECTIVE MEMORY

Collective memory is a concept developed by Maurice Halbwachs,[9] who saw memory as an agent of individual integration into society. From this point of view, the psychological function of memory lies in the reconstruction of a reliable identity image in a social group by an outsider. As to the state of alienation itself, it is due to a deterioration of the individual's social connection with others occasioned by a traumatic past. One finds an explicit example of this narrative pattern in Tourneur's film noir *Out of the Past* (1947), when the hero, who has succeeded in starting a new life in a distant city, is caught up by his past involvement with gangsters.[10]

In *Broken Lullaby*, Paul's tale about his pseudo-friendship with Walter in Paris can be interpreted as an attempt to integrate postwar German society. By borrowing a false identity that is grounded in an imitation of what he has learned about Walter from reading his last letter (DVD 05:56–07:00), Paul creates an acceptable role for himself in the family group – that of the dead man's friend. The fact that the rest of the population of the town is scandalized by his intimacy with Walter's family underscores the collective dimension of memory and the tension caused by his being an outsider. The choice of Schumann's piano piece "Träumerei" (Dreaming), to conclude the film and elucidate its title, *Broken Lullaby*, also confirms the collective nature of memory.

Music, as Halbwachs explains, is not only a semiotic system of codes and conventions but also a vehicle of memory for the listener. The example of piano music played by a violin – which is the case in Lubitsch's film – can be analyzed as an instance of memory at work:

> Where is the model which we recognize, if we do at all? It must be located in our memory and within the sound space. In our mind, as a disposition which we have acquired in our past to reproduce what we heard, but an incomplete disposition because we would have been unable to reproduce it. But the sounds now heard come forward and join the reproduction being performed.[11]

Though there is a great difference in the acoustic quality of the fluid continuity on the piano produced by the pedal sustaining the vibration of the strings and the essential fluidity of the violin's notes woven together by the musician's bow, we know the melody for having heard it in the past. In the film, the performance catches the attention of the characters, who show that they can identify it perfectly despite its being a transposition, as demonstrated by Elsa's sitting at the piano to play an accompaniment. The appeal to collective memory is dramatized by the performance since the experiences of playing and listening both involve memories of a common cultural artefact from the past. What is more, there is an emphasis on the present time of the shared moment, a merging of the temporality of the characters in the fiction and our own temporality as audio-viewers. The framing of Walter's parents sitting on a sofa symbolizes our fictive presence within the frame.

The narrative pattern of collective memory is therefore all the more significant as it sustains the dominant pacifist ideology of the film – that is, hatred being cured by love. When Elsa is told the truth by Paul, she first suffers from a fit of hatred, but she also wishes to protect her former prospective parents-in-law from having to endure this terrible truth. She accordingly decides that Paul will have to stay and live with them as the only way to atone for the killing, thus suggesting the necessary transformation of the passion of hatred into the passion of love.

In *Wings*, the individual experience of war also has a collective significance. Throughout the film, killing enemies is associated with collective recognition, as when Jack, David, and a third pilot are rewarded by a French officer for their military action by a Military Cross and the title of "aces."[12] The film also deals with collective memory in a long tracking-forward of the camera (VHS 1:08:42–1:09:60) between the tables of the Folies Bergère customers, framing each couple in a short pause before proceeding further on. We thus have portraits of different couples, such as two women clearly in love with one another, two Frenchmen in uniform entertaining women at two separate tables, until the camera finally singles out Jack with another French woman, blowing bubbles and enjoying himself. The unity of the sequence that the camerawork provides by the tracking forward shows that the effort to forget the war is a collective enterprise that Jack shares with other customers. Later, back at the front, waves of soldiers following one another in a general attack by the American Army against enemy lines are seen in long shots from a high angle, which emphasizes the collective nature of the action (VHS 1:45:23–1:46:50), even though Jack is also shown – in cross-cutting – hatching his secret plan to avenge David's death (VHS 2:10:52). Later still, when David's plane crashes into a farmhouse roof, French soldiers crowd into the farm and help carry David's injured body in a mise-en-scène of collective concern that is also relevant to the paradigm of collective memory. Significantly, in an earlier sequence, as David himself brings down a German plane, the dead pilot's body is recovered by his comrades and the camera lingers on the collective carrying away of his body (VHS 1:57:54). A careful viewer cannot miss the parallel between the two scenes, which provides another clue to the overall ideology of the film's condemnation of war. Finally, when the film focuses on Jack's return home, a street parade (VHS 2:11:00) emphasizes the collective significance of his having volunteered to join the French forces. He is seated in his plane, decorated with flowers on a triumphal chariot as the camera frames the cheering crowd.

According to Halbwachs, habit is yet another effect of the collective nature of memory, which has to do with education and custom and gives shape to the narrative of repetition already observed in the music scene. In *Broken Lullaby*, we see Walter's father Dr H. Hölderlin (Lionel Barrymore) having grown into the habit of hating the French although he says he has never actually seen any. He stares at Paul from behind his thick glasses in astonishment. The idea of an automatism is conveyed by conventional pantomime when Walter's father shows Paul the door and shouts, "Get out" (DVD 30:01–31:56). Walter's father initially shares this hatred for the French with the men who gather at the inn as a common act of mourning for their lost sons. However, having accepted Paul as a substitute for Walter, the Doctor analyzes such habits differently, rebuking fathers for the decision to send their sons to fight for them (DVD 48:41–50:06). Habit and collective memory are also tangible in the graveyard scene as Paul is discovered by Elsa putting flowers on Walter's grave, an action that is commented upon by the gravedigger. Habit is thus deconstructed and reversed by this unexpected gesture from an enemy: the hatred for the invader instilled by nationalist propaganda (the dialogue refers to Alsace-Lorraine, but from the German point of view [i.e., complaining their soil is now occupied by the enemy]) is unexpectedly effaced by Paul's gesture. Collective memory is shown in *Broken Lullaby* to be a matter of repetition, but the film also shows how an imaginative individual, such as Walter's father, is able to give up old habits and adopt new customs, such as his wife's mercy for the young.

But collective memory is necessarily dependent on a collection of private individual experiences, as the above study of forgetting and memorizing in the two films has shown. And individual memory is made perceptible to the audience by subjective camerawork, such as the main protagonist (an intradiegetic focalizer) gazing at a shocking scene just as we are doing so. The use of such a subjective camera has already been noted when Cadet White's "things" are framed in close-shots. However, the device is used for one particular visual motif that is found in the two films of our corpus as well as two others, as will now be discussed, since it is one of the mainsprings of the psychological war-film drama as a film genre.

MEMORY AS CONSCIOUSNESS SHARED BETWEEN PROTAGONIST AND AUDIENCE: THE CAIN/ABEL MOTIF[13]

Looking at these films as psychological dramas entails a study of the memory narrative within the paradigm of audience reception. As early as 1916, and therefore already of some use to Wellman's 1927 silent film – or rather to the earlier Vidor film *The Big Parade* (1925), which is the model of war iconography for the ensuing films – a study of audience psychology and film reception had been published by Hugo Münsterberg, *The Photoplay: A Psychological Study*.[14] In this treatise, the German specialist in early cognitivist sciences selects four main psychological functions of film viewers' reception – attention, memory, imagination, and emotion – all of which involve the audience's consciousness of experiencing the present, which also implies recalling each one's private history.[15] The subjective shot that enables us to share the character's gaze is a vehicle for the empowerment of these functions of film reception. Consequently, the emphasis on a significant visual motif by this device is a key element in the conversion of the viewer to a new ideology.

A specific image is found in both *Wings* and *Broken Lullaby*,[16] which I call, for the sake of argument, the visual Cain/Abel motif – that is, the moment of direct confrontation of the main protagonist with the man he killed, Jack and Paul having killed, respectively, David and Walter. This image is characterized by the summoning of the protagonists' whole attention, conquering their memory and imagination as they identify their victim as their "brother," these cognitive functions being accompanied by powerful emotions in which the shock of witnessing death is predominant. In *Broken Lullaby*, Paul kills Walter as they come across each other by accident in a deep shell-hole at night in "no man's land" between enemy trenches. Very similar encounters in such deep shell-holes are shot in *The Big Parade* (King Vidor 1925) and *All Quiet on the Western Front* (Lewis Milestone 1929). The confrontation between the two young men is dramatized by the eyes of the dying man, eyes that address the protagonist – and us spectators – with expressions of anguish and appeal to their enemy's mercy. In *Broken Lullaby* the dying man's hand points to a book lying on the ground, in which a letter appears that Paul helps him to sign "Walter," Paul's own very bloody hand grabbing Walter's white hand is emphasized by an insert – an emphasis intended solely for the viewer – of a high angle close up. The sense of touch is added here to the visual eye contact. In Vidor's film, Jim carefully slips his last cigarette into his enemy's lips, though unlit for fear of detection in the darkness. But the dying man drops it, and, after hesitating, Jim picks it up, once more joining touch to gaze in the recording of the memory image. In Milestone's film, Paul Baumer gives water to a dying Frenchman to relieve his thirst (DVD 1:15:00). Cigarette and water are tactile objects that are exchanged between the two men as emblems of their silent communication. Subjective camera close-shots of the dying man's gaze also contribute to our empathy with the hero's horror at eye-contact. It is in William Wellman's *Wings* that the Cain/Abel motif is given a more explicit biblical reference as the scene of the encounter between enemies takes place between actual brothers in arms: a close-shot shows us Jack's petrified face at recognizing the dying pilot's face as his friend David's (VHS 2:04:00–2:05:71) as he bends over his lips to hear his dying message of love.

The issue that the Cain/Abel motif hinges upon is all the more significant within the 1920s debate between moral conscience and consciousness of lived experience that can be traced back to important philosophers and novelists such as Henri Bergson (the concept of duration and memory, sometimes called "becoming") and William James (pragmatic theory), and, more generally, to the research in phenomenology. A crisis in consciousness of lived experience has brought about a moral crisis in which the justification for killing a man, called "duty" by society, suddenly appears in a new light as unacceptable. Cathy Caruth addresses the issue by drawing attention to a passage in Freud's *Beyond the Pleasure Principle* in which the hallucination of war trauma is questioned as being different from his theory of trauma and dream.[17] And, indeed, Freud's psychoanalytical theory of displacement and condensation of – mostly childhood – trauma into dreams does not appear at all in the visual Cain/Abel motif under discussion here. Another reading of the Cain/Abel scene in *Wings* might be that the emotional shock suffered is caused by the sudden disappearance of the official narrative justifying the killing of one's enemy, while a narrative that might explain the accident is felt to be missing.

THE MISSING NARRATIVE AND SURVIVAL AS THE ULTIMATE NARRATIVE

Caruth writes that, for Freud,[18] consciousness first arose as an attempt to protect the life of the organism from the imposing stimuli of a hostile world, and, more important for trauma analysis, the function of consciousness is to place stimuli from the outside world in an ordered experience of time. The cause of trauma, then, is an encounter with stimuli that are not directly perceived as a threat to the life of the organism but that occur as a break in the mind's experience of time. It is the lack of an appropriate narrative, which typically occurs in an accident, which would account for the shock of trauma.

In *Wings*, Jack is confronted with the shock of discovery when, after having proudly cut out the German cross from the plane that will officially prove his new exploit, he races to see the pilot whom the French soldiers have extracted from the plane. Jack has a narrative ready, which he has been instructed to believe in as a moral justification for the killing of another man, but he has no narrative to account for his unexpected confrontation with David's gaze. The dying David reminds him of this "official" narrative, "You didn't shoot me ... you brought down a Heinie, don't you see?" He then explains that he stole a German plane to get back across the lines, and, once he has died, the French officer who is present tells Jack, "*C'est la guerre*," which echoes Paul Baumer's "war is war" once he has told his story to his friend Kat in Milestone's *All Quiet on the West Front*. But Jack, when he meets David's mother, kneels before her in a posture suggesting a confession of guilt. She does make the effort of giving him a comforting, "It wasn't your fault, it was war" (VHS 2:14:00), but a close-shot of his weeping face shows us that the official narrative cannot atone for the killing of his "brother," a plain fact that remains without any explanation. A similar gap in narrative structure occurs in Lubitsch's film as Paul reads Walter's letter expressing his love for French culture. Again, as the priest reminds him, a narrative of war has been given to him as a moral justification for killing a man (DVD 08:02), but this implies an image of the latter as an enemy, not as a "brother."

And yet, both in *Wings* and in *Broken Lullaby*, the Cain/Abel motif is finally supplanted by a new visual and aural motif of an altogether different kind. In both films, official memorialization is presented as a collective ritual of glorifying remembrance, while its inefficiency is emphasized as far as individual consciousness is concerned. What Lubitsch's film offers instead is an act of creation, surmising that an artistic performance is always unique, so that the character's musical performance becomes a narrative of survival and remembrance. In a similar manner, looking at the concluding scene in *Wings* we are also shown a creative act with the image of the shooting star crossing the sky at night, which no longer means Jack's infatuation with speed, adventure, and David's girl Sylvia but, rather, his awakening to Mary's physical presence near him, confirmed by Jack's kiss, thus making an early bonding come true: "Kiss the girl you love when seeing a shooting star." Both images, Schumann's "Träumerei" and the shooting star, are thus used by the characters in order never to forget the obsessive Cain/Abel image and thus accomplish a new destiny, giving expression to a narrative of survival. To the audience, these two narratives appeal to our sensorium, visually and aurally, but an image of the sense of touch, already noted in the shell-hole "seminal scene," is also strikingly present on the screen.

The violin is a gift from Walter's father that passes from hand to hand and whose effectiveness as a sensory apparatus relies essentially on the sense of touch, complete with bow upon string, fingers, and shoulder. And Jack's kiss in *Wings* is also an explicit address to our sense of touch. Common to both images is a near erasure of the gaze, as if one's eyes had to close – Walter's parents close their eyes – and one's body had to submit to the sense of touch as a more reliable narrative of survival. Examining the other two films that have been referred to in this chapter confirms this intuition by an equally revealing appeal to our phenomenological body. In *The Big Parade*, Jim returns to the farm in France and embraces Melisande, and Milestone's *All Quiet on the Western Front* also offers such a narrative involving the sense of touch through a butterfly in no man's land between the two enemy lines, which Paul Baumer's outstretched fingers attempt to stroke. While it is by attempting to reach out for it with his hand that he attracts the attention of a French soldier who instantly kills him, it is the beauty of the butterfly's graceful life that creates a narrative of survival in the viewer's memory.

CONCLUSION

It might be seen as a common characteristic of these films that they deal with the issue of war memories in different narrative forms, whether the characters struggle between remembrance and forgetfulness, between individual and collective memories, or whether they create out of the "ashes of history," to quote Caruth again, a narrative of survival. The very absurdity of having survived is given an unprecedented significance by creative experiences, such as performing Schumann's music for a "new" family, and "kissing the girl you love when seeing a shooting star." As Caruth's analysis of the trauma of war memory suggests,[19] it is by achieving a re-enactment of the act of parting, which until then had remained unbearably inexplicable, that Paul and Jack endorse their true destiny as survivors. The post-First World War film then seems to have created a new film genre, the pacifist film. The other two films considered in this chapter as part of our corpus, King Vidor's *The Big Parade* (1924) and Lewis Milestone's *All Quiet on the Western Front* (1929), affirm the idea of a new genre since they share a common use of memory as a plastic raw material. In all four films, memory becomes a vehicle for the invention of a new narrative form and a new ethos – that of pacifist survivors.[20]

NOTES

1 *Wings*, directed by William A. Wellman (United States: Paramount, 1927). A spectacular tribute to the American flyers of the First World War, born of William Wellman's and John Monk Saunders's own experiences with the Lafayette Corps. Main cast: Gary Cooper (Cadet White), Richard Arlen (David Armstrong), Charles "Buddy" Rogers (Jack Powell), Clara Bow (Mary Preston). See William Wellman Jr, *The Man and His Wings – William A. Wellman and the Making of the First Best Picture* (Westport, CT: Praeger, 2006). I have been using my VHS 1989 Paramount Pictures copy. The timing indicated is that of the VHS tape, which lasts 219 minutes.
2 *Broken Lullaby*, directed by Ernst Lubitsch (United States: Paramount, 1932). Based on a French play by Maurice Rostand, original title: *The Man I Killed*, or *The Fifth Commandment*. Main cast: Lionel Barrymore (Dr H. Hölderlin), Phillips Holmes (Paul Renard), Nancy Carroll (Fräulein Elsa, Walter's fiancée), Louise Carter (Frau Hölderlin).
3 *The Big Parade*, directed by King Vidor (United States: Metro-Goldwyn-Mayer, 1925), adapted by Harry Behn from the play by Joseph Farnham and the autobiographical novel *Plumes* by Laurence Stallings. Main cast: John Gilbert (Jim), Renée Adorée (Mélisande), Karle Dane (Slim), Tom O'Brien (Bull). See Jean-Marie Lecomte, "King Vidor et l'Amérique en guerre: La déraison de l'Histoire," dans *King Vidor, Odyssée des inconnus*, *Cinémaction*, ed. Jean-Marie Lecomte et Gilles Menegaldo (Condé sur Noireau: Charles Corlet, 2014), 80–8.
4 *All Quiet on the Western Front*, directed by Lewis Milestone (United States: Universal Studios, 1930), based on Erich Maria Remarque, *All Quiet on the Western Front*, 1928–29. Main cast: Lew Ayres (Paul Baumer), Louis Wolheim (Stanislaus Katczinsky), John Wray (Himmelstoss), Arnold Lucy (Kantorek), Raymond Griffith (the French soldier Duval).
5 See my "Struggling for or against Memories: E.M. Remarque's 1928 Novel *All Quiet on the Western Front* and Its 1930 and 1980 Hollywood Transpositions," in "La guerre de 14 re–présentée: L'art comme réponse à la guerre – Representing World War I: Art's Response to War," *Caliban* 53 (2015), 135–52.
6 Wellman Jr, *The Man*, 110. This was what is now called "a cameo appearance" by Gary Cooper, but the author tells us that, on the contrary, it was the actor's first appearance on screen and he was so broke that Wellman kept him on the team for several weeks even though he no longer needed him. The very short and extremely effective sequence was what launched

his career, surfing on the successful wave of the film. See Wellman Jr, *The Man*, 123–4. While the text reads that White always carried a talisman and forgot it, it does seem that the copy of the film that I have (from TCM in the 1990s) says the contrary: he despises having one, and it is David who always has one, his teddy bear, but forgets it the day he is killed, after telling Jack he has a premonition he will not survive the day.

7 This is the narrative that Paul Baumer, the fictional narrator of the journal that is found after his death in Remarque's novel *All Quiet on the Western Front* (1928), finally agrees upon after having experienced a similar trauma as he kills a Frenchman named Duval, a printer, in a shell-hole: "War is war." In this similarly antimilitarist novel and Milestone's 1929 film adaptation, the protagonist is killed before the war ends, and so are many of his school friends. It is dedicated to those who, "though they may have escaped its shells, were destroyed by the war" (prologue to both novel and film).

8 "Träumerei" is one of the *Kinderszenen/Scenes from Childhood*, opus 15, no. 7, a set of thirteen pieces of music for piano written by Robert Schumann in 1838. Other music in the film is from Beethoven, Symphony no. 5 (uncredited); *Der Frohe Wandersmann*, music Theodor Fröhlich, lyrics Joseph Freiherr von Eichendorff; *Père la Victoire*, music Louis Ganne; *Quand Madelon*, music Camille Robert, lyrics Louis Bousquet.

9 Maurice Halbwachs, *La mémoire collective*, 1950 Ebook, https://www.puf.com/Auteur%3AMaurice_Halbwachs. See also "La mémoire collective chez les musiciens" (1939) on the same website. Since Lubitsch's film dates from 1932, perhaps the most relevant work on the subject is Maurice Halbwachs, *Les Cadres sociaux de la mémoire* (Paris: Alcan, 1925).

10 Anne-Laure Dubrac, "Out of the Past: Between the Past and Memory," in *Memory in/of English-Speaking Cinema: Le cinema comme vecteur de la mémoire dans le cinéma anglophone*, ed. Zeenat Saleh and Melvyn Stokes (Paris: Michel Houdiard Editeur, 2014), 361–7.

11 Maurice Halbwachs, "La mémoire collective chez les musiciens," trans. Lewis A. Coser, 1992, viewed 3 May 2020, http://classiques.uqac.ca/classiques/Halbwachs_maurice/memoire_coll_musiciens/memoire_coll_musiciens.html.

12 William A. Wellman, in Wellman Jr, *The Man*, 27. "For a pilot to receive the designation of ace, he must shoot down at least one enemy plane and have a 'kill' confirmed; … witnessed by at least two other flyers … or be seen from the ground by military observers … or the pilot must land his plane, tear off the black cross insignia from the downed aircraft and return it to his escadrille."

13 "The story of Cain and Abel is a 'primeval' story of punishment ... a story that occurs as a kind of foundation, rather than one that ... occurs within the terms of an already established law." See Cathy Caruth, *Literature in the Ashes of History* (Baltimore, MD: Johns Hopkins Press, 2013), 70–1. The interpretation of God's mark on Cain is one of a survival that is an absolute loss; however, within the pacifist ideology of reconstruction, these pacifist films displace God's mark on Cain within a duty of recollection, memorizing in order to avoid repetition or return.

14 Hugo Münsterberg, *The Photoplay – A Psychological Study* (New York: Bibliobazaar, 2007 [1916]). The book is grounded on German experimental psychology, a stimulating field of research that led William James to invite the author to Harvard. The echoes between his psychology of the audience's reception and Henri Bergson's *Matter and Memory* is a subject of research in itself. For an early discussion of this almost entirely forgotten book, see J. Dudley Andrew, *The Major Film Theories – An Introduction* (Oxford: Oxford University Press, 1976), 14–26.

15 As in the experience of reading a novel. See Marthe Robert, *Le roman des origines et les origines du roman* (Paris: Grasset, 1972).

16 But also in *The Big Parade* and in *All Quiet on The Western Front*. King Vidor's film is the source of this war metaphor, and it can be argued that it initiated an interest in war films among the American public. It was the main reference for the decision by Adolph Zukor and Jesse Lasky, the first "moguls" of the film industry, to make a war film on aerial combat. "The enormous success of *The Big Parade* (1925) and *What Price Glory* (1926) had fuelled new interest" in the war film (Wellman Jr., *The Man*, 103).

17 Caruth, *Literature*, 5.

18 Ibid.

19 Ibid., 13.

20 Though *Broken Lullaby* (1932) and the earlier *All Quiet on the Western Front* (1929) do involve Germany within this new process, the general state of mind of defeated Germany that led to the theory of a Weimar Republic left-wing plot has not been taken into account in this chapter. Maybe the difference is apparent after all: Remarque's hero dies, while in the American films he actually survives.

20

Peacekeeping Forces and Their Filmic Representations: The Case of Peter Kosminsky's *Warriors* (1999) and *The Promise* (2011)

Georges Fournier

Raphaëlle Costa de Beauregard has previously focused on a pacifist approach towards the tragic events of the First World War in Europe sustained by the psychological drama of memory, which opens a reflection on the apprehension of peace. In the chapters of part 4 of this volume, the authors scrutinize various wars, some of them more thoroughly documented than others. While the First and Second World Wars, along with the Vietnam War, gave birth to many books and films, there are conflicts that remain largely undocumented. Such is the case with recent conflicts, which have been denied this appellation, something from which soldiers have suffered. Among these conflicts are the ones fought by peacekeeping forces, a subject that is not very popular and one that is almost ignored by fiction and academics, which partly contributes to making it a blind spot in modern history. Peacekeeping operations, because of the complex nature of the responsibilities involved (political, military, and humanitarian), have redefined the role of modern armed forces, and sceptics might even worry about a move towards the "militarization of human security" or, conversely, towards the "humanization of the armed forces."

In recent decades, British military personnel have spent relatively little time fighting conventional battles; instead, they have participated in numerous peacekeeping operations. This chapter is structured around two films by the same director, Peter Kosminsky. *Warriors* and *The Promise* are films that he directed for British television and that testify to his status as a nonconformist in the television landscape. Kosminsky is a director who has chosen to work in television on the subject of conflicts, be they social, political, or military. His interest lies in the contrast between individual memories (e.g., those of soldiers) and history – namely collective memory, which Pierre Nora defines as, "The memory or set of memories, conscious or not, a living community has of identity, whether lived and/or mythified, and to which the past fully belongs."[1] More precisely, Kosminsky is interested in the opposition between individual memories and the absence of any narratives for some major events from recent times. His focus is on the protagonists of these events, whose experiences lie well beyond the realm of interest for almost everyone, and for historians in particular. Such is the case with the soldiers sent to Palestine under the 1923 League of Nations mandate and the soldiers of the British contingent sent to the former Yugoslavia under the 1992 United Nations mandate. His work, which now spans almost four decades, taps into the roots of British documentary with the re-examination of recent news and current events to promote the political awakening of citizens.

The Promise and *Warriors* stand as two instances of the fictional reprocessing of major conflicts from the twentieth century. They share a heavy reliance on testimonies, something Kosminsky is adamant about when it comes to endowing his work with all the required legitimacy. Although years apart, these conflicts are connected in more ways than one. First and foremost is the fact that the broadcasting of *Warriors* led to the inception of *The Promise*: *Warriors* is based on the testimonies of ninety-five former soldiers, mobilized in Bosnia between the autumn of 1992 and the spring of 1993. After watching *Warriors* on the conflict in the former Yugoslavia, a veteran with the British mandatory forces in Palestine contacted the director to let him know how much his fiction had echoed his own experience as a young soldier, an experience of years of unrelenting agony in buffer zones under a mandate to maintain peace between warring factions pending a resolution that, apart from civilians, nobody on the ground wanted. From this first contact sprang a collection of over sixty testimonies from veterans from Palestine, their stories forming the backbone of the filmmaker's work. Although diverse, these narratives stood out as variations on identical and ordinary human experiences that were easily amalgamated so as to provide a coherent story, thanks to which each soldier was allowed to relive the spirit of these years.

Although the soldiers sent to Palestine from 1945 to 1948 and those sent to the Balkans in the first half of the 1990s endured situations similar in many ways, this chapter draws clear distinctions between two radically different conflicts, from diplomatic and political points of view. While the British soldiers sent to Palestine in the post-Second World War years were charged with an exclusive mandate initiated in 1920 under the auspices of the League of Nations, the British troops sent to the Balkans under the aegis of the United Nations operated jointly with other national armies. While in the Middle East, the British Army soon appeared as the main opponent to be defeated in order to witness the advent of a Zionist state, in the Balkans the UNPROFOR was never seen as a belligerent;[2] rather, it remained all through the conflict as one of the forces to be reckoned with pending a diplomatic resolution.

An examination of the specificities of these conflicts begins with a special focus on places and the need, from a societal point of view, to keep traces of these military involvements in the history of Britain. Only then will it be relevant to pay attention to the impossible missions of the peacekeeping forces and the damage visited upon soldiers for lack of acknowledgment of the arduousness of their task and the enduring traumas that they suffered. A final part will be devoted to writing, particularly to fiction, as a means for both soldiers and nations to come to terms with the experience of war.

PLACES BLIGHTED BY HISTORY, PLACES FULL OF MEMORIES

Cursed Places

In both *Warriors* and *The Promise*, Kosminsky stages places steeped in history, places that haunt memories. While in Palestine the massive settlement of Jews from Europe after the Shoah initiated decades of violence, in the Balkans, the opposition between the Austro-Hungarian Empire and the Ottoman Empire degenerated into clashes between Catholics, Muslims, and Orthodox Christians, making this place, in the twentieth century, a "cursed place." As Elmira, one of the characters, explicitly states on several occasions: "This place is cursed."/ "Don't tell me that this place isn't cursed."

In *Warriors*, most of the plot takes place in Bosnia and Herzegovina, more precisely, in the Vitez region, and builds on the events that occurred there between the autumn of 1992 and the spring of 1993. Explicit references are made to the massacres of Bosnians by the Croatian Defence Council and to the displacement of Muslims from their original settlements to more welcoming areas in the Srebrenica region, the Croatian authorities hoping to reverse the demographic balance in their favour in the process and consequently proclaim their legitimacy over these territories. Many scenes hint at the forced exodus of Muslims, whose homes are burned down, though the reality of "ethnic cleansing" is never explicitly represented.

It is at an early stage in the film that the notions of displacement and ethnic cleansing are evoked; they then serve to explain the violence that prevailed at the time and the extent of the task that awaited the UNPROFOR soldiers. For instance, Almira Zec, one of the Bosnian protagonists, experienced a first exodus from Sarajevo, which, for her, was reminiscent of the displacement of Romani, Jews, and Serbs from 1941 onwards, and which eventually led to their deportation to the concentration camp of Jasenovac, in the Independent State of Croatia (1941–45), or to those of Nazi Germany.

The Promise, the film about Palestine, spans two separate periods, at the end of the British mandate in 1948 and in the twenty-first century, so that, once again, as in *Warriors*, viewers are confronted with places fraught with history, where the past can crop up at any time. Whenever it does, violence follows, and, from the reading of the very beginning of her grandfather's diary about his years in Palestine, Erin – the protagonist – is moved to tears. In Israel, confrontations between communities are numerous and each is instantly an opportunity for past feuds and sufferings to resurface, be it the Holocaust, the Nakba, the Six Day War, or other historic moments when the communities were pitted against each other.

Memory as a Cardinal Value

Memory has become a cardinal value of our time, a kind of norm designed to measure the democratic level of societies. Among its new assignments, there is the need to correct the past, to heal past wounds, to (re)write history in the name of the principles on which our present time rests. Films, and more particularly television films, contribute to this in their own way due to the impact mass media can have on viewers.

Likewise, the concept of memory from a social and not only cognitive point of view, is not recent. As members of groups, individuals have always shared collective memories, memories made of a series of exceptional events transmitted, more often than not, in the form of fictional narratives that serve as social bonds but also as foundations in the development of personalities.These fictional narratives relate the perennial memory of remarkable protagonists of events from a near or distant past.

What is valid for societies is also true for individuals: memory encapsulates the mysterious connections that link the past to the present, connections that bind each of us to the lives of those who made history. Moreover, it is through this process of appropriation of the experience of others that the sense of belonging to a community is built: as individuals, we are made of the history of the group to which we belong. Yet, while we praise collective memory, some are prevented from expressing themselves and so prevented from taking part in the history of their community, even though their stories rightly deserve to be recorded as part of their national history. This is the case with soldiers with peacekeeping forces. Their testimonies would be all the more legitimate as, in the twentieth century, the creed of objectivity was defeated by the idea that history ultimately rests on constructions developed from the narratives of the actors of the events rather than on the exhumation of archives.

Moreover, many written archives are testimonies, whether recorded when events were taking place or later, concealing an irreducible subjectivity that their status as archives does not diminish in any way. History originates in points of view, in diverse angles of observation: it springs from varied interpretations. For Henry Rousso, history is no longer one and indivisible, and "any source is an 'invented' source just like any historical figure is, in the sense that Max Weber used it, i.e., a construction, an ideal-type."[3] History would then be some form of fiction that "wanders through the words of others," namely, the actors of these tragedies.[4] Then, a national narrative would be made of the official version, which the stories of those who experienced the events from the inside would complement. Consequently, fiction itself becomes legitimate in its claim to testify to the past.

This is the approach adopted by Kosminsky, an approach that is nowadays echoed by the will of minorities to write their own history. Official narratives then collide with parallel accounts made of the testimonies of those whose experiences have been superseded, or overshadowed, by the official versions of the facts, as is the case of the soldiers who were sent to Palestine or to other missions abroad under the aegis of the United Nations.

PEACEKEEPING: IMPOSSIBLE MISSIONS

From Privates to Diplomats

Kosminsky carefully analyzes the manoeuvres involved in peacekeeping missions and the impacts they have on both military and civilians. "Outlawed wars," "peace through law,"[5] these expressions, well established from both political and diplomatic standpoints by the 1924 Geneva Protocol and by the August 1928 Briand-Kellogg Pact, sum up the spirit of these international military interventions. In such missions, the purpose is to impose a ceasefire agreed upon at a supranational level through the stationing of foreign armies at the heart of the conflict.

Even though officials may have given their consent to the deployment of independent peacekeeping forces, this does not imply that a unilateral consensus necessarily prevails at the local level, particularly if division endures on either side of the feuding parties or if military leaders have little control over the situation on the ground. This partly explains the constant breaching of ceasefires, the repeated attacks on civilians, and how frequently soldiers of peacekeeping forces face abuse and atrocities from militias. These are clashes Kosminsky stages in both films.

While in the Balkans, British soldiers have to cope with the contempt of the Serbian and Croatian militias, who scoff at their powerlessness; in Palestine, civilians throw the content of chamber pots at them during their patrols. Their assignment, which consists in behaving in a diplomatic and occasionally a humanitarian way (a mandate that contravenes the original tasks assigned to soldiers – i.e., waging war), makes them weak. Among their first and most mandatory constraints is the obligation never to shoot unless their lives are in danger, which de facto deprives them of their military status. Their attention is constantly drawn to this point from the very beginning of their time there. They are first told about it during a briefing session, and they are constantly reminded of it before leaving for an operation: "You are here to observe and nothing more!" Theoretically, they are asked to remain neutral, which is not tenable on the ground: being neutral means not interfering to set things right, although interfering is a constant necessity for soldiers sent to the heart of a conflict. Any act of interference is perceived as benefitting one group over another, and any action is interpreted by the camp that feels aggrieved as a breach of impartiality. Feuding parties know this all too well, and once peacekeeping forces, perceived as foreign battalions, reach the forefront, they are identified as military troops to be reckoned with in the preparation of future and decisive strategies.

Peacekeeping Soldiers: A Common Enemy

The obligation to remain neutral, almost indifferent, towards the dire plight of civilians is unbearable for UN soldiers, and violence can often explode. In *Warriors*, one such instance occurs when the Blue Helmets,[6] in order to protect Bosnians, decide not to warn their superiors about their displacement. Another takes place after Alan James, a soldier, pulls Almira Zec's husband out of a mass grave of almost a hundred corpses and verbally attacks a Serbian militiaman to force him to respond and so be rightfully justified in engaging in a fight. From the outset of conflicts, peacekeepers are the object of antagonism from at least one of the parties, and eventually they arouse the hostility of all the parties, even the civilians they protect, and this is the cause of many traumas for soldiers who are then held liable for whatever goes wrong. Gradually, the target is no longer the original enemy but the peacekeeping force, which then becomes the one that must be eliminated to achieve the initial intention: the removal of ethnic groups from a given place. They are gradually seen as the one and only illegal occupier.

This is particularly true in *The Promise*. The scenes of violence with which British troops have to cope are numerous. However, two are particularly noteworthy: the first is about the explosion at the King David Hotel, which was then the British administrative headquarters, and the second is the Deir Yassin massacre, both actions undertaken by Zionist paramilitary groups. These are pivotal moments in the history of the birth of Israel, promoted by the Balfour Declaration of 1917, and to which Irgun, the then main paramilitary group, largely contributed something a former militiaman sums up when he explains: "The British were blocking us. We dismissed them."

The soldier of the peacekeeping force as the real enemy is a trope that both films feature: he is the object of both retaliation and blackmail, which are instances of violence in the face of which he remains totally powerless and vulnerable. While in *The Promise*, only the violence from Zionist paramilitary groups is staged, in *Warriors*, it is the violence of both the Serbs and the Croats that is shown, not to mention the fact that Bosnians also take out their anger on UN soldiers due to their exasperation with the the latter's inability to ensure the protection of civilians, a situation that makes their position less and less tenable.

PEACEKEEPING FORCES: FROM HOSTAGES TO ACCOMPLICES

The trap in which UN soldiers are caught springs from their status, which prohibits them from using their weapons in an offensive manner, something which therefore disarms them. They even become the auxiliaries of the executioners in mass massacres, a situation that is bound to arouse feelings of guilt likely to lead to deep-seated traumas.

Soldiers as Hostages

Kosminsky's postulate is the untenability of the assignments given to peacekeeping soldiers: standing halfway between two warring factions and protecting civilians while never making any moves that might be understood as sign of allegiance to one particular camp. More precisely, what the filmmaker chooses to explore is the reality of these new forms of war and the status of these new soldiers, condemned to arouse anger and disapproval: they are the ones who are sent to war zones to try and impose pacification, something no one on the ground wants, and who are confronted with a situation that, with their arrival, radically changes shape.

In this new form of conflict, wounds are more psychological than physical, and the traumas experienced by the peacekeeping soldiers result both from the absence of any clearly defined militarized zones and from the presence of civilians as an extra factor to be taken into account while managing the delicate balance of forces on the ground. The main reason for their presence is to protect civilians who, for the warring factions, are part and parcel of the conflict. Ironically, although the main assignment of UN soldiers is to protect civilians, the latter are weakened by their presence. This is because, according to the strategies of modern warfare, civilians are no longer merely the targets of retaliation; rather, they stand at the centre of sophisticated tactics designed either to use peacekeeping forces against the rival faction or to get them to interfere and so be discredited.

UN forces find it hard to be accused of collaborating. This feeling of betrayal that they arouse on all sides originates in their proximity to the contending factions and to the civilians. In this context, they easily become the hostages of both militias and civilians. While the latter find protection in their vicinity, turning them into human shields in the process, the former seize their military equipment, first to neutralize them and then to use this very equipment against their opponents. In *Warriors*, several episodes show hundreds of civilians on the move suddenly gathering around the British tanks to achieve protection but rendering soldiers powerless and vulnerable in the process. For these episodes, Kosminsky tapped into actual events, in particular, direct references are made to the April 1994 hijacking of UN personnel by Serbs to use them as human shields against NATO bombings. The Serbs used the same strategy the following year against Atlantic Alliance raids on the city of Pale, where General Karadzick's headquarters was located.

The other major factor of violence in these films is the guilt with which soldiers eventually have to live: the soldiers were the witnesses of atrocities committed by the belligerents and, especially, of abuse against civilians that they failed to prevent – a breach of moral obligation that later becomes a major source of resentment towards their superiors and towards themselves. Guilt is all the greater when closeness with the civilians gradually sets in. This is particularly the case when soldiers happen to be stationed in the same place for months, even years, on end. The trauma comes from the proximity to these civilians whom they are supposed to protect and whom they see being massacred, without being able to prevent it. It can be even worse when they indirectly contribute to these massacres.

While contenders follow clearly defined lines of conduct, the foreign contingents act with no clear assignments, with unrealistic means, with no exact blueprints, which partly explains why, in the best of cases, they are taken by surprise, and, in the worst of cases, they are used by one of the factions, against their will, in order to serve warmongering purposes.

PEACEMAKERS AS EXECUTIONERS

Strangely, peacekeepers weaken the people they are supposed to protect. This, for example, is the case with British troops in Palestine and in the Balkans. This is the meaning of a scene from *Warriors* in which an elderly couple is killed by Croatian militias in retaliation for the intervention by British soldiers who had, some time before, prevented the ransacking of their home. By interfering, the British soldiers only made things worse, and what was first intended as a sequence in a looting spree turned into a massacre: Dario Kordic's men take revenge on the couple and on their dog.[7] Before leaving, they take pains to nail the old man to the façade of his house in a posture reminiscent of Jesus Christ on the cross and hide explosives intended to harm or even kill anyone who might attempt to take him down. Likewise, when Serb militiamen find out that British soldiers have rescued a young boy and sheltered him at the back of their warrior,[8] they threaten to report the incident, which they take as an outright instance of interference, and they take the young boy with them to kill him, we are led to fear. These scenes serve as examples for other similar situations that occurred at that time in the Balkans.

The Promise raises the question of commitments, political but also human – in particular non-contractual, but nonetheless binding, commitments – that one makes to others. This issue stands at the centre of the script, for Len not only fails to bring Mohamed's son back to his father but also exposes him to the bullets of his enemies. It is difficult for soldiers not to join in and take sides when they are sentimentally committed, as is the case with Len, who, in *The Promise*, falls in love with Clara, who, as we learn in the course of the film, is a prominent agent of Irgun. Although lovers, neither considers their sentimental pledge as superior to their political commitment. Just as Len refuses to become an informer, Clara is, above all, a soldier highly devoted to the Zionist cause. Late in the film, when they meet by chance at Deir Yassin in an episode that evokes the massacre of the population of this small town on 9 April 1948 by paramilitary groups (an action made possible by the murky atmosphere caused by the progressive withdrawal of the British contingent), Clara has changed in every way – a metamorphosis that makes their relationship impossible. If life is meaningful for Clara, who, as an active agent of Irgun, can make sense of each event that occurs, such is not the case with Len, who is always taken unawares. To that extent, Len and Clara embody the positions of two forces in war-zones: (1) the natives – often militants – who are moved by a cause they have to defend, and (2) foreigners sent there to impose an undesired ceasefire. There is a huge chasm between young soldiers who recently left their safe and secure homeland and people who, at the front, are ready to lay down their lives for a national project that far exceeds their personal interests.

Len is sentimentally committed to Clara, a position that blinds him and prevents him from being fully aware of his lover's political involvement. He is naïve, which leads him to betray and expose friends and colleagues; he is even indirectly responsible for the death of one of them. Clara scrupulously follows the orders given by Irgun and, even though there is no doubt about her love for Len, she deceives him every time they meet: she manages to extract sensitive information from him, and even goes so far as to pretend that she has been tortured by Lehi, a Zionist paramilitary group, on account of what it believed to be acts of treason on her part. Len falls into all the traps she lays.

In *Warriors*, soldiers who are nervous, exhausted, and left to their own devices, try to make sense of their lives by forming fortuitous bonds with the locals. Lieutenant Feeley falls in love with a young Bosnian, Almira Zec, and Lieutenant Loughrey flirts with the camp interpreter. It is in these circumstances, and because they happen to get close to one of the warring camps, often the weakest one, that they become biased and take sides. These relationships highlight the harshness of the reality endured by the victims. Soldiers discover the violence levelled at the Bosnian civilians: families are murdered, the old are savagely slaughtered, and the local population is hunted down and displaced in an attempt to achieve ethnically safe zones.

"I wouldn't give this job to a monkey," says Sergeant Andre Sochanik. As they are confronted with instances of extreme pain and suffering on a daily basis without being able to bring the sufferers any relief and, because they are subject to orders that contradict evidence, soldiers gradually lose all sense of morality. Before being mobilized, Lieutenant Neil Loughrey was about to get married; yet, once in the Balkans, he becomes the interpreter's lover and when back home beats his pregnant wife. Likewise, John Feeley, after being invited by a young Bosnian couple, becomes the wife's lover as soon as the husband leaves to join Bosnian paramilitary groups. Finally, once back in Britain, he commits suicide. Private Alan James, for his part, becomes violent while still in service and, when back in his local town, he commits acts of incivility and, for example, smashes bus shelters.

UNKNOWN SOLDIERS IN UNKNOWN CONFLICTS

Post-traumatic stress disorders remain largely undocumented, and few people are aware of what lies at their origin and of the many forms they take. To that extent, *Warriors* and *The Promise* stand as rare instances of dramatization of the consequences of PTSDs on ordinary young men who, as soldiers, are sent to conflict zones where they are confronted with atrocities that strongly affect their mental health.

The special conditions in which the missions of the peacekeeping forces are undertaken raise the question of the absence of any collective memory of what soldiers experienced, of what they went through during their mandates. The conflicts in which the UN soldiers intervene are external conflicts: they do not directly concern their fellow citizens. Media coverage is occasional, episodic. If today everyone in the United Kingdom is still well aware of the Falklands War, which happened in 1982, few remember anything about the conflict in the former Yugoslavia, which raged between 1991 and 2001. Fewer still are those who know anything about Palestine in the first half of the twentieth century, even though the British Army was the only foreign party there.

The specificities of these conflicts largely account for the invisibility of the soldiers who fought battles on foreign territories for stakes that exceeded the interest and understanding of the person in the street. These missions, largely undocumented as far as most British citizens are concerned, turn these soldiers into unknown soldiers or, even worse, into soldiers of wars who are denied this status due to lack of prominence in recent history and are consequently seen as regional disputes. The indifference that surrounds these interventions is particularly frustrating for soldiers whose memories need to be heard in order not to foster traumas.

From Helplessness to Silence

Due to their relative invisibility, these military missions have failed to achieve the place they deserve in recent history. Those who took part in them are consequently deprived of the recognition their perilous assignments deserved. The lack of relevant acknowledgment, in the form of gratitude towards them for having put their lives at risk to fight for the values that their country promotes, is a great source of trauma. While among their minor traumas is the denial of their existence as combatants and as helpless witnesses of the exactions of the rival parties, among their major traumas is the guilt of being responsible for the deaths of civilians.

The soldier in the peacekeeping force is often associated with humanitarian personnel and their altruistic functions. His role is then seriously diminished since people associate his missions with those of the humanitarians, which brings forth images of camps where food is distributed to the population – positive images that seriously contradict the sufferings the soldiers must endure. They are perceived as good Samaritans more than as fighters in a regular army. Yet, unlike regular humanitarians, they are on the operational front and are witnesses, and occasionally the auxiliaries, of genocidal situations. Because they lack the ability to provide remedies they are confined to enforced helplessness. Their forced passivity in the Balkans had devastating psychological effects on them. Their daily patrols were often carried out among lines of dead bodies that they were able neither to remove nor to bury for fear of being accused of interfering. All these corpses, left to rot in the open air, remind one of Polynice's body and Creon's refusal to allow Antigone to proceed with the mourning and burial of her brother. This analogy is an apt reminder of the tragic dimension of the soldiers' missions.

Veterans of the armed forces in the Balkans were condemned to indefinite silence after returning home without glory and following a withdrawal perceived, by some, as a defeat and, by others, as the final stage of a humanitarian expedition. Likewise, in 1948, silence was imposed on the less glorious episodes in the history of the British Army and, in particular, the mandate in Palestine in the context of a country ruined by the Second World War and undermined by the collapse of its empire.

Repressed Memories

The experience of war ensures that soldiers mature rapidly, and, even though they manage to cope with atrocities on the ground, they return home with deep traumas. Regarding Len, Kosminsky explains:

> When you meet him, in 1945, he has already made the most extraordinary of emotional journeys. He is a paratrooper, he participated in the landing of D-Day, in the disaster of Operation Market Garden [the attempt to retake the Arnhem Bridge, which

> saw 30% of British paratroopers die in a single attack], in the Battle of the Ardennes, and in the liberation of the Bergen-Belsen concentration camp. He has matured at full speed, his character is already well-drawn.[9]

Hell for soldiers is also the fact that they are trapped in a world in which words no longer have the same meaning as they did before: when they call their relatives and are told about "cleansing," for them this can only mean "ethnic cleansing," which makes communication all but impossible. This episode foretells the hardship and dissociative disorders they will experience when back home when they are constantly at cross-purposes with their nearest and dearest and "normality" being almost unbearable.

Once demobilized, soldiers in the peacekeeping forces feel incapacitated, trapped in a kind of limbo borne out of a silence that results from society's general lack of interest in them – something that they experience as an obstacle to testifying. This feeling is increased by the vacuity that characterizes the return to the West, where the futilities of everyday life are totally disconnected from the realities of the war zone. The trauma is all the greater as the passage is quick and the UN soldiers, once back home, are haunted by the spirit of the place they were sent to – the curse of the Balkans, for instance – and by images of their recent descent into hell. Kosminsky refers to this as a lift into adulthood and maturity from which all find it hard to recover.

Official decisions taken at supranational levels often make things worse. In June 2017, the Hague International Court of Justice gave full legitimacy to the guilt felt by the UN Dutch soldiers by blaming them for the Srebrenica massacre. Originally, in 2007, a complaint had been lodged by Van Diepen and Van Der Kroef, a law firm, on behalf of the survivors and relatives of the victims of Srebrenica against the Netherlands for non-compliance with contractual obligations, "failure to prevent a genocide," and "non-declaration of war crimes." The 450 Dutch peacekeepers, stationed near the enclave they were supposed to protect, were found guilty by the Court for not having stepped in to stop the attacks of about a thousand Serb soldiers.

On 16 July 2014, the Hague tribunal estimated that the Netherlands was civilly liable for the three hundred deaths in Srebrenica: the Dutch soldiers should not have evacuated these men from the base where they had taken refuge. The decision was confirmed on 17 June 2017, and the Netherlands was condemned to pay financial compensations to the families of the victims of the Srebrenica massacre. This decision has had serious consequences for UN missions and has not helped assuage the guilt felt by soldiers who acted in earnest but who were manipulated by the Serbian militias. It imposed a final verdict that discredited and silenced alternative approaches to the issue.

Unwilling Actors of Tragedies

The verdict regarding the role of the Blue Helmets in the Srebrenica massacre has turned peacemakers, who had been sent to protect civilians and to keep contending parties at bay, into murderers, making the path to recovery particularly arduous for them. De facto, Dutch soldiers were deprived of the attributes ascribed to victims even though they were twice subjected to traumatic experiences themselves: first, when they were forced to be the non-committed witnesses of the deportation of male Bosnians, second, when they were obliged to accept the accusation according to which they were responsible for the genocide.

Being on the executioners' side, they were condemned to remain silent about the situation as it happened and about their understanding of the matter. Kominsky aptly defines the mixture of guilt, frustration, and resentment in the concluding part of the films, in which he describes the soldiers' return to their homes. Private James sums it up when he answers a passerby's compliment by saying: "It was shit we made there." From then onwards, we see how soldiers who strictly abided by the order of non-intervention, and who were later accused of having contributed to a massacre committed by others, were denied justice. The same judgment was passed on the British soldiers in Palestine, who, it is widely believed, helped the Zionist cause by condoning the displacement of Palestinian populations, in particular in 1948: the Nakba.

This conflict between memory and repression, resulting from the impossibility for soldiers to testify, resembles a struggle between darkness and light. It is as though these soldiers were to remain forever in the shadow of history, their military involvement never acquiring the status of a national victory. Repressed memories and enforced silence foster traumas, which, because the soldiers experiencing them are not offered a means of release, turn into violence. In the absence of recognition of what they endured, these soldiers are led to commit acts of violence towards society at large, towards their loved ones, and towards themselves. How can society prevent these soldiers from being eternal victims and eternal victimizers? *The Promise* offers a positive resolution, presenting writing as salvation, while *Warriors*, with its irresolute conclusion, ends on a note of sadness.

FICTION

From Woes to Words

The Promise shows the path to salvation through writing. Writing a diary helps victims overcome traumas. It offers future and potential recognition – the diary being intended to be read, to be staged, to be brought to light. Nearly half a century after it was written, Len's diary is read, and this gives it the status of undisclosed evidence, like the testimonies of the soldiers that need to be heard, to be acknowledged. Dramatizing Len's diary means the broadcasting of an episode of history that up until then had been largely ignored. Its primary function is cathartic and, with this diary, written day after day during the years he spent in Palestine, Len shows the way to salvation. The diary, which records his feelings and the events of his daily life, helped the young soldier he was make sense of what he had to go through. It also gave meaning to the absurdity of war.

A diary recounts the fresh memory of events as they unfolded. It is not a fictionalized account that the passage of time either prettifies or darkens but, rather, the strict recording of the highlights of a soldier's traumatic experience in the heart of darkness. Len's diary, like any diary, has an important cathartic dimension: it is a true testimony, the kind that could be given by the witnesses and the actors of events. By writing down detailed records of his daily life, Len managed to maintain a sense of normalcy, something that other soldiers were not able to do – an inability that condemned them, when they least expected it, to be the victims of repressed memory. What is true at an individual level is also true for society, and this is probably what explains, in the Western world, the commemoration of past atrocities, as if by collectively commemorating events such as the Holocaust the West were sending victims signals of atonement in order to prevent them from falling into violence or madness. The idea is that collective commemoration brings general appeasement.

The Fictionalization of History/His Story

By drawing inspiration from the stories of veterans, Kosminsky shows that history is more than memorable deeds and dates: it is also the stories of the protagonists. What matters here is the link between individual stories and collective history, their respective places and their differences, what Maurice Halbwachs calls "the existence of frameworks of collective memory on which individual memory relies."[10]

With the fictionalization of the testimonies of more than 150 veterans, the point of view of rank-and-file members is heard several years after the facts: the human aspect of war is brought to light. As a consequence, Kosminsky's fictional narrative is a legitimate form meant to transmit soldiers' memories, as are films on the Second World War and/or the Vietnam War. The relevance of fiction with regard to collective memory is extremely important as, unlike in the past, when national memory would fall within the province of the military only, today history is composed of the stories of people who actually took part in it.

For the soldiers, testimonies mean the sharing of individual memories; for society at large, it means having individual experiences inscribed in the official history. This is what films attempt to do by forming a parallel narrative that complements official war records. According to Paul Ricoeur: "to remember we need others."[11] This is what Kosminsky endeavours to do by highlighting the need to share the experiences and traumas of those who are ready to lay down their lives. Experiences and traumas then become part and parcel of the collective memory. This is what the director clearly understood by opting for a televized fiction film, one that addresses the broadest possible audience.

CONCLUSION

As the main mass media channel, television has a bardic function,[12] and this predisposes it to national debates and to the release of the voices of those who are neglected and misunderstood. An example of these neglected voices is provided by peacekeeping forces, who may feel resentment and develop post-traumatic stress disorders due to the indifference fellow citizens show towards the hardship they endured.

Filmmakers produce knowledge that is widely disseminated in the public space and that contributes to the building up of a common memory; for the protagonists of historic events, the dramatization of their hardship represents a form of acknowledgment. That is what Kosminsky achieves when he offers, through extremely innovative films, interrogations of today's military interventions abroad from perspectives that are both political and human.

Carefully avoiding any Manichean opposition, Kosminsky refuses to take sides. For him, all the actors on the ground are victims, and he blames higher authorities for the conflicts in which they were involved. On several occasions in *Warriors*, he attacks those who set up military peacekeeping interventions that proved to be catch-22 situations for both the soldiers and the civilian populations. In *The Promise*, he is particularly hard on the British authorities. For him, the Palestinian conflict began when the United Kingdom was the primary colonial power in the world and was therefore responsible for what happened in countries under its authority.

The testimonial dimension of fiction films that present themselves as being historical is crucial and should not be underestimated, particularly when it comes to work dealing with recent history. Epistemologically speaking, this is what Kosminsky does by explaining his approach and fighting the doctrine that documentary is "truer" than fiction – something that other directors, such as Costa Gavras and Ken Loach, also do in their own way. This assignment is crucial if fiction film directors are to assume their responsibility as participants in the building up of a national history – something Peter Kosminsky tries to achieve by espousing controversial views on recent events.

NOTES

1 My translation of "La mémoire collective est le souvenir ou l'ensemble des souvenirs, conscients ou non, d'une expérience vécue et/ou mythifiée par la collectivité vivante de l'identité de laquelle le sentiment du passé fait partie intégrante." See Pierre Nora, *Présent, nation, mémoire* (Paris: Gallimard, 2011), 300. All other unacknowledged translations are mine.

2 United Nations Protection Force, Richard C. Hall, *War in the Balkans: An Encyclopaedic History from the Fall of the Ottoman Empire to the Breakup of Yugoslavia* (Oxford: ABC-Clio, 2014), 316.

3 "Toute source est une source 'inventée' au même titre que tout 'individu historique', au sens où l'employait Max Weber, est une construction, un

idéal type." See Henry Rousso, *Face au passé. Essai sur la mémoire contemporaine* (Paris: Belin, 2016), 303.

4 Arlette Farge, *Le Goût de l'archive* (Paris: Seuil, 1989), 147.

5 Hans Kelsen, *Peace through Law* (New Jersey: The Lawbook Exchange, 2000), 71.

6 Another word for peacekeepers because of their light blue helmets.

7 Between 1992 and 1994, Dario Kordić was a Bosnian Croat politician and military commander of the Croatian Defence Council (HVO). He was sentenced to twenty-five years in prison for war crimes committed against Bosniacs and was released in June 2014.

8 "Warrior" was the name of the British tanks used at the time.

9 Sophie Bourdais, "Peter Kosminsky: La Grande-Bretagne a une responsabilité dans l'actuel conflit palestinien," *Télérama*, 17 mars 2011.

10 Maurice Halbwachs, *On Collective Memory* (Chicago: University of Chicago Press, 1992), 174.

11 Paul Ricoeur, *Memory, History, Forgetting* (Chicago: University of Chicago Press, 2004), 624.

12 John Fiske and John Hartley, *Reading Television* (London: Routledge, 2004), 200.

5

Intimate Memories of War

21

Requiem for a Tommy: Impersonality and Subjectivity in Stuart Cooper's *Overlord* (1975)

Nicole Cloarec

Both historians and film critics concur that when people think of war films, the Second World War is the war that has shaped the epitome of the genre.[1] Conversely, cinema has been the main vehicle for perpetuating its memory.[2] Unlike the Great War, which soon came to epitomize the absurd futility of modern conflicts, the Second World War remains in British collective memory what comes closest to a just and necessary war and the perpetuation of collective myths such as the "People's war," which have been well served by the cinema.[3] Few films have challenged this view, but Stuart Cooper's *Overlord* is one of them. Although it was conceived in the early 1970s, it appears as the antithesis of the cycle of war epics that was spawned by the success of *The Longest Day* (1962), all restaging major campaigns of the Second World War,[4] filmed with big budgets, displaying high-production values and a galaxy of stars, all focusing on spectacular action sequences in colour and widescreen. In stark contrast, although its very title conjures up such epic war films, Cooper's *Overlord* is a low-budget, low-key, and intimate film shot in black and white with no well-known actors.[5] What is more, despite the expectations raised by its title, very little of the actual D-Day landing or its subsequent military campaign is shown. While in the famous movie *The Longest Day*, D-Day is undoubtedly the climactic moment of the film, in *Overlord* the military operation appears more as an anti-climax since, in a bitterly ironic twist, the main protagonist is killed even before landing on the beach at the end.

However, Cooper's film is also a far cry from the cynical view adopted in satiric films that started appearing in the 1960s, films such as Britain's *How I Won the War* (1967) or the United States's *Catch-22* (1970),[6] in which exposing the absurdity of the Second World War was inevitably perceived as an indirect indictment of the American involvement in Vietnam. Indeed, if *Overlord* can be construed as an anti-war film, its tone and aesthetics are unique, not least because the film was produced by the Imperial War Museum and pays tribute to the newsreel and service film units who recorded the war,[7] and of whose archival footage the film makes extensive use.[8]

I first analyze how this original use of archival footage makes *Overlord* a hybrid feature film, in which the blending of fiction and documentary creates tensions and discrepancies between different fields of perception that disrupt both time and space. What is more, these disruptions raise the critical question of the different points of view that the film adopts. As they dramatize the incompatible articulation between the impersonal and the infra-subjective, their interplay ultimately questions both the position of the viewers and the ability of all images to depict the "real" and convey any truths.

FROM THE "OVERLORD EMBROIDERY" TO *OVERLORD*

The film originated in 1972 when Stuart Cooper was hired to make a documentary about the "Overlord Embroidery" that was commissioned by Lord Dulverton in 1968 to commemorate the Normandy landings of 1944.[9] This led the filmmaker to researching the film archives of the Imperial War Museum, spending more than three thousand hours in the vaults. Cooper never actually completed the documentary but, instead, conceived a feature film, shot in black and white and combining a fictional narrative with about one-third of archival footage.

In the foreword to the novel that Stuart Cooper wrote with Christopher Hudson after the film, the authors insist on the unprecedented truthfulness of the film's depiction of war experience, based on authentic historical sources: "The outcome of the last great war has shaped the lives of a whole generation – but it is a generation which has only experienced the chaos, cruelty and uncertainty of those years through the media. Direct knowledge would be impossible, but even now books and films continue to hack out the same old fantasies: blood and guts, excitement and lantern-jawed heroes. Here at last, is a story that describes what it was really like. *Overlord* – the allied code name for the D-Day landings – uses unique and authentic sources from the Imperial War Museum. Based on detailed research, incidents depicted actually happened."[10]

Such a claim, however, is fairly common – all war films have claimed to be "the real thing." Nor is the insertion of archival footage unusual as many war films have been using it to give dramatic authenticity and heighten realistic effects,[11] even though the results are not always as successful as expected since archival footage is often so powerful that it sets off the artificiality of the reconstitution.

What makes *Overlord* unique, though, is the extent of its use of archival footage, which amounts to a third of the movie and,[12] even more crucially, its radically different approach. Whereas in all other cases archival footage is meant to substantiate the film's storyline, *Overlord* proceeds the other way around: the storyline was actually written after the archival footage was selected – thus it was conceived to fit and serve the footage, which remains the main driving force.[13] Moreover, in parallel with the use of the visual archive, most of the storyline itself is also based on period testimony, chosen from the diaries, letters, and postcards of ordinary soldiers.[14]

BLENDING ARCHIVAL FOOTAGE AND LIVE ACTION

In this regard, the main storyline appears definitely loose and tenuous. Contrary to what the title might suggest, very little action occurs, and, even more strikingly, very little dialogue is used. All the usual tropes from war films are thwarted. If most of the film follows Private Tom Beddows through basic military training up to his first assignment on D-Day, it eschews any sociological portrayal of recruits who stand for a cross-section of the whole society or the equally classical narrative structure depicting a motley crew of awkward army conscripts who are eventually moulded into an effective fighting force.[15] Nor is the film a satiric indictment of inhumane drilling, as in *Full Metal Jacket* (1987), for example.[16] Likewise, if Tom meets a woman one night in a NAAFI, their romance soon fizzles out.

What prevails is a series of impressionistic tableaux, scenes that alternate parades, drills, long periods of emptiness in which soldiers are whiling away time, travelling on trains, or in trucks from camp to camp. This creates an overall feeling of disconnection and isolation from any meaningful structuring principle. Many scenes include long silent takes of empty space, devoid of any human figure. When on screen, Tom, the main protagonist, is often filmed alone as a diminutive figure isolated in a large landscape or decentred within the frame.[17] Except for a few interactions with the woman he encounters and his two mates Arthur and Jack, there are remarkably few counter-shots: most of the orders – and the people shouting them – remain off screen. Tom's face is filmed in close shot facing the camera, facing the viewer in a virtual counter-shot. In the film, improbable counter-shots are provided through editing with the intercutting of archival footage, which is often presented as if it were taking place in Tom's mind, most specifically while he is shown dozing off in transitory states induced by his journeys.

Indeed, one of *Overlord*'s remarkable features is the way the film blends the archival footage with live action. Contrary to most other feature films that use already-edited compounds devised for documentaries or newsreels,[18] and, with the exception of the propaganda film that Tom watches in a film theatre,[19] Cooper actually used non-edited original nitrate negative for *Overlord*, which accounts for its high, pristine quality. Conjointly, cinematographer John Alcott was able to find two sets of 1936 and 1938 German Goetz and Schneider lenses, which he rebuilt and remounted so as to aesthetically match the look of the 1940s archive. Thus, archival footage and live action are stylistically matched by the technical quality of lenses and lighting. The use of lap dissolves, and some graphic editing and cutting on action (as when Tom is filmed running forward at the beginning), also contribute to providing smooth editing between the heterogeneous sequences.

Nevertheless, for all this elaborate stylistic continuity, the film's narrative structure is characterized by discontinuity, marked by numerous ellipses and ruptures. No explicit link between the archives and storyline is made. When Tom is first filmed leaving home after he has been called up, his train journey is intercut with images of the Blitz. In the next scene, Tom is alone on a station platform after he has missed his connecting train to the camp. From a hint from the station master we then surmise that Tom's train was held up because of bombing, but this remains the only connection that is made explicitly. Many scenes appear utterly disconnected: the baptism scene in the vault of Saint Paul's Cathedral, for example, is not only filmed from afar but offers no direct link to any characters or events from the storyline. Later, montage sequences of archival footage are only loosely connected through the soundtrack, through either a song playing or Tom's voice reading the letter he is writing home.

While in most other war films the insertion of archival footage into live action is used to provide a larger, more comprehensive vision of the theatre of operations, thus allowing some meaningful structure to emerge, here, what prevails is discontinuity, creating acute tension and discrepancies between macro and micro vision, impersonality and subjectification.

DISJOINTED TIME

Most significantly, except for D-Day, no temporal landmark is clearly stated. Although most of the archival images refer to emblematic events that have shaped the collective memory of the war – the arrival of the German troops in Dunkirk, the London Fire Brigade fighting the blazes, the large-scale evacuation of children, the bombing raids over Essen – they seem to evolve in a parallel time line to Tom's story. Following the landmarks, Tom joins the army in 1940 but then has no assignment until D-Day, which is rather unlikely. After the opening of the film on a montage sequence of archival documents to which I will return, the first live action scene occurs on D-Day, showing soldiers waiting in a landing craft. The camera dolly slowly moves from left to right detailing the anxious face of each soldier. Just before the end, the same scene reoccurs, but this time the camera dolly moves from right to left and includes the main character, Tom, at the end of the line. The whole film comes full circle, framed within this moment fraught with tension and foreboding. Time appears disjointed, marked by repetitions and entrapment, echoing the feeling of disorientation evoked by the soldiers themselves as they constantly move from one camp to another. As one of them comments: "I've lost all sense of direction." His remark is followed by performer Nick Curtis singing "We Don't Know Where We're Going (Until We're There!)"[20]

Tom's "journey" is thus marked by a compelling sense of entrapment and doom, as the scene of soldiers waiting in the craft occurs three times with slight variations – at the beginning without him, then as part of his training, then just before the end. But this recurring realistically shot scene is accompanied by dream-like scenes of premonition occurring three times as well. Right at the beginning, even before the main character is introduced, a blurred silhouette is filmed in long shot as he is running, facing the camera, before being suddenly shot.[21] The same leitmotiv recurs in slow motion when Tom tumbles down a hill during training (after twenty-one minutes). Then, the first scene is being re-enacted as part of a dream in the first half of the film (after thirty minutes) and at the end, but this time it is filmed in an extreme close shot focused on the eye of the character.

However, this premonition might as well be the ultimate irony in thwarting Tom's destiny and the film's climactic ending since it is dramatized only as a dream sequence. In the actual narrative, Tom never lands and is actually hit within the craft by a stray bullet. Then for us viewers, where we might have expected a scene evoking Robert Capa's legendary photographs of the D-Day landing, we are instead reminded of his iconic photograph of the "Falling Soldier" taken during the Spanish Civil War. What is striking in Cooper's opting for a different angle is the tension between the iconic model capturing a "decisive moment" that condenses all action before and after,[22] and its unfolding in filmic time. Along with the use of slow motion and soft focus, this creates an uncanny effect of time expanding implausibly, highlighting the unearthly feeling of agony.

In addition to these recurring flash-forwards, the film's timeline is further disrupted by dream sequences or sequences that appear as flashbacks.[23] Tom meets a dead German soldier in a dream,[24] and he takes out his wallet to peruse his personal items and photographs. The next shot shows a photograph of Tom himself, suggesting the dead soldier was his doppelganger. Likewise, Tom meets the woman again only in dreams while he is on the landing craft, saying goodbye to her during a parade and meeting in a large empty shed where she shows him how "to prepare the dead." Then even what appears as memory is endowed with dream-like quality, as when Tom and his mate Arthur slip away on their last night before landing and come across a little girl forced to sing in an empty theatre.[25] The whole scene is filmed in strikingly dramatic lighting that conveys a powerfully surreal effect, turning an anecdote into an uncanny, slightly unnerving episode. Tom's storyline is thus both utterly mundane, reduced to the routine steps of soldiers whiling away time, and endowed with unrealistic poetic effects, which are underscored by the musical leitmotiv associated with him. Paul Glass's music is melancholic and hauntingly moving, at times dissonant but mostly reminiscent of the romantic style of Vaughan Williams's pastoral symphony, which itself was conceived as an elegiac meditation on the First World War.[26]

CHALLENGING HUMAN SCALES OF PERCEPTION

As Tom's fluxes of consciousness intermingle with archival footage, an extreme tension is created between the impersonal and the infra-subjective. In most classical movies that purport to subjectivize war, History becomes a meaningful background against which personal ordeals are played out, where, however much they embody the "common man," the main protagonists eventually fulfill exceptional fates. Here, by contrast, not only does Tom remain everyman, but History is depicted as an impersonal machine, a huge juggernaut like the Great Panjandrum, the wheeled infernal-like contraption that is tested on a Devon beach alongside similar monstrous mechanisms.

In *Overlord*, the experience of war is properly dehumanizing not so much because of the extreme feelings of fear and horror it brings about but because it altogether challenges all human scales of perception. The ellipses and ruptures between the different storylines, the discontinuities between their levels of reality within the narrative, expose the unfathomable discrepancies between incompatible fields of perception in what Paul Virilio refers to as the war's "logistics of perception."[27] As Tom writes in the letter he will never send: "It's like being part of a machine which gets bigger and bigger while we grow smaller and smaller until there's nothing left." As he speaks, a slow tracking out underscores his words.

In this respect, archival footage may first appear to stand in contrast to the dream-like sequences of the staged fiction, and there is no denying it provides a most powerful sense of immediacy and connection to the past. And yet, its intermingling with the flashbacks, flashforwards, and dream sequences questions the very status of the images, which becomes unclear. Some of them, although they have long been familiar icons of the war, become uncannily unfamiliar: London's East End engulfed under the blaze and heavy smoke regains a surreal, properly haunting quality; night aerial bombings appear like abstract fireworks, generating a startling beauty out of sheer horror.[28]

In *Overlord*, the use of found footage acknowledges the essential ambiguity of the film image. While undoubtedly retaining its referential potency, it becomes a figuration of the past rather than a simple index of it. As Catherine Russell aptly points out: "The found image doubles the historical real as both truth and fiction … Its intertextuality is always also an allegory of history, a montage of memory traces by which the filmmaker engages with the past through recall, retrieval and recycling."[29]

Significantly, as the film starts with a sequence of archival footage, it opens on a very lengthy black screen. For more than one minute, the only sound that is heard is the pounding of horses' hooves, marching feet, and engines revving, increasing in intensity – all the sound effects of an army on the move. It is no small paradox to start introducing a sequence of realistic documents with only sound effects, thus conjuring up the viewer's inner representations and subjectivity. From the outset, seeing is thus exposed as problematic. It conveys a paradoxical reminder that all archival footage is actually silent and any sound effect comes as an addition. This also reminds the viewer of the production context, which from the start was characterized by some loss of senses as the camera was capturing only a segment of reality.[30]

Throughout the film recurrent disjunctions occur between image and sound, with other fades to black that linger while sound starts anew, or, conversely, abrupt cuts to silent archival shots. Life in the camp unfolds in silence, derealizing the trivial routines of the soldiers. At the end, the landing scene is introduced once more by a lengthy black screen with only the sound of distant explosion that becomes louder and louder; when the image appears again, the sound of intense bombing has become almost deafening but then stops abruptly, although explosions are still visible on the image track. Even within the staged fiction, some scenes unfold with no synchronous sound, in dream sequences as well as realistic ones. When the prostitute solicits Tom in the film theatre, they are filmed talking to each other in close shots but no dialogue is heard; at the end, after Tom has been hit by a stray bullet, all sound effects stop in favour of the music off.

A PERCEPTUAL EXPERIENCE OF THE IMPERSONAL

Because war challenges the limits of filmic representation, it raises the question of the strategies film adopts to depict war not only as a representation but also as an experience. In *Overlord*, this questioning is conveyed through the multiple disjunctions produced by the constant interplay between archival footage and staged drama as well as between the visual and sound tracks. While the former questions the respective ability of images to depict the "real" and convey any truths – be they impersonal or subjective, the latter interrogates perception through what escapes it – whether aurally or visually. Ultimately, as the film plays on different levels of reality involving discrepancies of scale and perspectives, it raises the question of the point of view that is assigned to the viewer.

Indeed, what is striking in the use of so much archival footage is its denial of any defined perspective. In archival footage, the viewer is made to align with the impersonal eye of the camera without any mediating characters. And although optical perspective is always acutely specific, close to point-of-view shots, this point of view remains anonymous, producing a disturbing tension between filming techniques that highlight subjectivity and its depersonalizing effects within the narrative. In other words, as the viewers lose their position of omniscient observers, they are effectively trapped in the perspective of the camera, which paradoxically induces both direct immersion into a perceptual experience and its distancing through foregrounding the technical apparatus of the camera replacing the eye.[31] James Chapman points out that this insistence on optical subjectivity has long been a feature of war films, and he traces its origin back to the cinema of attraction, "when 'phantom rides' were a favourite genre of filmmakers exploring the potential of camera subjectivity,"[32] whereby a camera was fixed on the front of a moving vehicle, usually a train, tram, or omnibus, which gave the illusion of movement. Chapman comments further: "A similar effect was achieved during the Second World War with operational footage of combat operations shot from cameras fixed to the nose or wings of combat aircraft. It soon became a convention of aerial combat films, both fictional and documentary, to include point-of-view shots through the bombsight and overhead shots of bombs falling into space ... The point-of-view shot through the bombsight, in particular, is another form of spectacle: the shot always has the effect of drawing attention to itself due to its extreme subjectivity."[33]

One striking feature of *Overlord* is the impressive number of aerial shots coming from archival documents, footage taken from the cockpit, through the bombsight or the trapdoor, which places the spectators in the position of both observers and participants, however unwilling they may be. What is more, aerial shots appear to challenge our perception of reality, erasing familiar markers for distance, surface, and depths, therefore questioning the possibility of their interpretation. As Rosalind Krauss writes about aerial photography:

> What is striking is that, unlike other types of photographs, aerial views raise the question of their interpretation and decoding. This is due not simply because, seen from far above, objects are difficult to recognize – which they are – but more specifically because the sculptural dimensions of reality become highly ambiguous: all difference between cavities and protrusions, between convex and concave, disappears. Aerial photographs make us face reality which is transformed into something that requires decoding. There is a hiatus between the angle of vision under which the photograph was taken and the one required to understand it. Aerial photographs disclose a breach in the fabric of reality, a breach that most ground photographers endeavour to conceal.[34]

Consequently, these impersonal images act as a double source of disturbance, both from a perceptual and from an ethical point of view. And unattributed point-of-view shots are not limited to archival footage sequences but also abound in the staged drama, as when scenes are filmed from the truck driver's seat or follow the gliding of the Lancaster bomber over the south coast of England and Corfe Castle. This aerial shot of Corfe Castle is the very last shot of the film, after intercutting between archival footage of the D-Day landing and of casualties carried back home.[35] Although it represents an echo of Tom's visit to the castle earlier in the film, such a distanced shot of a historical monument appears as a rather unexpected way to conclude such a subdued and intimate film.

Once again the human scale is discarded, this time in favour of a symbolic war-torn place that has stoically endured the ordeals of British history.[36] If the scene may be interpreted as some sort of transcendence of the sum of memories into History, what is more cogent still is the image of the ruins as a figuration of History. What these ruins reveal is that the past can only be experienced as a loss while perpetuating the trace of this loss. And so can the memory of all Tommies be experienced through archival footage. In this respect *Overlord* does not simply present a war narrative set in the past: it also reflects on the possibilities and limits within which the experience of this past war can be represented and retrieved.

CONCLUSION

Overlord offers a surprisingly bleak view of one of the most famous episodes of the war that made the British so proud and rightly so – yet, in a way, it remains very much in tune with the British spirit of quiet, stoic endurance that characterizes so many classical British war films.[37] Although any action is thwarted, the character's nobility lies in enduring his fate while being aware of its absurdity.

However, *Overlord* radically differs from most other war films in refusing to adopt a visceral way of shooting, meant to immerse the viewers in the experiences of the main characters. Instead, the film offers a poetic meditation on war and a reflection on how its experience can be represented and retrieved. Its aesthetic hybridity, blending archival footage with staged drama, builds up a series of contrasts and discrepancies that highlight the intractable tension between the different scales and "fields of perceptions" that war produces: impersonal on the one hand, infra-subjective on the other.

If the main tension between the impersonal and the infra-subjective conveys a heartfelt indictment of warfare depicted as a large-scale impersonal machine crushing humans, it also questions the respective ability of images to depict the "real" and convey any truths – be they impersonal or subjective. Ultimately, both modes of representation are equally subject to surreal, almost abstract effects, rewriting one of the most famous episodes of the Second World War into a stark elegiac film poem.

NOTES

1 In her pioneering study of war films as a genre, Jeanine Basinger writes: "Triggered by a catastrophic historical event and the resulting social upheaval, this genre [the Second World War combat film] filled the needs of the wartime public for information placed in a narrative, and thus more personal, context. Juxtaposed as it was with actual newsreel and documentary film, as well as with numerous newspapers and magazine photographs of 'reality,' this genre provided comparison, contrast and emotional relief. The World War II combat genre existed for the period of the war, but by virtue of its popularity has remained a genre (or accepted story pattern for films) until the present day. Furthermore, once established, the combat film influenced the entire concept of the war film. The pattern of the World War II combat movie is now the most common pattern for all

combat movies." See Jeanine Basinger, *The World War II Combat Film: Anatomy of a Genre* (Middleton, CT: Wesleyan University Press, 2003), 8–9. English historians and film critics concur: "The Second World War is a visual war above anything else. In the popular mind the definition of a war film is one made during or about the Second World War." See Mark Connelly, *We Can Take It! Britain and the Memory of the Second World War* (Harlow: Pearson Longman, 2004), 6; "As the last war with set piece battles and almost universal involvement, the Second World War is likely to continue to dominate the way in which the war is represented in the cinema." See Robert Murphy, "War. The Triumphs and Tragedies of Battle," *screenonline*, http://www.screenonline.org.uk/film/id/446224/.

2 "War films – including anti-war films – have established the prevailing public images of war in the twentieth century … The image of combat in World War II remains shaped for American audiences by films like *Sands of Iwo Jima* (US, 1949) and *The Longest Day* (US, 1962)." "The photographic and celluloid images of World War II are so frequently used and reused that they have begun to supersede experience and memory." See John Whiteclay Chambers II and David Culbert, eds, *World War II, Film, and History* (Oxford, New York: Oxford University Press, 1996), 2, 155; "At the same time, however, film has determined popular impressions of what war is like. From *The Battle of the Somme* to the reporting of the Gulf, Bosnia and Iraq wars, film has purported to show the reality of war even while it has resorted to artifice in doing so … The popular response to films such as *Saving Private Ryan*, acclaimed for their realism and authenticity, suggests that the public at large have become conditioned to accepting the filmic representation of war as being like the real thing." See James Chapman, *War and Film* (London: Reaction Books, 2008), 246.

3 For an analysis of the part cinema played in upholding "the British myth of the Second World War," see Mark Connelly: "The essentials of the British Second World War we know today were put together between 1939 and 1960. Our contemporary conception of the war is the result of a strange fusion of images produced in wartime and reactions to a glut of post-war remembrances. The most potent mediator of the British version of the war was the cinema, in particular the on-screen reworking of the Second World War that occurred during the 1950s" (Connelly, *We Can Take It*, 198).

4 *Battle of the Bulge* (1965), *Anzio* (1968), *Battle of Britain* (1969), *Tora! Tora! Tora!* (1970), *Midway* (1976), and *A Bridge Too Far* (1977).

5 In a 2008 interview for the *Guardian*, Stuart Cooper explains the final certified cost of *Overlord* was £89,951. See Stuart Cooper, "A Camera

Instead of a Rifle," *Guardian*, 18 January 2008. The main part was interpreted by Brian Stirner, who was playing his first role in cinema. The low cost was partly made possible thanks to the contribution of the Imperial War Museum, whose staff designed all sets, props, and costumes. The crew also benefited from the help of the Ministry of Defence, who provided a lot of facilities such as the main barracks and training camp, and a Lancaster bomber (the last operational one in the UK at the time) for aerial shots.

6 Although such films are not numerous, British films include *The Long and the Short and the Tall* (1961), *How I Won the War* (1967), *The Long Day's Dying* (1968), and *Play Dirty* (1968). Similar tendencies can be seen in *The Dirty Dozen* (1967) and *Too Late the Hero* (1969), which have some British input but are essentially American films.

7 The novel that Stuart Cooper wrote with Christopher Hudson after the film is dedicated to "the cameramen and photographers of the second World War."

8 Notwithstanding the prize it won at the 1975 Berlin Film Festival, the film was not released or broadcast in the UK until 2014. It also failed to find an American distributor until 2006 but was broadcast on Z Channel in the 1980s.

9 Since 1984, the embroidery has been housed in the D-Day Museum in Southsea, Portsmouth.

10 Stuart Cooper and Christopher Hudson, *Overlord: Codename: D-Day, June 6, 1944* (New York, Lincoln, Shanghai: Authors Choice Press, 2007 [1975]).

11 Classical tropes include the use of archival footage as authentic stamp in the opening credits (e.g., in *The Victors* [1963] or *The Eagle Has Landed* [1977]), the insertion of newsreels within the films (often watched by the characters in film theatres), or the intermingling of archival footage with live action and reconstruction through intercutting or even back screening (to name but a few examples: *Bombardier* [1943], *Iwo Jima* [1945], *Theirs Is the Glory* [1946], *They Were Not Divided* [1950]; *The Cruel Sea* [1953]; *The Dam Busters* [1955].

12 *Overlord* makes extensive use of archival footage: most originate from the Army Film and Photographic Unit (the scenes of drills and training at sea, of soldiers whiling away the time at the training camp, footage of D-day landing and the bringing back of casualties) and the RAF (the night bombing raids over Hamburg and Essen, day air footage of shooting at trains and other transports), but some footage also comes from the captured German film collection (footage of Dunkirk with the arrival of the

victorious German troops, the aerial shots of Paris's Arc de Triomphe and of Rome's Coliseum, shots of Hitler on a plane, which were part of the German propaganda film *The Baptism of Fire* documenting the invasion of Poland), the London Fire Brigade (images of the Blitz), and British railway footage (the children's evacuation).

13 Significantly, the live action scenes were shot in two weeks as opposed to seven months spent in the cutting room.

14 In particular, Stuart Cooper and Christopher Hudson explain they used the diaries of two Scotsmen, Sergeant Edward Robert McCosh and Sergeant Finlay Campbell, who were both D-Day soldiers. Their diaries inspired the scenes of the filling-in of the will form, the birthday anecdote, the bonfire, Arthur and Tom slipping away on their last night before landing, and coming across the little girl forced to rehearse in the empty theatre.

15 One classical instance is Carol Reed's *The Way Ahead* (1944).

16 Similarly, *Overlord* eschews traditional stereotyping and has no "whipping boy" or "chaps' favourite chum" among the army boys.

17 For example, after Tom arrives at the camp as he is told he is late and during parade reviews, which are filmed in lateral tracking shots.

18 As some critics have noted, one notable exception is Jeremy Isaacs's epic twenty-six episode TV documentary *The World at War* (1973), for which Isaacs and his crew also did thorough archival research. James Chapman thus notes: "*The World at War* differed from previous television documentaries, such as the American produced *Victory at Sea* and *Crusade in Europe*, in that it was compiled from unedited actuality film held by the archives rather than using only the commercial newsreel libraries. The production team embarked upon what, at the time, was the most extensive research in the film archives (though they were denied access to Soviet and East German sources) and, led by the producer, Jeremy Isaacs, were at pains to ensure that they used only actuality film and not reconstruction" (Chapman, *War and Film*, 58).

19 One exception in the film is the propaganda film *Lambeth Walk*, also known as *Germany Calling*, which Tom watches in a film theatre. The film was produced by the Ministry of Information and edited by Charles A. Ridley in 1941. It cleverly recuts shots from Leni Riefenstahl's film *Triumph of the Will* to the rhythm of the popular 1937 tune "The Lambeth Walk" so as to ridicule the Nazis shown goose-stepping back and fro. The film was very popular at the time.

20 The song was written by Noel Gay (music) and Ralph Butler (lyrics) and published in 1944. Interestingly, Nick Curtis was to become famous for performing the signature tune of the series *Goodnight Sweetheart* (BBC1,

1993–99), in which the main protagonist travels back in time to the Second World War.

21 Significantly, the first subjective shots are introduced with no specific relation to any character since his face remains unrecognizable.

22 The term was used by Henri Cartier-Bresson in the keynote text of his 1952 book *Images à la sauvette,* whose English-language edition was titled *The Decisive Moment.* The photographer explains: "Photographier: c'est dans un même instant et en une fraction de seconde reconnaître un fait et l'organisation rigoureuse de formes perçues visuellement qui expriment et signifient ce fait." [To me, photography is the simultaneous recognition, in a fraction of a second, of the significance of an event as well as of a precise organization of forms that give that event its proper expression.] My translation. See Henri Cartier-Bresson, *The Decisive Moment* (New York: Simon and Schuster, 1952), 1–14.

23 In this respect, one may see an ironical comment on what happens to the film's narrative timeline in the insertion of "The Lambeth Walk" and its contrivance of editing the same shots to and fro.

24 The dream encounter of a German soldier is strongly reminiscent of Wilfred Owen's poem "Strange Meeting" (1919), in which the narrator has killed a sleeping German soldier who is also a mirrored self, foreshadowing his own encounter with death.

25 The song is the folk song "Let Him Go, Let Him Tarry," which ironically features in the 1945 war film *The Way to the Stars.* The couplet can be heard distinctly:
"Let him go, let him tarry, let him sink or let him swim
He doesn't care for me and I don't care for him
He can go and get another that I hope he will enjoy
For I'm goin' to marry a far nicer boy."

26 It is interesting to note that Vaughan Williams's Symphony No. 3, published as *A Pastoral Symphony* in 1922, as well as his later Symphony No. 5 in D Major written between 1938 and 1943, were inspired by the First World War and Second World War, respectively. Both are elegiac meditations on war and peace and a tribute to the dead.

27 Paul Virilio, *War and Cinema: The Logistics of Perception*, trans. Patrick Camiller (London: Verso, 2009), 8. In *War and Cinema* Paul Virilio analyzes the interconnections between military and visual techniques, from aerial photography to film propaganda to modern simulators. Indeed, Virilio's premise is that there is no war without representation. From this perspective, cinema is defined not so much as the mere production or reproduction of images but as their manipulation through time and space,

which, most specifically, produces new dimensions in movement and depth of field.

28 Cooper actually intercuts these shots of abstract pyrotechnic battle with ground shots of devastated buildings and casualties to remind the viewer of the actual effects.

29 Catherine Russell, *Experimental Ethnography: The Work of Film in the Age of Video* (Durham, ND: Duke University Press, 1999), 238.

30 In an essay about the relationship between cinema and history, Jean-Louis Comolli asserts: "Une bien curieuse demande insiste aujourd'hui quant aux archives filmées. Il est exigé qu'elles fournissent une preuve irréfutable de l'existence de ce qui a été, alors qu'elles ne peuvent porter trace que de la conjonction particulière qui les a permises et fabriquées, rencontre circonstancielle d'un référent et d'un procès de production, conjonction remise en jeu dans une seconde rencontre non moins circonstancielle, celle de cette trace et du regard qui se dit capable de l'interpréter." [Today archive footage is expected to act in a very curious way. One demands it provides irrefutable evidence of what has been, whereas it can only testify to the particular conjunction which produced it, namely, the circumstantial meeting between a referent and a production process. This conjunction is itself brought again into play in a second – no less circumstantial – meeting between these traces and the interpreting gaze.] My translation. See Jean-Louis Comolli, *Arrêt sur histoire* (Paris: Centre Georges Pompidou, 1997), 38–9.

31 François Niney expounds on this paradox: "La caméra subjective aura alors produit l'effet inverse de la fascination qu'on lui prête d'habitude: elle aura rendu visible la mise en scène, provoqué le recul du spectateur, sa 'désidentification.'" . [The subjective camera will then have produced the reverse effect of fascination that it is usually supposed to create: it will have exposed the *mise en scene*, provoking the viewer's distancing and 'de-identification.'"] My translation. See François Niney, *L'Epreuve du réel à l'écran: Essai sur le principe de réalité documentaire* (Brussels: De Boeck Université, 2000), 213.

32 Chapman, *War and Film*, 96.

33 Ibid.

34 "Ce qui frappe c'est que, contrairement à la plupart des autres photographies, la vue aérienne soulève la question de l'interprétation, de la lecture. Il ne s'agit pas simplement du fait que, vus de très haut, les objets sont difficiles à reconnaitre – ils le sont effectivement – mais plus spécialement de ceci que les dimensions sculpturales de la réalité sont rendues très ambiguës: la différence entre creux et saillies, convexe et concave, s'efface.

La photographie aérienne nous met en face d'une 'réalité' transformée en quelque chose qui nécessite un décodage. Il y a césure entre l'angle de vision sous lequel la photo a été prise et cet autre angle de vision qui est requis pour la comprendre. La photographie aérienne dévoile donc une déchirure dans le tissu de la réalité, une déchirure que la plupart des photographes au sol tentent ardemment de masquer." See Rosalind Krauss, "La photographie comme texte: Le cas Namuth/Pollock," *Le Photographique: Pour une Théorie des Écarts*, trans. Marc Bloch and Jean Kempf (Paris: Macula, 1990), 96–7. My translation.

35 The cross-cutting between the different types of landings – of D-Day soldiers disembarking on Sword Beach and of causalities carried back home – provides an unusual demonstrative comment, standing for a cause and effect relationship of war.

36 Corfe Castle stands above the village of the same name on the Isle of Purbeck in Dorset. It was built by William the Conqueror and is well known for being twice besieged by the Parliamentarian army during the English Civil War. Its National Trust website stresses its eventful history and endurance: "Thousand-year-old royal castle shaped by warfare"; "Corfe Castle is a battle-scarred survivor with a turbulent past." See "Corfe Castle," http://www.nationaltrust.org.uk/corfe-castle.

37 "The popularity of British as against American war films in the 1950s indicates that it was the myths around Britain's achievement in the Second World War rather than a celebration of war as such that attracted audiences. Few of the films can be accused of romanticizing or glamorizing war, and the predominant tone is one of subdued realism – the stiff-upper lip ethos acting as a convenient shorthand for necessarily suppressed pain, bitterness and fear." See Robert Murphy, *British Cinema and the Second World War* (London: Continuum, 2000), 233.

22

"Our Visit to Waterloo": Representing the Battlefield in the Memoirs of Charlotte Eaton and Elizabeth Butler

Nathalie Saudo-Welby

In June 1815, hundreds of civilians followed the British troops to Belgium and thus found themselves some thirty kilometres away from Waterloo when Napoleon came back from Elba. Among the British non-combatants who were present in Brussels at the time, there were officers' wives, such as Lady de Lancey and Madame d'Arblay (Fanny Burney), as well as busybodies who had anticipated the conflict and were eager to experience the social events that preceded the military operations. The war having broken out earlier than expected, the residents in Brussels describe scenes of confusion and panic. They were unable to tell whether the chariots rambling past their windows were moving forward to the front or retreating in flight. They were caught between injunctions to flee and the desire to stay, between contradictory rumours of victory and imminent defeat. The well-off ladies staying around Brussels's Parc spoke of one another in their memoirs. They visited one another and constantly ran to the window to try and make sense of the tumult outside. Madame d'Arblay writes in her journal: "The individuals … only resided at the windows: so that the whole population of the city seemed constantly in public view."[1] On the pretext of gaining privileged access to what was happening outside, people placed themselves in a position of forced visibility and, hence, publicity.

One such inquisitive and visible witness was Charlotte Eaton, born Charlotte Waldie. In June 1815, she went to Brussels with her brother and younger sister Jane, an accomplished painter who soon after painted a watercolor panorama of the site of the battle. A first version of Charlotte's narrative of their experience in Brussels was published anonymously under the title *The Battle of Waterloo* as early as August 1815, along with a plan of the battle and Jane's watercolor panorama, a wide fold-out page, designed to give the reader an idea of the topography of the place. This bucolic landscape is empty of people or animal, except for two tiny human figures standing in front of the door of La Haye Sainte. Charlotte Eaton then rewrote and considerably expanded her text into her memoir, which came out in 1817 as *A Narrative of a Residence in Belgium during the Campaign of 1815*.[2] This longer version contains a vivid description of her visit to the battlefield. The immediacy and intensity of her personal experience are so marked that they obscure the mediated character of her experience of the battle, as it was reported by the eye-witnesses she met.

Fifty years later, on her nineteenth birthday, Elizabeth Thompson, later Lady Butler, toured the sites around Waterloo. The account is one of the only youthful memories she retained in the autobiography she published in 1922, and it is inseparable from her choice, even at this early age, to be a military painter, "a type of work rare at the time in England but one associated with the prestige of history painting."[3] The account of her visit is directly followed by a chapter narrating her experiences as a student at the South Kensington School of Art and at the Female School of Art in London. The extract from her diaries included within the account of her visit to Waterloo is nourished by the thorough history education she was given by her father and by references to painterly and literary renderings of Waterloo.

Charlotte Eaton's and Lady Butler's narratives were thus written almost a century apart. Being present in Brussels in June 1815, and on the battlefield some three weeks after the events, Charlotte Eaton was both an immediate witness and a not very distant non-combatant. Her decentred position with respect to the battlefield was an intensified version of that of the British public. As often in British history, the fighting had not occurred at home,[4] and the British public were avid to bridge the gap that separated them from this victory. John Scott, who visited the battlefield a few weeks after Charlotte Eaton, noted that "multitudes were on the road following [his] footsteps."[5] It is my contention that Eaton built upon her decentred position to identify with the British population as a whole. Since hers was also a privileged position, her account creates and sustains the illusion that she can give her British readers the direct access to the events they were yearning for.

Lady Butler's account has greater literary merit than, and it throws a contrasting light on, Charlotte Eaton's text. It valorizes the notion of distance as an essential element of historical appreciation and as a prerequisite for artistic representation.

CHARLOTTE EATON'S NARRATIVE

The initial forty-page version of Eaton's narrative, published anonymously in 1815, was told in the first person plural, which contributed to depersonalizing it. It hardly resembles the second version, which is intensely personal and presents her as a first-hand witness.[6] Charlotte is indeed eager to remind her reader of how close she was to the events and boasts of her position as a privileged observer. The British tourists who rushed to Flanders in the months following the battle also hoped that they had been quick enough to have an immediate experience of the events before time had taken away their freshness. "I trust that I have not come too late to experience a portion of the advantage which has thus been enjoyed by the many writers who have taken Waterloo for their subject,"[7] John Scott wrote in his account of his visit in August 1815.[8] In her 1817 preface, Charlotte Eaton calls her narrative:

> the simple and faithful account of one who was *a spectator of the scenes she describes, and a witness of the events she relates*, during those days of desperate conflict and unparalleled victory, which must be forever memorable in British history, and interesting to every British heart. It was written whilst the impression of those eventful scenes was yet *fresh upon the mind* ... The Author must be permitted most earnestly to disclaim all idea of entering into competition with the writers whose talents and genius have been so well employed in describing the battle and the field of Waterloo. They were not, however, like the Author, *on the spot at the time*; they were pilgrims who afterwards visited the memorable scenes of these glorious events, and wrote from report: they related the past – she described the present.[9]

Contrary to a few first-hand women witnesses,[10] Charlotte Eaton was a non-participant in the battle and did not approach the battlefield until three weeks after the events. Her constant efforts to minimize the distance between herself and the events make her memoir seem overwritten: hers is an intensely subjective and jingoistic account, which today's readers might find misplaced and naïve. Her boast of being "on the spot at the time" may result from her need, as a woman, to legitimize her authority to narrate the events she describes. Her position as a non-combatant observer overlapped with that of the British public, who had little experience of combat on their own ground. The observant Charlotte could justifiably pretend to stand for the majority of the British people, who regarded Waterloo as a distant, foreign place that they could now claim as partly theirs. This may also justify the triumphalist and partial tone of her narrative. One can feel that Charlotte's exuberant display of emotions is often motivated by a desire to experience what the situation requires from her, as a British Woman.

Even though she calls her head "unmilitary,"[11] her unproblematic display of authority concerning the events she narrates might make her readers uneasy. "But I am relating the history of the battle, forgetful that I am only describing the field,"[12] she writes. Eaton often appears as the direct source of information and interpretation of what she sees, even though she was shown around the sites by officers and local farmers. The presence of these participants is sometimes suggested by incidental clauses, such as "I understood" or "we were told,"[13] but their voices generally disappear behind hers. Even though she gives pride of place to Baptiste la Coste, Napoleon's guide,[14] the discursiveness and individuality of what she is being told is downplayed.[15] Discursiveness was often foregrounded by the authors and artists who represented Waterloo. Sir Edwin Landseer's *A Dialogue at Waterloo* (1850) depicts the duke of Wellington and his daughter-in-law on horseback on the battlefield of Waterloo. They are talking to a peasant girl who is trying to sell them prints of the battle as well as shreds from uniforms or flags. The duke's massive presence and his downward-pointing finger, as if to say, "this is the exact spot ...," indicate his authoritative knowledge of the events and his right to narrate them. His authority seems to be accepted by the two attentive female listeners. The three dialogists seem to be silent and in accord at that precise moment, but the painting remains dialogic in so far as it builds upon a preceding dialogue and juxtaposes different and possibly differing perceptions of the same event. In the left-hand half of the painting, a picnic scene demonstrates, along with the girl's wares, how present-day life is making its own claims on what Waterloo is and has become. A hunter is thus using a strategic depression in the ground to court a countrywoman, unseen by her husband who is tending the horses and looking admiringly at the duke.

Eaton has no difficulty in adopting an authoritative posture similar to that of Wellington on Landseer's painting. At one point, she unobtrusively quotes her earlier, less personal account of the battle, crediting herself in a note.[16] Her comments are often judgmental: "in consequence of the cowardice of the Belgians and baggage-men," she writes, "the last and most dreadful alarm of Sunday night was spread over the whole country."[17]

Eaton's position of authority as the legitimate source of information is strengthened by her direct contact with the concrete remains of the fighting. The impression of immediacy is reinforced by references to the offensive smell of the corpses, which she describes as "in some places scarcely bearable. Deep stagnant pools of red putrid water, mingled with mortal remains, betrayed the spot where the bodies of men and horses had mingled together in death."[18] Such vivid descriptions give a sensory quality to her already graphic and emotional account, a technique frequently used in early visitors' narratives. With her eyes riveted to the ground, Eaton gives the reader long lists of the vestiges of war:

> The ground was ploughed up in several places with the charge of the cavalry, and the whole field was literally covered with soldiers' caps, shoes, gloves, belts and scabbards; broken feathers battered into the mud, remnants of tattered scarlet or blue cloth, bits of fur and leather, black socks and haversacks, belonging to the French soldiers, buckles, packs of cards, books, and innumerable papers of every description.
>
> I picked up a volume of *Candide*; a few sheets of sentimental love-letters, evidently belonging to some French novel; and many other pages of the same publication were flying over the field in much too muddy a state to be touched. One German Testament, not quite so dirty as many that were lying about, I carried with me nearly the whole day; printed French military returns, muster rolls, love-letters, and washing-bills; illegible songs, scattered sheets of military music, epistles without number in praise of "*l'Empereur, le Grand Napoleon* [sic]," and filled with the most confident anticipations of victory under

> his command, were strewed over the field which had been the scene of his defeat. The quantities of letters and of blank sheets of dirty writing paper were so great that they literally whitened the surface of the earth.[19]

There is some poetry in representing the battlefield as strewn with blood and paper, as if the words of the dead were waiting to be collected by the historian so as to become living words again. Most of the visitors who streamed over to Waterloo in the summer and autumn of 1815 noted the mixture of bodies, weaponry, papers, and personal objects that covered the ground. When John Scott visited the place a few weeks later, many of the relics were for sale.[20] The muddy mixture of blood and paper symbolically represents the difficulty of sorting out the subjective from the objective in historical narratives, which Eaton claims to leave to "[a]bler pens than [hers]."[21]

However, Charlotte Eaton's keen interest in the material vestiges of war raises questions related to the respect due to the battlefield and to the dead and their possessions. As she rummages around, "searching among the corn for some relics worthy of preservation," she comes upon a human face,[22] as well as "a human hand, almost reduced to a skeleton, outstretched above the ground as if it had raised itself from the grave."[23] Her attempts to connect to the battlefield in concrete terms result in a frantic collection of memorabilia, which today's readers are likely to find distasteful. Besides the papers mentioned above, she collects wild broom seeds in Hougoumont and gathers ashes, with the idea that they might become "all that is now to be found upon earth of the thousands who fell upon this fatal field!"[24] She also obtains from a poor woman "the broken sword of a British officer of infantry."[25] At the end of her visit, she hires a young girl to carry for her "for half a franc" the cuirass that she wants to bring back to England.[26]

Her taste for vestiges brings her very close to war looters and later shoppers for souvenirs alike. Ironically, she condemns the female looters without seeming to notice the similarity with herself: "there was still plenty of rubbish to be picked up upon the field, for those who had a taste for it like me – though the greatest part of it was in a most horrible state. It was astonishing with what dreadful haste the bodies of the dead had been pillaged. The work of plunder was carried on even during the battle ... The most daring and atrocious of these marauders were women."[27] The unavoidable echoes between these marauders and the author confirm the problematic nature of Charlotte Eaton's acquisitiveness and appropriation of the events. This acquisitiveness and appropriation resonates with her narrative method of taking the emotional power of the place upon herself rather than letting others' voices and emotions speak.[28]

Well satisfied with her personal examination of the place, Charlotte Eaton joins her brother and sister who have been quietly "taking sketches at a little distance."[29] By contrast, these appear to be more suitable and respectful witnesses. Their unobtrusive presence announces the coming of a later visitor, the Victorian painter Lady Butler.

LADY'S BUTLER'S ACCOUNT

Elizabeth Thompson joined what her father called the "tremendous ruck" of battle artists very early in her career.[30] She comments modestly in her *Autobiography* that there were so few battle artists at the time that her choice of this vocation may have allowed her to distinguish herself. Painting historical scenes for the Royal Academy made her part of a masculine universe, as did her marriage in 1877 to Major William Butler, a Catholic Irishman who had made a good career in the army and written several accounts of his campaigns.[31] Lady Butler described an official dinner with table companions, "of whom only six were ladies."[32] Her paintings feature very few women,[33] and her work was famously described by John Ruskin as "Amazon's work":[34] "I was very pleased to see myself in the character of an Amazon,"[35] she commented. The self-portrait she made in 1869 depicts her with boyish features, wearing a hat resembling a service cap, boldly facing the viewer and deliberately avoiding seduction.

Lady Butler's account of her visit begins with a long extract from her teenage diary, reporting in inverted commas how she visited the battlefield of Waterloo on her nineteenth birthday. Having been taught history very intensively by her father, Elizabeth Thompson was well informed about the facts and she drank in "with avidity" "the stirring narrative" of Sergeant Mundy, the veteran officer who served as a guide.[36]

The reported nature of what she hears is constantly foregrounded by verbs of discourse and by references to their guide, whom she variously calls "a fine old man," "the truthful old sergeant," the "positive old soldier," the "man of discipline," and "our mentor."[37] Her narrative also involves the man's daughter and an eighty-eight-year-old woman who nursed the wounded after the battle. The inclusion of these survivors confirms how receptive she is to the human side of the conflict as well as to the weight of the personal testimonies and memories of ordinary participants. These survivors have, in a sense, become living relics. By contrast, two paintings by two battle artists of her day are blamed for their inaccuracy: she calls upon Sir Edwin Landseer to hear old Mundy's declaration that the duke of Wellington never visited Waterloo again.[38] On Sergeant Mundy's authority, she also blames Daniel Maclise for mistakenly locating *The Meeting of Wellington and Blücher* at the farm of "La Belle Alliance" in his 1861 canvas. Since she claims to have written these pages immediately after their visit, the young Elizabeth appears remarkably familiar with the productions of her contemporaries.

The battlefield is no longer strewn with paper, but it calls up literary memories: "Ah! the wall with the loopholes. I knew all about it and hastened to look at it. Again all the wonderful stratagems and deeds of valour, etc. were related, and I have learnt the importance, not only of a little hedge, but of the slightest depression on a battlefield."[39] Elizabeth is here taking a strategy lesson, like a soldier. Her concrete involvement with the battlefield involves performing the movements of the soldiers so as to understand them:

> There was the field, planted with turnips, where our Guards lay down, and I could not believe that the seemingly insignificant little bank of the road, which sloped down to it, could have served to hide all those men until I went down and stooped, and then I understood, for only just the blades of the grass near me could I see against the sky. Our Guards must indeed have seemed to start out of the ground to the bewildered French, who were, by the by, just then deploying.[40]

Elizabeth is experiencing physically what she will later turn into the careful anatomical representation of movement.

However, her concrete involvement does not include any interest in souvenirs of the battle. She despises the vendors of war vestiges and expresses her relief at being protected from them by Sergeant Mundy. "He says they have sold enough bullets to supply a dozen battles,"[41] she writes. The vestiges no longer speak of Waterloo but of a morbid taste for collecting relics. A few years later, she would probably have found such objects useful as models to make her pictures more authentic,[42] and her memoir explains how she sometimes had to have veterans and their uniforms brought to her for posing sessions.[43] As a young girl, however, she was more sensitive to the memories of the living than to the remains of the dead. She describes the content of Sergeant Cotton's Museum as "a collection of the most pathetic old shakos and casques and blundering muskets, with pans and flints, belonging to friend and foe; rusty bullets and cannon balls, mouldering bits of accoutrements of men and horses, evil-smelling bits of uniforms and even hair, under glass cases; skulls perforated with balls, leg and arm bones in a heap in a wooden box; extracts from newspapers of that sensational time, most interesting; rusty swords and breastplates; medals and crosses, etc., etc., a dismal collection of relics of the dead and gone. Those mouldy relics! Let us get out into the sunshine."[44]

Lady Butler respected and maintained the distance that separated her from the time and place of combat. She constantly expressed distaste for the depiction of actual combat. Her preference went to the moments right before or after the battle.[45] Those of her pictures that depict moments of onslaught, such as *The Defence of Rorke's Drift*, *Scotland Forever* and *Floreat Etona!*, were not among her greatest artistic successes.[46] Being suspicious of sensationalism and conventional depictions of war heroism, she generally preferred to build on her decentred position as a woman to question standard representations of heroic fighting and the artists' preference for officers as subjects at the expense of rank-and-file soldiers. This approach was radicalized after her marriage, when she came to share her husband's anti-jingoistic attitude and his critical stance towards British imperial policy.

In her autobiography, her creed of distance is precisely borrowed from the duke of Wellington's answer to those who nagged him for details of the action at Waterloo:

> My own reading of war – that mysteriously inevitable recurrence throughout the sorrowful history of our world – is that it calls forth the noblest and the basest impulses of human nature. The painter should be careful to keep himself at a distance, lest the ignoble and vile details under his eyes should blind him irretrievably to the noble things that rise beyond. To see the mountain tops we must not approach the base, where the foot-hills mask

> the summits. Wellington's answer to enthusiastic artists and writers seeking information concerning the details of his crowning victory was full of meaning: "The best thing you can do for the Battle of Waterloo is to leave it alone." He had passed along the dreadful foothills which blocked his vision of the Alps.[47]

The details of the action of Waterloo are likened to foothills blocking vision, potentially turning the observer into a failed climber who remains level with the "ignoble and vile." By contrast, the summit represents the lofty purposes, which motivated the war, and the heroism and pathos that it generated. The metaphor provides an apt illustration of Eaton's and Butler's differing perceptions of the battles. Eaton had her eyes riveted to the ground, where she found an endless list of bloodstained objects, which were coveted by relatives, looters, and tourists alike. Lady Butler's conception of Waterloo is nourished by her historical and literary readings as well as by her experience of war painting and of coping with her two sons' participation in the First World War. These filters are so clearly foregrounded in the contents and organization of the text that they clarify our vision.

Lady Butler's critical and historical distance towards "Waterloo" is sufficient for it to gain a philosophical dimension. For Lady Butler, Waterloo is not a victory, it is not even a battle but, rather, the archetype of a human situation, fraught with complexities and paradoxes, generating ambivalent feelings of relief and grief, pride and disgust. It becomes an instrument of analysis, which allows her to adjust her response to later events. As she leaves the site along the Charleroi road, filled with sadness, she is reminded of her father quoting Wellington's words to her when leaving this place of "carnage": "A defeat is the only thing sadder than a victory."[48] As in the extract quoted above, Wellington's philosophical pronouncement has aesthetic implications: what truly matters in a battle is not the heroes, or the victors, but the human cost that her military paintings strive to show. Wellington's words are cited again, although in a slightly different form,[49] when Lady Butler comments on her 1877 canvas, *The Return from Inkerman*, the fourth of a series of paintings depicting episodes from the Crimean War: "No doubt for some the subject of this work is too sad, but my dominant feeling in painting it was that which Wellington gave expression to in those memorable words on leaving the field of battle at Waterloo: 'There is nothing sadder than a victory, except a defeat.' It shows the remnants of the Guards and the 20th Regiment and odds and ends of infantry returning in the grey of a November evening from the "Soldiers' Battle," most of the men very weary."[50]

With its heavy losses, Waterloo is the archetype of a sad victory and a model for later representations of victorious fighting. The violent fight on the heights of Inkerman cost the lives of twenty-five hundred British soldiers and twelve thousand Russians. Lady Butler's picture is particularly representative of her compositional technique: a few foregrounded figures are projecting forward, in a triangular shape, as if they were going to walk out of the canvas into the viewer's space. The massed troops behind them are passing dead bodies and their disregarded remains. They give the canvas its monumental proportions and a collective, national scope. The line of crows in the sky, which was initially flying from right to left (a bad omen), is shown to swerve back to the right (a good omen) in the foreground. The straggling column of wounded and exhausted men and the meditative and grave expression on the face of the officer mounted on a horse give no indication that they have just been victorious. Butler's paintings and titles rarely indicate whether the fight was successful or not.

Lady Butler's account of her visit to Waterloo concludes on a paragraph in which she speaks as a battle artist and as the mother of two sons who fought in the First World War. The narrator is now the remembering and writing "I," and there are no inverted commas: "About this same Battle of Waterloo. Before the Great War it always loomed large to me, as it were from the very summit of military history, indeed of all history. During the terrible years of the late War I thought my Waterloo would diminish in grandeur by comparison, and that the awful glamour so peculiar to it would be obliterated in the fumes of a later terror. But no, there it remains, that lurid glamour glows around it as before, and for the writer and for the painter its colour, its great form, its deep tones, remain."[51]

Having to face war "for real" carried a sombre irony for the battle artist. It also relativized her artistic achievement by challenging the authenticity of her work. The war broke out just as the "one-man" exhibition of her Waterloo watercolours was opening: "To think that I have lived to see it!"[52] But a few pages later, she wonders: "Who will look at my 'Waterloos' now? I have but one more of that series to do. Then I shall stop and turn all my attention and energy to this stupendous war."[53] Waterloo, which she once called "that inexhaustible battle,"[54] is preceded by the adjective "my" as in the preceding quote, but the plural form turns the idealism of "my Waterloo" into a metonymy for her numerous and inadequate representations of the battle. The occurrence of real war makes their relevance and suitability problematic. She takes a sudden dislike to the sparkle of uniforms and guns of the Napoleonic wars. She is now curious to learn what real preparation for war means, as she observes men learning to entrench themselves at Lyndhurst.

CONCLUSION

Lady Butler wrote, rather paradoxically, that "it was against [her] principles to paint a conflict."[55] While this assertion applies to one of her later depictions of Waterloo, *Dawn of Waterloo. The "Reveille"* (1895), it does not quite apply to *The 28th Regiment at Quatre Bras* (1875), where the action is so imminent that the guns appear to be already smoking, or to *Scotland Forever* (1881), which represents a frozen moment of action. Yet, her statement confirms that distance and displacements were part of her creed as an artist. Lady Butler's representations favour post-battle scenes, they also adopt a narrow perspective, focusing as they do on a few isolated figures. Lady Butler never adopts a panoramic, global view of the events, as in conventional military painting.[56] Her focus appears as deliberately selective and narrow. The same is true of her literary depiction of the battlefield in her autobiography, which is mediated by a specific voice, by discursive and artistic references, and by personal experiences that are specifically mentioned. This narrowness of vision is not to be understood in a negative sense, and it can be linked to her serene acceptance of her decentred position with respect to the conflicts.

This shows up very clearly the paradox or the weakness in Charlotte Eaton's narrative of her visit to the battlefield: in her readiness to demonstrate her proximity to the events, she responds to the battlefield by giving pride of place to her own emotions and sensations. The disproportionate tone of her response distances her readers further from the reality of the events. While trying to position herself as a privileged first-hand witness, she only succeeds in appearing as a prefiguration of the battlefield tourist.

While Charlotte Eaton strives to close the gap that separates her from the male preserve of war by emotionalizing her perception, insisting on the authenticity of her private feelings and striving to display the right emotions, Lady Butler chooses the opposite strategy, relying on the narrative of Sergeant Mundy, the "positive" guide and "mentor,"[57] who took her and her family around the sites of Hougoumont, La Haye Sainte, and Mont St Jean. Lady Butler's account uses her decentred female position to gain the right perspective on the battle, which she viewed through as many filters as she could find, without ever trying to remove them. On the contrary, she tends to exhibit these intermediaries and filters that separate her from the battle. This attitude turned into an aesthetic principle later in her career, when she opted for a representation of war that focused on its observable human effects.

NOTES

1 Fanny Burney, *Diary and Letters of Madame d'Arblay*, vol. 4/4 (London: 1840), 321.
2 It was republished in 1853 as *The Days of Battle* and in 1888 as *Waterloo Days*.
3 Penny Dunford, *A Biographical Dictionary of Women Artists in Europe and America since 1850–1990* (London: Harvester Wheatsheaf, 1990), 51.
4 These are the first lines of chapter 32 of *Vanity Fair* (1848): "We of peaceful London City have never beheld – and please God never shall witness – such a scene of hurry and alarm, as that which Brussels presented." See William Thackeray, *Vanity Fair* (Harmondsworth: Penguin, 1968), 369.
5 John Scott, *Paris Revisited in 1815 by way of Brussels, including a walk over the field of battle at Waterloo* (London: Longman, 1816), 170.
6 A note by Charlotte Eaton confirms that she was the author of the earlier version. I quote from an anthology that contains a re-edition of the 1888 edition by Edward Bell, which itself takes up the text of 1852. See Charlotte Eaton, *Ladies of Waterloo* (Oakpast: Leonaur, 2009), 141n31.
7 Scott, *Paris Revisited*, 223.
8 Eaton writes that, in the days following the battle, "there were those of my own country, and even of my own sex, whom I heard express a longing wish to visit this very morning the fatal field of Waterloo!" See Eaton, *Ladies of Waterloo*, 80.
9 Eaton, *Ladies of Waterloo*, 15, emphasis added. The preface of the 1817 edition is even more candid than that of 1852: "This little Narrative has, however, one claim on its attention which no other possesses, in being the simple and faithful account of one who was herself a spectator of the scenes she describes … and to whom every little circumstance was related

with all the freedom and egotism of colloquial intercourse." See Charlotte Eaton, *Narrative of a Residence in Belgium during the Campaign of 1815* (London: Murray, 1817), iv.

10 These include Juana Smith, who rode across the battlefield to join her husband, and Lady de Lancey, who spent a week at Mont-Saint-Jean where her husband, one of Wellington's close officers, lay dying.

11 Eaton, *Ladies of Waterloo*, 116.

12 Ibid., 127.

13 Ibid., 116, 124.

14 Ibid., 134–5. He is also Paul's guide in Walter Scott's *Paul's Letters to his Kinsfolk. The Prose Works,* vol. 5 (Paris: Galignani, 1827). His name is spelled in various ways: known today as Jean-Baptiste Decoster, he was called "Baptiste la Coste" by Eaton, and "John Lacoste" by Walter Scott.

15 There are several exceptions: "the ghastly spectacle … was described to me by an eye-witness of this scene of horror" (Eaton, *Ladies of Waterloo*, 125) and "a countryman, who belonged either to La Belle-Alliance, or to some of the neighbouring cottages, told me…" (139).

16 Eaton, *Ladies of Waterloo*, 140–1.

17 Ibid., 118.

18 Ibid.

19 Ibid., 127–8.

20 "Almost every house in the hamlet of Mont St Jean poured forth women and old men, to every fresh arrival of visitors, – who eagerly offered relics of the battle for sale. From the complete cuirass, the valuable sabre, carbine, and case of pistols, down to the buttons that had been torn from the jackets of the slain, – all the wreck of the field had been industriously collected, and each article found ready purchasers. Letters taken from the pockets of the dead, were frequently offered, and were always eagerly bought" (Scott, *Paris Revisited*, 207).

21 Eaton, *Ladies of Waterloo*, 146.

22 Ibid., 124.

23 Ibid., 129.

24 Ibid., 133.

25 Ibid., 132.

26 Ibid., 147–8.

27 Ibid., 142–3.

28 For instance, she condemns the "dreadful effects of the unrestrained indulgence of th[e] passionate and heart-breaking grief" of a widow and a sister looking for the corpse of "a brave and lamented British officer" (Eaton, *Ladies of Waterloo*, 144).

29 Eaton, *Ladies of Waterloo*, 147.
30 Elizabeth Butler, *Elizabeth Butler, Battle Artist: Autobiography* (Santa Fe, NM: Fisher Press 1993), 75.
31 Once Lady Butler entered her life as "a soldier's wife" (Butler, *Elizabeth Butler*, 133), she no longer devoted so much time and energy to her painting, and the latter part of her autobiography shows her to be absorbed by her numerous official duties and travel, which made her a familiar of Empress Eugenie and a frequenter of Queen Victoria.
32 Butler, *Elizabeth Butler*, 186.
33 One major exception must be made for *Evicted* (1890), which depicts a lonely Irish woman standing desperate in desolate surroundings as her house is being consumed by the fire. A crowd of women bystanders can be seen in *To the Front: French Cavalry leaving a Breton city on the declaration of war* (1888–89).
34 John Ruskin, *Notes on some of the Principal Pictures Exhibited in the Rooms of the Royal Academy: 1875*, (London: Allen 1875), 57.
35 Butler, *Elizabeth Butler*, 117.
36 Ibid., 27.
37 Ibid., 25, 26, 26, 27, 29.
38 Ibid., 26.
39 Ibid., 28.
40 Ibid., 27.
41 Ibid., 25.
42 Lady Butler's oil paintings on the subject are *The 28th Regiment at Quatre Bras* (1875) (National Gallery of Australia, Melbourne), *Scotland Forever* (1881) (Leeds Art Gallery), and *Dawn of Waterloo. The "Reveille" in the bivouac of the Scots Greys on the morning of the battle* (1895) (Palace of Falkland, Fifeshire). A fourth canvas, entitled *On the Morning of Waterloo* (1915), was probably destroyed after her husband's death.
43 "Because detail was crucial as a means of confirming the historical, and by implication, the moral veracity of the picture, Butler devoted much time to getting her subjects right." See Claire Bowen, "Lady Butler: The Reinvention of Military History," *Revue LISA/LISA e-journal*, 1, no. 1 (2003): 127–37, §9, https://doi.org/10.4000/lisa.3128
44 Butler, *Elizabeth Butler*, 26.
45 "Three of Butler's major successes, then, were characterized by the choice of a historical military subject, by the decision to paint a post-battle scene and by a desire to depict the effects of battle on ordinary soldiers" (Butler, *Elizabeth Lady Butler*, §8).

46 Ibid., 6.
47 Butler, *Elizabeth Butler*, 37.
48 Ibid., 29.
49 Charlotte Eaton cites yet another version of the same quote: "there is nothing more melancholy than a victory – except a defeat" (*Ladies of Waterloo*, 140).
50 Butler, *Elizabeth Butler*, 133.
51 Ibid., 29.
52 Ibid., 253.
53 Ibid., 258.
54 Ibid., 252.
55 Ibid., 148.
56 One can compare the last three previous paintings with H.F.E. Phillipoteaux's *La Charge des Cuirassiers français à Waterloo* (1844), as the authors of the catalogue of Lady Butler's works invite us to do. See Paul Usherwood and Jenny Spencer-Smith, *Lady Butler, Battle Artist, 1846–1933* (Gloucester: Alan Sutton, the National Army Museum, 1987), 163–4.
57 Butler, *Elizabeth Butler*, 26, 29.

23

Historically Estranged Generations: Memorials and the Relevance Effect in Nigel Farndale's *The Blasphemer* and Tatiana de Rosnay's *Sarah's Key*

Marzena Sokołowska-Paryż

Historical estrangement is a condition of those generations who have access to the past only through *lieux de mémoire* (sites of memory) that "originate with the sense that there is no spontaneous memory, that we must deliberately create archives, maintain anniversaries, [and] organize celebrations."[1] Sites of memory are an inevitable substitute for real memory, yet, in the words of James E. Young, "in this age of mass memory production and consumption, ... there seems to be an inverse proportion between the memorialization of the past and it contemplation and study."[2] For historically estranged generations, sites of memory may have as little significance as the past they were intended to commemorate. Memorials and museums tend all too easily to become mere tourist attractions, or – more often than not – they become "unremembered," that is, present but not noticed. The "temporal distance" unavoidably written into memorialscapes may nonetheless be "diminished" by the adoption of strategies that effectively shape people's "commitments and responses."[3] It is possible to "[make] past moments close and pressing"[4] by means of reducing "formal, affective, ideological, and cognitive distances"[5] and thus achieve the "relevance effect" to be understood as a deliberate construction of a "contemporary" significance for historical events.

The relevance effect is strongly connected to the concept of "formal distance," defined by Mark Salber Phillips as "the wide variety of textual or other representational devices that shape the reader's experience of the text, or – to change the scene – the visitor's tour of a monument or exhibition."[6]

Both Nigel Farndale's *The Blasphemer* (2010) and Tatiana de Rosnay's *Sarah's Key* (2007) use parallel plots to establish the relevance of the past of the present. *The Blasphemer* opens with a prologue set in July 1917, introducing Private Andrew Kennedy who has just arrived at the front prior to the oncoming battle of Passchendaele. The following chapter shifts time to present-day London, beset by threats of terrorist attacks, where Andrew's great grandson, Daniel Kennedy, a specialist in nematology, is preparing for a journey to the Galapagos Islands with his partner. There are altogether fourteen chapters set between July 1917 and September 1918, and thirty-one chapters set in the present day. Though the contemporary plot is obviously the structurally dominant one, the importance of both the framing and intercutting Great War plot is underscored by the use of present tense narration, which creates a stark contrast with the past tense narration in the contemporary plot. The use of the present tense stylistically eradicates the "pastness" of the past by producing an almost cinematic "now." The sudden shifts from present-day London to wartime Flanders and France augment the relevance of the Great War plot for understanding the contemporary one by spotlighting the reasons the story of Andrew Kennedy is consistently interrupting the story of Daniel Kennedy. Daniel Kennedy's panic-driven swim away from the crashed plane, where he leaves the mother of his child behind, is a mirror image of his great-grandfather's desertion from the Passchendaele battlefield, the parallel experiences underscoring an inherent similarity between possible cowardly conduct under extreme circumstances.

Sarah's Key adopts a similar strategy, with the first two chapters putting forth the leading characters of the novel, the (yet) unnamed Jewish girl taken away with her family by the French police in July 1942, and Julia Jarmond, an American journalist who is given the task to write about the infamous round-up of Jews in Paris on the occasion of its sixtieth commemoration. Julia discovers that she and her family are about to move into a flat where a boy died waiting in vain for his sister, Sarah Starzynski, to return for him. The novel comprises twenty chapters set during the Second World War and sixty-two chapters set predominantly in present-day Paris. The initial chapters are very short, ranging between two and four pages, with each chapter about the Jewish girl immediately followed by a chapter about the contemporary journalist. This structural pattern interlaces the two plots, concomitantly linking Sarah and Julia more strongly. While the Second World War plot and the contemporary plot intertwine only in the first part of novel, they merge when Sarah returns to her apartment for her brother and when Julia's father-in-law tells her about his meeting with the little Jewish girl. At that moment in the novel, Julia has become so obsessed with Sarah that her story is also Sarah's. The past and present are shown to be one and the same, hence there is no longer a need for two separate plots.

DIMINISHING "COGNITIVE DISTANCE": HISTORICAL KNOWLEDGE

Historically estranged generations need knowledge before they can feel a connection to the past. Mark Salber Phillips defines diminishing "cognitive distance" as "a widespread critique of long accepted models of knowledge and their replacement by new strategies of investigation."[7] For a historian, the knowledge of the past is the essential (and obvious) foundation on the basis of which he/she can indulge in "striking new themes of historical research," "a new politics," and "experimentation with a new historical form."[8] In *The Blasphemer*, Daniel Kennedy's knowledge of the Great War encompasses no more than the catch phrase "lions led by donkeys."[9] Daniel's father, himself a war veteran of campaigns in Northern Ireland, the Falklands, and the First Gulf War, is annoyed by his son's ignorance:

> [Daniel] knew his great grandfather had died on the first day of a big battle, but he couldn't remember which one. He wanted to say Somme but that didn't sound right. He read the caption again … "Passchendaele was the one with all the mud, wasn't it?" … "And we won it, right?" If Philip was irritated by this comment, he chose not to show it. "The British gained five miles of ground. Four months later they withdrew, leaving the ruins of Passchendaele to its ghosts. If this was a victory, it was a pyrrhic one."[10]

The meaning of the battle of Passchendaele has been consistently interpreted by historians within the ideological framework of the futility myth: "On the Somme [Haig] had sent the flower of British youth to death or mutilation; at Passchendaele he had tipped the survivors into the slough of despond,"[11] "In spite of the monumental losses … Haig remained confident in his strategy. … [The] men were … irrevocably committed to the toil, the agony, the weary crucifixion of the long slog through the mud to Passchendaele."[12]

As Farndale points out with the example of his character Daniel, it is wrong to assume that there exists a collective memory of the war shared by generations across time, with collective memory to be understood as "not a remembering but a stipulating: that this is important, and this is the story about how it happened."[13] The importance of Passchendaele is totally lost on Daniel. Though he remembers how "as a child he had sat on [his father's] knee and listened in awe to tales about the world wars in which his grandfather and great-grandfather had fought," as a young man he started deliberately to forget while emotionally severing himself from his father, mostly out of guilt, for "it was obvious that Philip had hoped his only son would follow him into the Medical Corps."[14]

John R. Gillis writes that "in the past decades memory has become … more global … Events and places with international meaning such as Hiroshima, Chernobyl, Auschwitz, and Nanjing capture the world's attention even when the nation's responsible may wish to forget them."[15] This is true to some extent, and yet one can hardly speak of a global historical consciousness, particularly when education in schools is predominantly nation-oriented. De Rosnay's Julia is an American, and it is her "Americanness" that she blames for lacking knowledge about wartime France: "Listening to Joshua, I realized how little I knew about what happened in Paris in July 1942. I hadn't learned about it in class back in Boston."[16] Considering the years she spent in Paris, married to a Frenchman, her ignorance cannot be totally exonerated for she should have taken the trouble to know:

> "Julia," said Joshua, looking up at me over his glasses, "this is up your alley. Sixtieth commemoration of the Vel' d'Hiv." I cleared my throat. What had he said? It sounded like "the veldeef." My mind went blank … Joshua continued. "The great round-up at the Vélodrome d'Hiver. That's what Vel' d'Hiv is short for. A famous indoor stadium where biking races were held. Thousands of Jewish families, locked up there for days, in appalling conditions. Then sent to Auschwitz. And gassed." It did ring a bell. Only faintly.[17]

Significantly, in both novels, the contemporary protagonists meet up with characters whose role is chiefly that of a history teacher. In *The Blasphemer*, the battlefield tour guide and amateur archeologist Clive is a character introduced only to provide Daniel with a summary of the realities of the battle of Passchendaele, whereas in *Sarah's Key*, the purpose of the character of Franck Levy is that of a source of information about the Vel' d'Hiv children. It is beyond doubt that both Farndale and de Rosnay intended their fictions to also feature a didactic content, on the assumption that their readers need this historical knowledge as much as their fictive characters.

RESTORING THE MEMORIALSCAPE

According to Pierre Nora, "our relation to the past is now formed in a subtle play between its intractability and its disappearance, a question of representation – in the original sense of the word – radically different from the old ideal of resurrecting the past."[18] Farndale's and de Rosnay's contemporary protagonists embark on a tour of the places marked by the historical events that had affected their families. This is to be their symbolic journey back in time leading through the existing commemorative space, to underscore, and concomitantly "undo," the profound disconnectedness between the past and the present. In *The Blasphemer*, Daniel's father takes him to the exact place in Belgium where Andrew Kennedy went missing in 1917. Yet there is no Passchendaele anymore: "the village that gave the battle its name ha[d] been wiped off the map by British artillery during the First World War. In its place were pastureland and fields of maize, as well as warehouses and a sewage treatment plant."[19] Daniel discovers, however, that the past cannot be obliterated altogether, and this battle – as many other battles of the Great War – had written itself firmly into the soil of the country. Not only are unexploded shells and bodies of soldiers still occasionally unearthed, but also the frequent disappearances of farm animals down into the shafts left by tunnels are, as Daniel comments, literally instances of "the present collapsing into the past."[20]

Though the Menin Gate Memorial makes an impression on Daniel, its history and meaning must be explained to him:

> "They built it here," Philip said, "because it was the route every British soldier would have taken on his way to the front." "Through the gate?" "There was no gate. Not even an arch. It was a gap in the ramparts which encircled the town, a bridge across a moat" ... Daniel stared up the single span of stone above him ... Every surface had been carved with chiseled capitals over leagues of whites stones ... "So, this is everyone killed

> along the Ypres Salient during the war?'" Daniel asked, "Lots of names." "No, these are the ones who have no known grave."[21]

Daniel's reaction to the Menin Gate Memorial is comparable to that of Elizabeth Benson in Sebastian Faulks's *Birdsong* (1993):

> She peered at the stone ... Every grain of the surface had been carved with British names, their chiselled capitals rose from the level of her ankles to the height of the great arch itself ... "Who are these, these ..." She gestured with her hand ... "The lost, the ones they did not find. The others are in the cemeteries" ... When she could speak again, she said, "From the whole war?" The man shook his head. "Just these fields." He gestured with his arm.[22]

In both Faulks's and Farndale's novels the emphasis is laid not just on the protagonists' lack of knowledge about the Great War but also on their ignorance of the memorialscape: "on a more general level, we might ask of all memorials what meanings are generated when the temporal realm is converted to material form, when time collapses into space, a trope by which it is then measured and grasped."[23] The importance of the Menin Gate Memorial and the Thiepval Memorial has been underscored in many memorial studies as well as memoirs and literature,[24] yet Farndale and Faulks clearly indicate that there is a discrepancy between what a memorial should mean and what it actually does (not) mean for historically estranged generations. Hence the imperative to restore knowledge of Great War memorials for "we might also remind ourselves that public memory is constructed, that understanding of events depends on memory's construction, and that there are worldly consequences in the kinds of historical understanding generated by monuments."[25]

In *Sarah's Key*, Julia wants to retrace the journey of Sarah Starzynski. However, the Vélodrome d'Hiver no longer exists:

> The rue Nelaton was dark and silent. It obviously never got much sunshine. On one side, bourgeois stone buildings built in the late nineteenth century. On the other, where the Vélodrome d'Hiver used to be, a large brownish construction, typically early sixties, hideous in both colour and proportion, MINISTÈRE DE L'INTERIEUR, read the sign above the revolving glass doors.[26]

Though there is a commemorative plaque in the vicinity of the one-time Vélodrome d'Hiver, its inscription reading "Passerby, never forget!" contrasts all too strongly with the obscure location: "I wondered if anyone ever glanced at it."[27] Next, Julia goes to Drancy from which "over sixty trains had left ... to Poland during the war." There is a "a large modern sculpture" erected for the purposes of commemoration as well as "the Drancy Memorial Museum ... [where] it was the first time [Julia] had seen a real [yellow star]." There are all too few visitors to the museum: "Schools sent their classes ... and sometimes tourists came." What shocks Julia the most is the fact that "the camp had barely changed in the last sixty years. The huge U-shaped concrete construction, built in the late 1930s as an innovative residential project and requisitioned in 1941 by the Vichy government for deporting Jews, now housed four hundred families in tiny apartments, and had been doing so since 1947." The answer to the question as to "how could anyone live within these walls," is as easy as it is disheartening: "Drancy had the cheapest rents one could find in the vicinity."[28] The next place on Julia's commemorative itinerary is Beaune-la-Rolande: "a sad empty place ... a technical school had been built over it in the sixties. The camp used to be a couple of miles away from the station ... The train station was no longer in use. It had been renovated and transformed into a day-care centre,"[29] and "the technical school was a grim," with "an unkempt square of grass in front of the school ... [with] strange, curving sculptures with figures carved into them. On one of them we read, 'They must act with and for each other, in a spirit of fraternity.'"[30]

British children are taught about the Great War in school, and part and parcel of the school curriculum is a battlefield tour. This does not ensure, however, a connectedness with the past, as exemplified by the behaviour of the pupils at the memorial sites: "The locals get upset because the children don't know how to behave. They jump over gravestones, drop litter, swear."[31] The "Tommy's Gift Shop" is a telling example of the commodification of the Great War, human experience reduced to an array of cheap replicas and tacky souvenirs: "[it] was festooned with Union Jacks and poppies. It sold poppy-patterned umbrellas, replica Vickers guns and helmets, mugs and spoons emblazoned with images of British Tommies."[32] Even worse, however, is when historically estranged generations do not wish even to inquire about a past written into the places in which they live and raise their children. The people who have chosen to live in Drancy live in the now, ignorant of the past: "I asked … if the residents … had any idea where they were living … Most of the people here were young. They didn't know, and they didn't care,"[33] and the same situation occurs in Beaune-la-Rolande:

> A woman in her late twenties carrying a toddler in her arms came out to ask me if I needed anything … She had never heard of a camp in the area … She didn't know. She was too young anyway.[34] … I asked one of the students if the sculptures had anything to do with the camp. He asked, "What camp?" A fellow student tittered. I explained the nature of the camp. It seemed to sober him a little.[35]

The emphasis in these citations is on the young age of the people whom Julia meets, just as Farndale underscores the reaction of school children, for childhood and youth is the time of a natural self-centred focus on the here and now. Historical knowledge is necessary but insufficient to reconnect one to events of an all-too-distant past.

DIMINISHING "AFFECTIVE DISTANCE": EMPATHY THROUGH IMAGINATIVE PLACEMENT

According to Mark Salber Phillips, "the idea of distance offers us a wider range of positions to work with and a more complex cross-hatching of form, affect and ideology," and he emphasizes the importance of "biography, memoir, or autobiography, but also … the *emotional* resonances of films, museums, and heritage sites, all of which have become important media for the representation of the past."[36] "All memory is individual, unreproducible – it dies with its person," Susan Sontag writes, strongly asserting that "there is no such thing as collective memory."[37] There is a reason, however, that the term "memory" tends to be used for what is really a demand for historical knowledge. The opposition between history and memory derives from the assumption of their different positioning in relation to the past. History – as knowledge – is assumed to be a detached awareness of events having taken place in a past ruptured from the present, whereas the concept of collective memory underscores an empathetic connectedness to the past and the conviction of the continuing relevance of the past for the present.

In *The Blasphemer* and *Sarah's Key*, the contemporary protagonists must imaginatively place themselves in historical circumstances so apparently alien (?) to their own lives. Without the workings of the imagination empathetic identification is not possible: "empathy is the ability to 'imaginatively' enter into and participate in the world of the cultural Other cognitively, affectively, and behaviorally,"[38] and without empathetic identification there can be no understanding of the experiences of others in the past: "Empathy is a form of receptivity that provides input to further processing which results in (empathic) knowledge of another individual. An individual comes to know what another is feeling because she or he feels it too."[39] For Daniel to be able to imagine what his great-grandfather had gone through when "[the] whole salient ... was a quagmire. Liquid mud," he is shown a panoramic photograph of "a charred and jagged landscape that was almost featureless apart from the barbed wire and a few splintered tree stumps over on Passchendaele Ridge."[40] It is, however, at the exhibition of war artists that Daniel, looking at one of Paul Nash's paintings, discovers in a moment of epiphany the connection between himself treading over his wife in order to escape from the crashed plane and his great-grandfather leaving his comrades during the battle of Passchendaele. In *Sarah's Key*, Julia imaginatively "re-lives" the tragic fate of Jews taken to concentration camps:

> It was as if the trenches were entering his own memory, his own consciousness ... His great-grandfather had been there ... these were visions of hell ... It was also producing a sense memory ... a feeling of being back in the plane as it was falling to earth. He had been certain that he was about to die and the imminent prospect of his death, of the nothingness, it represented, the hell of not being, terrified him."[41]

> We were then shown into a single cattle wagon that stood in the middle of the lawn, just outside the [Drancy] museum ... I tried to imagine the wagon filled up with masses of people, squashed against each other, small children, grandparents, middle-aged parents, adolescents, on their way to death."[42]

By "re-living" Sarah's harrowing experiences, Julia brings her back to life, saving the Jewish girl from oblivion:

> I stopped by the rue de Bretigne. The garage was still there ... I felt I could almost see Sarah coming down the rue de Saintonge on that hot July morning, with her mother and her father and the

> policemen. Yes, I could see it all, I could see them being pushed into the garage, right here, where I stood. I could see the sweet heart-shaped face, the incomprehension, the fear.[43]

The essential similarity between the characters of Daniel Kennedy and Julia Jarmond is that their historical estrangement is largely due to the non-existence of family memory, which David Gross defines as "a form of collective memory since the family is, after all, a kind of Intimgruppe that mediates between the individual and the larger social whole."[44] For family memory to be formed there must be "a continuous interaction between an older generation that wishes to transmit memories and a younger one that is present and willing to receive them."[45] In *The Blasphemer*, the shame of having had a deserter sentenced to death in the family is shrouded in silence: "sometimes it is best not to disturb the past."[46] In *Sarah's Key*, no member of the family talks about the acquisition of a new apartment by the Tezac family during the war. The fact that Sarah had returned for her brother remains the shameful secret hidden in the minds and hearts of those who met the girl, for they had gained much better living conditions by taking over the home of the evicted Jews. It is only when Julia – an outsider in the Tezac family – probes the subject, that her father-in-law feels forced to tell her the truth. Julia is warned, however, that "truth is harder than ignorance."[47]

UNWANTED HISTORIES

The Blasphemer and *Sarah's Key* are novels not only about, respectively, the Great War and the 1942 Round Up, but also about the changing historical politics with regard to "unwanted histories." *The Blasphemer* raises the issue of cowardice on the battlefield and in peacetime within the framing context of the attitudes towards the "Shot at Dawn" in British commemorative practice. A memorial commemorating British soldiers executed for alleged cowardice during war was unveiled at the National Memorial Arboretum in Staffordshire no earlier than in 2001,[48] not to mention that the relatives of the condemned soldiers could not take part in the official ceremonies of Remembrance Day until 2000.[49] In 1989, a campaign for blanket pardon for the executed men was launched,[50] ultimately forcing politicians to take action in 2006: "All 306 soldiers of the First World War who were shot at dawn for cowardice or desertion will be granted posthumous pardons, the

Ministry of Defence [Des Browne] said last night."[51] Published in 2010, *The Blasphemer* explicitly addresses the controversies raised by the campaign. Daniel's father, Philip Kennedy, takes part in the Remembrance Day ceremony. Thinking of his own father's heroic death during the Second World War, he is not sympathetic towards the supporters for blanket pardon:

> There were no First World War veterans left to take place in the march past. Although he had noticed, with a disapproving eye, a Wren in her forties, possibly holding a wreath of poppies on behalf of the SAD campaign. Shot at Dawn … Philip had been asked to support the campaign but had declined. He had also argued at the War Graves Commission that the SAD campaigners shouldn't be part of the parade. They made mockery of the men – men like his father and grandfather – who had given their lives gallantly in battle.[52]

Philip, who himself had an unblemished career in the Medical Corps, decides to investigate his grandfather's past, only to discover that he had deserted from the battlefield. Yet, it is his son's confession of his own cowardice after the plane crash and his redemption when endangering his life to save his daughter from a kidnapper that makes Philip change his mind and, in a symbolic manner, throw away his medals. Likewise, Daniel finds a very personal connection with the past:

> Well, I wonder if there is a gene for cowardice. Obviously, it skipped two generations in the case of our family … My great-grandfather and me, the yellow streak … I saved myself and left her to drown … She called me a coward. I'm surprised she didn't present me with a white feather. She has every right to.[53]

Brian Lickel, Toni Schmader, and Marchelle Barquissau state that, though "it is possible for people to feel both shame and guilt in response to a particular event," "the two emotions are likely to stem from very different interpretations of the event, and their differing implications for people's self concept."[54] In the case of Daniel, his is "a self-caused shame [that] occur[s] when people feel that a flawed aspect of their personal identity has been revealed,"[55] whereas in the case of Philip, his anger at the presence of the SAD campaigners during an official ceremony may be seen as representative of a "collective shame [that] stems from perceiving that the actions of the ingroup confirm or reveal a flawed aspect of one's social identity."[56] It is not a coincidence that Philip served with distinction in the Medical Corps and that his father was a hero of the Second World War. In constructing such characters, Farndale points to the existence of a national pride that perceives the military effort of one's country as a continuity of the men's readiness to sacrifice. There may be thousands of names of the dead listed on the Menin Gate Memorial and the Thipeval Memorial to the Missing of the Somme, yet one should also consider the magnitude of these architectural designs – these are also monuments in honour of a victorious British Empire. In his novel, Farndale strongly advocates the need to "re-see" the alleged "cowards" of the Great War not as a shameful blemish on the nation's past but as a source of the nation's collective guilt (executions of soldiers during the war and the exclusion of these men from commemorative practice), which would allow people to "take responsibility for and make reparations for their group's mistreatment of another group."[57]

In *Sarah's Key*, the emphasis is placed on the wilful "amnesia" of the French people about the historical fact that it was the French policemen, and not the occupying Nazis, who were responsible for sending thousands of their countrymen of Jewish origin to their death: "We don't talk about it. We don't want to remember. Some people here don't even know,"[58] "Nobody cares anymore. Nobody remembers … Tell Joshua the Vel' d' Hiv is a mistake. No one will read it. They'll yawn and turn to the next column."[59] This collective forgetting derived from an acute sense of shame. An old woman, a witness to the round-up of 1942, tells Julia:

> We French had been told for years that Jews were the enemies of our country, that's why in '41 or '42, there was an exhibit, at the Palais Berlitz, if I remember correctly … called "The Jew and France." The Germans made sure it went on for months. A big success with the Parisian population. And what was it? A shocking display of anti-Semitism … Shame on all of us for not having stopped it … Nobody remembers the Vel' d'Hiv children, you know. Nobody is interested … Nothing has changed. Nobody remembers, Why should they? Those were the darkest days of our country.[60]

When people feel shame, "they feel that wrongs committed by their ingroup implicate something about the very nature of who they are."[61] And, in France, "under direct Nazi occupation and under the Vichy puppet regime – the authorities and a key section of the population cooperated enthusiastically in the transport for mass execution of the Jew."[62] It was not until 1995 that the French responsibility for the death of thousands of the country's Jewish citizens was officially recognized by the French president, Jacques Chirac, who emphasized the need to undo the nation's shame by acknowledging the nation's guilt: "There are moments that hurt the memory and the idea we have of the country through a nation's lifetime ... It is difficult to speak of them too, because these dark times soil its history forever, and are an insult to our past and traditions." It does not suffice, however, to admit that "Fifty-three years ago, on July 16th 1942, 450 French policeman and gendarmes, under the authority of their leaders, answered Nazi orders."[63] For, as Tatiana de Rosnay underscores in her novel, there is also an urgent need for France to both acknowledge and take moral responsibility for those of its members who benefited from the eviction of the Jews.

CONCLUSION

Living in their own present, historically estranged generations are often unaware that they are surrounded by the past in the form of existing memorialscapes. Memorials, museums – or the lack of them – bespeak primarily of the way these events have been written (or unwritten) into the national identity. Film and literature may play an equally important role in either remembering or forgetting. Nigel Farndale's *The Blasphemer* and Tatiana de Rosnay's *Sarah's Key* both belong to the category of commemorative fiction that, by re-remembering the memorialscapes contemporary generations tend to ignore, "create a subtle interplay between [their] historical and contemporary meanings and, by doing so, ... disclose the manner in which our understanding of the past is constructed by forms of commemorative practice."[64]

NOTES

1 Pierre Nora, "Between Memory and History: *Les Lieux de Mémoire*," *Representations* 26 (Spring 1989): 12.
2 James E. Young, *The Texture of Memory: Holocaust Memorials and Meaning* (New Haven: Yale University Press, 1993), 5.
3 Mark Salber Phillips, "History, Memory, and Historical Distance," in *Theorizing Historical Consciousness*, ed. Peter Seixas (Toronto: University of Toronto Press, 2006), 95.
4 Ibid., 96.
5 Ibid., 97.
6 Phillips, "History, Memory, and Historical Distance," 97.
7 Ibid., 100.
8 Ibid., 99–100.
9 Nigel Farndale, *The Blasphemer* (London: Doubleday, 2010), 27.
10 Ibid.
11 John Keegan, *The First World War* (London: Pimlico, 1999), 395.
12 Lyn Macdonald, *They Called It Passchendaele* (London: Penguin Books, 1993), 153.
13 Susan Sontag, *Regarding the Pain of Others* (London: Penguin Books, 2003), 76.
14 Farndale, *Blasphemer*, 24.
15 Gillis, "Introduction," in *Commemorations: The Politics of National Identity*, ed. John R. Gillis (Princeton, NJ: Princeton University Press, 1996), 14.
16 Tatiana de Rosnay, *Sarah's Key* (London: John Murray, 2008), 28.
17 Ibid., 27.
18 Nora, "Between Memory and History," 17.
19 Farndale, *Blasphemer*, 355.
20 Ibid., 57.
21 Ibid., 335–6.
22 Sebastian Faulks, *Birdsong* (London: Vintage, 1994 [1993]), 264.
23 Young, *Texture of Memory*, 15.
24 For an analysis of the historical and cultural meanings of the Menin Gate Memorial, see Marzena Sokołowska-Paryż, *Reimagining the War Memorial, Reinterpreting the Great War: The Formats of British Commemorative Fiction* (Newcastle upon Tyne: Cambridge Scholars Publishing, 2012), 25–30.
25 Young, *Texture of Memory*, 15.
26 de Rosnay, *Sarah's Key*, 39.

27 Ibid., 60.
28 Ibid., 134.
29 Ibid., 138.
30 Ibid., 145.
31 Farndale, *Blasphemer*, 225.
32 Ibid., 337.
33 de Rosnay, *Sarah's Key*, 134.
34 Ibid., 138.
35 Ibid., 145.
36 Phillips, "History, Memory, and Historical Distance," 99, emphasis added.
37 Sontag, *Regarding the Pain of Others*, 76.
38 Carolyn Calloway-Thomas, *Empathy in the Global World: An Intercultural Perspective* (Los Angeles: Sage, 2010), 8.
39 Lou Agosta, *Empathy in the Context of Philosophy* (New York: Palgrave Macmillan, 2010), 11.
40 Farndale, *Blasphemer*, 360.
41 Ibid., 303–4.
42 de Rosnay, *Sarah's Key*, 135
43 Ibid., 133
44 David Gross, *Lost Time: On Remembering and Forgetting in Late Modern Culture* (Amherst: University of Massachusetts Press, 2000), 112.
45 Gross, *Lost Time*, 112.
46 Farndale, *Blasphemer*, 28
47 de Rosnay, *Sarah's Key*, 124
48 For an analysis of the meanings of the Shot at Dawn memorial, see Sokołowska-Paryż, *Reimagining the War Memorial*, 55–6.
49 Jenny Edkins, *Trauma and the Memory of Politics* (Cambridge: Cambridge University Press, 2003), 72.
50 See David Johnson, *Executed at Dawn: British Firing Squads on the Western Front, 1914–1918* (Stroud, Gloucestershire: Spellmount, 2015), 129–153.
51 Ben Fenton, "Pardoned: The 306 Soldiers Shot at Dawn for 'Cowardice,'" *Telegraph*, 16 August 2006.
52 Farndale, *Blasphemer*, 99.
53 Ibid., 343.
54 Brian Lickel, Toni Schmader, and Marchelle Barquissau, "The Evocation of Moral Emotions in Intergroup Contexts," in *Collective Guilt: International Perspectives*, ed. Nyla R. Branscombe and Bertjan Doosje (Cambridge: Cambridge University Press, 2004), 43.
55 Lickel et al., "Evocation of Moral Emotions," 42.

56 Ibid., 42–3.
57 Ibid., 35.
58 de Rosnay, *Sarah's Key*, 144.
59 Ibid., 51.
60 Ibid., 69.
61 Lickel et al., "Evocation of Moral Emotions," 43.
62 Adam Jones, *Genocide: A Comprehensive Introduction*, 2nd ed. (London: Routledge, 2011), 258.
63 Jacques Chirac's 1995 Vel d'Hiv speech: "Il est, dans la vie d'une nation, des moments qui blessent la mémoire, et l'idée que l'on se fait de son pays. … Il est difficile de les évoquer, aussi, parce que ces heures noires souillent à jamais notre histoire, et sont une injure à notre passé et à nos traditions. … Il y a cinquante-trois ans, le 16 juillet 1942, 450 policiers et gendarmes français, sous l'autorité de leurs chefs, répondaient aux exigences des Nazis," "Le discours de Jacques Chirac au Vel d'hiv en 1995" (*Le Figaro*), http://www.lefigaro.fr/politique/le-scan/2014/03/27/25001-20140327ARTFIG00092-le-discours-de-jacques-chirac-au-vel-d-hiv-en-1995.php.
64 Sokołowska-Paryż, *Reimagining the War Memorial*, 2.

24

An "Abominable Epoch": An Australian Woman's Perception of Occupied France

Sylvie Pomiès-Maréchal

> The "Abominable Epoch" is 1940–41 in France. It is late to tell stories of that time, and mine are not such as to lead to sensational discoveries. They are extremely humble and personal, and yet, whenever I look through my diary, I think their interest is wider than the purely personal, if only I can tell them properly. There was a beautiful spring and then a wretched year, which had, in spite of wretchedness, as much beauty in it as the spring. If I can show that, the book is justified. I dedicate it, with affectionate gratitude, to the persons, not all still living, who befriended me in France during the "Abominable Epoch."[1]
>
> Christine Morrow

As demonstrated in the previous chapters, the past is "never dead" because it is kept alive through our personal recollections and the work of historians. The connection between history and memory constitutes a central point of debate within the historical community. Sometimes regarded as complementary, sometimes seen as contradictory, the two concepts refer to the question of the representation of the past. History is commonly presented as "dispassionate," more distant and objectifying than memory, perceived to be subjective, if only in its dimensions of oblivion and silence.

Building upon the theories of Paul Ricœur,[2] François Dosse states: "The only knowledge of the past we can get is through a narrative ... This is the path within the collective memory of experiences which has continuously enriched history in its reflexive and historiographical stage ... Memory first gives rise to history as writing. It then forms the basis of the new appropriation of the historical past as memory enlightened by transmitted history."[3]

The role and status of narratives in the shaping of collective memory and of history is clearly reflected in Christine Morrow's writing process. Christine Morrow was an Australian doctoral student caught up in the upheaval of the debacle of Occupation from 1940 to 1941. *Abominable Epoch* thus traces her epic journey through occupied France as her nationality forced her to retreat to the unoccupied zone while attempting to return to Australia.

Published after her death in 1971, this war diary sheds light on the way narratives can convey war experiences. Though issued thirty years after the events, this account constitutes invaluable material in that it combines both memory-related history and immediate history. Christine Morrow mostly recounts her experiences through her personal notebooks. However, some of the notes taken at the time of the events appear word for word, without any redrafting, especially at the end of the book. Both a confidant and an outlet for strong emotions, this diary traces the trials of daily life: the exodus and the bitter struggle to find accommodation, the deprivation, the endless lineups, the way refugees were perceived; but also, making up for all this, the tremendous feeling of solidarity that is gradually growing.

Through this account, we primarily focus on the shaping of identity in wartime, whether individual or collective. After a few biographical and contextual elements, we examine Christine Morrow's perception of the enemy and her relation to the mother country, as much in her belonging to the Commonwealth as in the way in which she espoused the ills and anxieties of an adopted homeland hurt in its national pride. We then consider the dynamics of solidarity and mutual assistance that were put in motion through such networks as the AFDU (Association Française des Femmes Diplômées de l'Université).

BIOGRAPHICAL AND CONTEXTUAL ELEMENTS

Christine Morrow was born in 1902 in Kalgoorlie, Western Australia, and grew up in Perth. After her studies, she became a French teacher. In 1935, she was awarded a grant and went to England and France to undertake doctoral studies. She was about to complete and defend her doctoral dissertation at the Sorbonne when Marshall Pétain signed the Armistice that put France under the Nazi yoke. She was then on holiday in Agon, a small town in the Cotentin. Here is how she announces the defeat of France:

> There was one other lodger, Mme V. At first, we were served at separate tables, bowing politely at each other, but after the bombing of Paris we ate together and became more friendly as events grew more alarming … The Prime Minister, Reynaud, said frequently on the radio "*Nous vaincrons parce que nous sommes les plus forts.*" (We shall conquer because we are the strongest.) … One evening at dinner time Reynaud spoke on the air again, this time in a different tone. He did not say "*Nous vaincrons.*" He spoke to America, imploring help. There was no word about the British. Probably there was silence then everywhere in France. Mme V. and I broke it in whispers as we went upstairs to our rooms. I said "Can this mean that…" Mme V. said, "No, not …" When we shook hands to say goodnight, we clung to each other's hands, doubtless hysterically, trying to bridge a rift between our nations.[4]

Though undated, this radio announcement likely occurred on 15 June 1940. The next day, French prime minister (Président du Conseil) Paul Reynaud resigned, having failed to obtain a promise of military engagement from the United States and his cabinet having rejected the British proposal to unite France and the United Kingdom, a proposal that he had fully backed.[5]

As the "rift" between nations and individuals implies, the Armistice marks a drastic change of identity for Christine Morrow who, as a member of the Commonwealth, turned from an Australian national into an illegal refugee in enemy territory. As early as May 1940, France had been faced with a large-scale flow of Belgian and Dutch refugees. According to a Red Cross telegram: "Over 2 million French people, 2 million Belgians, 70,000 Luxembourgers and 50,000 Dutch nationals [were] in France, living in a state of overwhelming deprivation."[6] Christine gives a powerful depiction of the first Belgian refugees who arrived in Agon:

> They arrived … as all Belgium seemed to arrive, in Agon … The old ladies wore clothing so ill-fitting that I thought it must have been borrowed, but apparently, it fitted before the journey made them thin … They went on foot, getting lifts sometimes, were

> often under shelling, slept where they could, on the ground, in forests sometimes, were urged to go on whenever they thought of resting, because the enemy was always close … The children were very frightened, and did not realize immediately that in Agon, they need not throw themselves on the ground or run to cover at the slightest noise. Soon, however, it was they who made the noise, playing at war when temporarily out of it, ambushing in imaginary trenches, calling "Bang, bang!" realistically.[7]

The French capitulation considerably exacerbated these movements, this time primarily displacing French nationals. When the Germans entered Paris, then declared an open city on 14 June, less than one-third of the population remained. Conversely, Beaune-la-Rolande, in the Loiret department, saw its population increase from seventeen hundred to forty thousand inhabitants.[8] Altogether, an estimated 10 million people, almost-one quarter of the French population, fled during the 1940 exodus.[9] Here again, Christine Morrow's account is consistent with the numerous descriptions and photographs of the exodus that have become engraved in collective memory:

> There they all were, the usual motley crew: … refugees from everywhere going everywhere, since no direction seemed better than another. There they all were, a mixture of selfishness and generosity, panic and fortitude, former wealth and former poverty, old and young, strong and feeble, good and bad, promenading mattresses, perambulators, blistered feet, with their families or without them, having in common weariness and sorrow, in the civilized country of France, in the year of our Lord 1940.[10]

The narrative brings to light the role of writing in terms of identity, putting wartime experiences into words. At the time of writing, Christine already includes herself within the community of the displaced and the oppressed – a community that she is soon to officially join.

WARTIME IDENTITY

This identity-building process primarily expresses itself through a sense of belonging to a particular group, but it also expresses itself through the staunch rejection of the enemy and the categorization and extreme polarization of individuals along lines of nationality:

> We began to see Germans often. Accounts said Coutances was full of them. Two frequented the bar of my hotel … They were absorbed in two objects, to drink as much as possible without taking their attention off the truck, and to fraternize, according to instructions, with the French in the bar … The Germans said, between drinks, *'Français Kamerad!', 'Anglais pan pan!'*, and made a shooting gesture to show that "pan pan" meant "bang bang." Sometimes, instead of "pan pan," they made a different gesture, finger drawn across the throat, – a throat that I qualified to possess more than the others at the hotel … I began to resemble a mouse waiting for the cat.[11]

Agon had indeed turned into a mousetrap. Being naturally inclined to take action, Christine decided to undertake her "walk across France" as she humorously called it.[12] Christine Morrow perceived the German military presence as an "infestation," a recurring term that literally associates the enemy with vermin and parasites. When Christine began her diary, she agreed with Gisèle (a fellow traveller) to write the word "infested" for "Germans," a word that transcribed the prevailing feeling at the time "until there was no sense in it, because everything was 'infested.'"[13] Thus, when stopping at Fougères, a Breton town on their way south, Christine writes: "the town was 'infested' and, not inured to it yet, we would not eat in a room where we saw the green uniform."[14] Just like the "infested" leitmotiv, the loathing, even the hatred, inspired by this colour comes up frequently: "I should be loath for a word of mine to advocate hatred of Germans. I hated them only for wielding power where they had no right, but may hate them again if they wear a certain green in France."[15]

This echoes another account, one of the letters addressed to the BBC and retranscribed by Aurélie Luneau in her book *Je vous écris de France*. This is a letter written in February 1941: "The Jerries are everywhere, our little village has turned green with them and our roads tremble. As to the warbling of our bell tower, they have also forbidden it, probably because this quivering, this crystal flight disturbed those souls which are neither pure nor noble-minded."[16]

Military occupation is bound to arouse hatred. This deep-rooted abhorrence of the occupying Germans appears quite blatantly in Christine Morrow's narrative. Although she feels some aversion at this emotion, she considers it to be justified by the harm and humiliation inflicted on an adopted homeland for which she professes a love equalled only by her hatred of the enemy. War reduces people to "simpler terms and [makes them] somehow more real,"[17] to echo Christine's words. Likewise, it crystallizes the deepest feelings:

> Hatred is wrongful of course, and hatred of a people in a lump perhaps worse than the rest, but it is bound to occur when one people comes among another by force. I saw no atrocities on the part of the Germans but, loving France, I hated them for their presence there. Their stolen authority above all was unendurable, and efficiency, patronage, even kindliness, made it worse … In some of the villages there was writing on the doors, "*Gute Leute hier*" – "Good people here," in German script. This "protection" seemed more insulting than bullying, and to see the invaders fondling children roused black enough feelings in Gisèle and me to suggest that it is difficult for occupiers to ingratiate themselves.[18]

REFUGEES IN TOULOUSE

The unoccupied zone had to absorb a massive influx of refugees. Christine Morrow recounts her arrival in Bordeaux where she had no other choice than to sleep on a bench of the railway station. The station was a natural point of convergence for the refugees who were lying on the ground along the platforms. If Bordeaux was a sorry sight, what about Toulouse? The capital of Languedoc had already absorbed a massive flow of Spanish refugees and, within a few weeks, had become the "terminus station" of human misery, with about 210,000 refugees in a single month:[19] "The Town Hall facade is plastered thick with notices, all the breadth of the building, up too high to read, down too low to read with ease, hundreds of little pieces of paper; they dance, get jumbled; no-one ever read them all. – New addresses of half of France; enquiries about the other half. No, not half of France, only the first wave of refugees."[20]

Christine gives a lucid description of the evolving feelings of the residents of Toulouse towards these newcomers. Though not espousing their views, she speaks for them, illustrating the mind shift from benevolence to sheer animosity:

> Refugees move the heart to pity, – which endures in inverse proportion to habitual proximity. Towns bombed, inhabitants on the roads, destitute, homeless, suffering, – old people sleeping

in the forests, children crying; hunger, fatigue, sorrow. Who does not grieve for these stories? One gives clothing one does not need, money. The refugees come among us, relatives, colleagues, other unhappy humans … How tired of it we grow! They don't go away. More come, requests still when there's nothing more to give. Congestion everywhere, home, street, restaurant, no air to breathe. No ordinary conveniences left in shops, food getting scarce, hygiene not so good. These refugees have not nice habits; they are locusts stripping the town bare. This influx is beyond measure of decency. Surely, they ought to stop coming! Why cannot they stay at home?[21]

According to Philippe Nivet, this kind of tension, so woefully contemporary, "is symptomatic of how communities can cut themselves from one another in wartime."[22] The biblical allegory of the locust invasion is the domestic side of the other "infestation" mentioned above, that of the occupier who requisitions, loots granaries, or seizes livestock. The reception of refugees is characterized by suspicion. Not only must they withstand the loss of property and the separation from their relatives, they must also endure the hostility of local populations and the anguish of finding accommodation, most likely something wretched, with rents verging on extortion. Christine narrates her first night in a "dirty café, with a dirty cellar, with a dirty mattress on its dirty floor":

The dirt was not very visible at first, for the cellar had no electric light … On my mackintosh spread on the end of the mattress, I sat and waited for the morning, electric torch in one hand, a pair of scissors in the other, a weapon of defence, since the door would not lock. But the only molesters were fleas, thick as hail in the light of my torch.[23]

Yet, simultaneously, Christine describes a profound humanity and a spirit of resistance to the Vichy government and the occupying forces' attempts to sow division, a spirit manifesting itself in altruistic and disinterested attitudes and in a commitment to human values:

Those were strange days. How good it would be if anything but dread and sorrow could make people of one spirit as we all were then! From then on began for everyone an abominable time. In

> a book given me during the year that ensued [a book offered by Jules Puech, soon to be mentioned], there is an inscription: – "*En souvenir de quelques bons moments dans une époque abominable.*" [In memory of a few good moments in an abominable epoch.] The strange thing is that there were "*bons moments.*" Their memory is more precious than that of any other *bons moments* at any other time. They had no other source than human goodness. We all seemed to become more simply ourselves. I met of course not only goodness. What happened from then on is that it became easier to know people. They seemed reduced by misery to simpler terms and to become somehow more real. I met certainly much more goodwill than illwill [sic] and much true kindness out of which good moments were made.[24]

THE ROLE OF ACADEMIC COOPERATION DURING THE SECOND WORLD WAR: THE CASE OF THE AFDU

Before continuing the story of Christine Morrow's stay in Toulouse, and to illustrate the spirit of mutual assistance that gradually developed, it is necessary to introduce a couple without whom Christine Morrow's war experience may have turned sour. Jules and Marie-Louise Puech were intellectuals committed to pacifism and feminism and who figured among the ranks of the first *Résistants*. Jules Puech was a Proudhon specialist[25] as well as the author of a biography on Flora Tristan that remains authoritative.[26] He was also editor-in-chief of *La Paix par le Droit*. When he was called up for the army during the First World War, his wife Marie-Louise ensured the publication of the journal. After the war, Jules Puech quite naturally supported the League of Nations, to which he devoted a book in 1921.[27] Marie-Louise had a similar background: a distinguished scholar, she spoke three languages fluently and joined McGill University in Montreal, where she taught French literature from 1900 to 1908, an experience that enabled her to build a significant international academic network.[28]

In the immediate postwar period, Marie-Louise became one of the officials of the AFDU (Association des Françaises Diplômées des Universités),[29] the statutes of which aimed at "the intellectual and moral reconstruction of a new world thanks to professional cooperation and mutual understanding."[30] This feminist and supportive professional network attained its utmost importance during the

Second World War, when it assisted student refugees who had fled to the unoccupied zone. The mission that Marie-Louise Puech assigned herself was to help these isolated and destitute refugees to overcome the hurdles of everyday life and help them complete their PhD studies. Though many of them were Jewish, they all shared the goal of attempting to escape the trap in which they had been caught after the fall of France. Every possible effort was made to help them find accommodation and work as well as books. This implied the setting up of a network of members who became "wartime godmothers," a practice dating back to the First World War.[31] The epicentre of this network was located in Borieblanque (Tarn),[32] in the Puech family estate, where the couple had retreated, consequently transferring the administration of the AFDU to the unoccupied zone. Rémy Cazals, professor emeritus of history at the University of Toulouse, collected these papers and devoted himself to the publication of *Lettres de réfugiées*, a "treasure" that included Marie-Louise Puech's correspondence with these refugees. Among them is Christine Morrow, herself a member of the Australian section of the International Federation of University Women. Christine Morrow refers to Marie-Louise Puech as "a grain of hope in a bag of something quite different."[33] In an unrevised excerpt of the diary dated 27 October 1940, Christine evokes the clandestine activities of the Puech couple:

> Mme Puech says why on earth didn't I enquire for the federation when I was so destitute there [Toulouse]? My hosts are on the Nazi black list. Their flat in Paris has been searched more and more. It would be dangerous for them to go back to Paris. They are delighted. It shows they have annoyed the enemy.[34]

Though Christine Morrow often related her story in an almost light-hearted way, she faced considerable perils, among which figured internment and deportation.[35] Fortunately, she did not have to rely on luck alone and could count on the assistance of her protectors. The semblance of stability she had found in Toulouse was indeed threatened. As an Australian, she was officially under house arrest in Grenade, near Toulouse. This was mentioned in red ink on her residence permit, but, thanks to Marie-Louise and Jules Puech, she was granted a one-month exemption on medical grounds, and this immunity was renewed twice.[36] In addition to material help, the couple helped Christine complete her thesis on *le Roman irréaliste*.[37]

As Christine liked to say, it was an ad hoc thesis and, just like her, an illegal refugee at the University of Toulouse.[38] Jules and Marie-Louise financed the publication of the book through the AFDU funds as well as through their own resources. According to Christine, it was "probably the only thesis to be published and defended by a British subject in France between the Armistice and the Liberation."[39] The very fact that this defence took place was a bold challenge to the established order. Christine and her dissertation were "illegal as the blackest crime,"[40] and attention was not to be drawn to them. Consequently, a single notice was "placed in a very dark corner" to avoid any embarrassing anglophile outbursts.[41] Two of the jurors were members of the resistance, and one of them, who was to die in deportation, declared: "I'm glad to see Australians defend themselves so well."[42] The defence had indeed taken place not long after the Australian victory in Benghazi.

EVERYDAY ACTS OF RESISTANCE

This natural and humanist resistance to the spirit of division fostered by the enemy also took more active forms. In a sub-chapter ironically entitled "*Vive Pétain!*" (long live Pétain!), Christine Morrow writes:

> On the walls all over the place someone began to chalk, "Down with the Jews!" In the same handwriting, there was also "*Vive Pétain!*" I rubbed out "Down with the Jews" but let Pétain live if he could … [It was] a gesture, a trifle, a tiny declaration. I made mine with scraps of paper dipped in the dirty gutters. As the paper wore out quickly, I began by erasing the word "Jews." People clustered round sometimes. "Down with what, Madame?" – "Down with Jews, Madame. I erase it for questions of Christianity." … My Jewish friends said "Why bother?" My French non-Jewish friends said, "Leave that sort of thing to us, and try not to get yourself noticed." But I liked my wall cleaning activities.[43]

Though described as "trifling," these acts of resistance doubly endangered her, first through her rejection of the ideology advocated by the occupant but also through her status as enemy of the nation. There again, we can find an echo of Christine's activities in the accounts sent to the BBC by anonymous French people. In a letter dated 16 January 1941, a Parisian woman thus describes the fight against German propaganda:

> Dear Sir, one day as I was going for a stroll, I saw a poster, such a poster! Judge for yourself. A Fritz soldier holding a French child in his arms. The child was eating a huge slice of bread and below, there was this inscription: "Trust the Germans." My goodness! Another poster: in the background, there is a heap of stones and in the foreground, one can see a woman, sitting with her two children and beholding what must have been her home. Above the scene, one reads "The English did that." But some French hands wrote this: "No! Adolf did it." As you can see, we have reestablished the facts. Another poster again. This time, it shows a French sailor sinking at sea and in a last effort before disappearing, he's brandishing the French flag. German inscription: "Remember Oran."[44] But Oran had disappeared as if by magic. "Oran" had been erased or torn away. On one of the posters I read "Remember…. 1870, 1914–18."[45]

IDENTITY AS SYMBOL

The experiences narrated by Christine Morrow investigate the nature of identity in wartime, an identity that here expresses itself in a commitment to humanist values. This identity also turns out to be dual and fluctuating: as an illegal refugee, Christine has no choice but to hide her nationality. Yet this very nationality can also convey hope. As an ambassador of the Allied forces, most people greet her as a beacon of liberty. Such scenes recur quite frequently in Christine's narrative, showing the intricate nature of Franco-British relations under the Occupation. As an illustration, let us quote the reaction of this old woman from Blagnac, in the suburbs of Toulouse, who offered half of her bed when there was nothing else to offer. She thus introduced Christine to the other "pensioners": "With a fine sense of drama she cried as we approached; 'I bring you good news; *Mademoiselle est Anglaise!*' Was I dreaming, hallucinating from fatigue? There in the gloaming, they all stood around me, shaking my hands till the bags fell away from them, asking about England, asking about Churchill, as if I was a messenger straight from him or from Heaven."[46] In another passage, she assumes symbolic status for her guilt-ridden French benefactors:

> It seemed sometimes as if we were a team to help me in my difficulties. This is no megalomania. It often was Anglophilia. There were Oran moments doubtless, and Dakar moments, but other moments too. There were bombing of London moments. Never did I see such glances from France towards England as in that time when England fought alone and France was supposed to be letting her down. In inverse ratio to its waning official value, the British passport increased in sentimental value. When I was officially nearly an enemy alien, individual kindness, from French and others, knew no bounds ... When I apologised for importuning and they said it was a pleasure, I knew it was, because I knew I was a symbol.[47]

The narrative similarly throws light on unconscious, imposed, or revealed identity, such as in the case of the Noufflard family. The Noufflards are friends of Christine Morrow's. They happen to rediscover their Jewishness at the onset of persecution, with all the tragic consequences this entailed: "They had been French for so many generations that, until the persecution began, they hardly ever thought of being Jewish. They were French and artists, and being Jewish seemed quite unimportant ... Mme N.'s mother had the order of the Legion of Honour for services to France in the First World War. She was 80 now, and summoned to the police commissariat to declare her race and possessions."[48]

CONCLUSION

Beyond Christine Morrow's individual experience and singular personality, this narrative takes on a universal dimension and indirectly pays tribute to the inner strength and resilience shown by women during the Second World War. The status of first-eye witness interacts with the role of participant in the events related, which makes this account even more significant.

Whether written at the time of the events or long after they occurred, such narratives bring a touch of authenticity that the historian, detached from these life experiences, is hard pressed to recreate. This is particularly true for contemporaneous notes: if they might have shortcomings, such as lack of distance, they can create a real sense of immediacy that literally takes the reader to the very heart of the events. This is how feelings of fear, despair, and the sense of emergency acquire substance, as in the telegraphic style of some *Abominable Epoch* extracts.

Christine Morrow's diary offers a tale of contrasts, whereby wretchedness, deprivation, the disruption of communities, and the humiliation of German occupation are largely outweighed by mutual assistance, friendship, and resistance to the Vichy government. These contrasts also reflect shifting identities, or "identities in movement," with polarized emotions such as the wartime rejection of the enemy or allegiance to a community and/or a nation. The symbolic nature of identity as exemplified in Christine Morrow's status as an Australian citizen also allows us to explore Franco-British connections under Occupation. The account finally offers a valuable insight into the dynamics of international academic cooperation in wartime.

After a year of waiting for her visas, Christine Morrow went back to England via Lisbon. She then returned to Australia, transiting through Canada and Rio de Janeiro on a very perilous journey. She then joined the University of Western Australia, where she became a dedicated French teacher.[49] As for the manuscript, Christine had the opportunity to submit her diaries to her wartime friends after a brief stay in Toulouse in 1955. This was to be the expression of her gratitude towards the people of France. Nevertheless, the book was not published during her lifetime but later, at the initiative of her colleagues from the University of Western Australia. The publication was also linked to the creation of the Christine Morrow Memorial Prize, which is still awarded annually to a student from the University of Western Australia's French department.[50] All proceeds from sales of the book still go towards funding the grant.

NOTES

1 Preface to Christine Morrow, *Abominable Epoch* (Western Australia: s.n., 1972).

2 Paul Ricœur, *La Mémoire, l'histoire, l'oubli* (Paris: Le Seuil, 2000), 676.

3 François Dosse, "L'histoire à l'épreuve de la guerre des mémoires," *Cités* 33 (2008) 36–7: "On ne peut avoir de connaissance du passé que médiatisé par un récit ... C'est ce parcours dans la mémoire collective des expériences qui a enrichi constamment l'histoire en son stade réflexif et historiographique. ... La mémoire est d'abord matrice de l'histoire en tant qu'écriture et en second lieu, elle est à la base de la réappropriation du

passé historique en tant que mémoire instruite par l'histoire transmise et lue."

4 Morrow, *Abominable Epoch*, 3.

5 For further information and biographical elements on Paul Reynaud, see Encyclopaedia Britannica, viewed 15 June 2021, https://www.britannica.com/biography/Paul-Reynaud.

6 Philippe Nivet, "Les réfugiés de guerre dans la société française (1914–1946)," *Histoire, économie et société*, 23: 2 ; Philippe Nivet, *La société, la guerre, la paix, 1911–1946*, (2004), 254: "plus de 2 millions de Français, 2 millions de Belges, 70 000 Luxembourgeois et 50 000 Hollandais se trouvent en France dans un état de dénuement accablant."

7 Morrow, *Abominable Epoch*, 2–3.

8 Nivet, " Les réfugiés de guerre dans la société française (1914–1946)," 255.

9 Aurélie Luneau, *Je vous écris de France, Lettres inédites à la* BBC *1940–1944* (Paris: L'iconoclaste 2014), 47.

10 Morrow, *Abominable Epoch*, 4.

11 Ibid., 6–7.

12 Humour is an obvious feature of Christine Morrow's writing. Even in the bleakest moments of her journey, she always succeeded in seeing the funny side of situations.

13 Morrow, *Abominable Epoch*, 8.

14 Ibid., 10.

15 Ibid., 14.

16 Luneau, *Je vous écris de France*, 88: "Les Boches sont partout, notre petit village en est vert, nos routes en tremblent et le chant de notre clocher, ils nous l'ont interdit aussi sans doute parce que ce frémissement de l'air, cette envolée de cristal troublait ces âmes qui ne sont ni pures, ni belles."

17 Luneau, *Je vous écris de France*, 4.

18 Ibid., 13–14.

19 Sébastien Ambit, "Entre le 14 mai et le 10 juin 1940, 200.000 réfugiés transitent par Matabiau: il y a 60 ans, l'exode," *La dépêche du Midi*, 18 June 2000, http://www.ladepeche.fr/article/2000/06/18/76272-entre-14-mai-10-juin-1940-200-000-refugies-transitent.html.

20 Morrow, *Abominable Epoch*, 22.

21 Ibid.

22 Nivet, *Les Réfugiés de guerre dans la société française (1914–1946)*, 247: " des tensions révélatrices des replis communautaires au moment des guerres. "

23 Morrow, *Abominable Epoch*, 25.
24 Ibid., 4–5.
25 Jules Puech, *Le Proudhonisme dans l'Association Internationale des Travailleurs* (Paris: Félix Alcan 1907), 285.
26 Jules Puech, *La vie et l'œuvre de Flora Tristan* (Paris: Marcel Rivière 1925), 514.
27 Rémy Cazals, *Lettres de réfugiées* (Paris: Tallandier 2004), 27.
28 Ibid., 31.
29 Now AFFDU for Association Française des Femmes Diplômées des Universités. The AFFDU is the French section of the IFUW (International Federation of University Women).
30 Cazals, *Lettres de réfugiées*, 35: "la reconstruction intellectuelle et morale d'un monde nouveau grâce à l'entraide professionnelle et à la compréhension mutuelle."
31 Cazals, *Lettres de réfugiées*, 15.
32 The Tarn is a French department in South Western France.
33 Morrow, *Abominable Epoch*, 40.
34 Ibid.
35 See the reply of the Secretary of State for Foreign Affairs at the House of Commons on 20 February 1941, House of Commons, *Hansard*, http://hansard.millbanksystems.com/commons/1941/feb/20/occupied-france-british-subjects.
36 Cazals, *Lettres de réfugiées*, 131–2.
37 Christine Morrow, *Le Roman irréaliste dans les littératures contemporaines de langue française et anglaise* (s.l.: Didier, 1941), 332.
38 The University of Toulouse had become a real place of refuge for many displaced students, especially for Belgian students (there were more than thirty thousand Belgians in Toulouse), which prompted the dean of the university to say that the Belgian University had been transported to the banks of the Garonne [the Garonne is a river in southwestern France] and that Toulouse had become the capital city of Belgium. See Ambit, "Entre le 14 mai et le 10 juin 1940."
39 Morrow, *Abominable Epoch*, 83.
40 Ibid., 85.
41 Ibid..
42 Ibid., 86.
43 Ibid., 69.
44 "Oran" is a reference to the Battle of Mers-el-Kébir, a naval base on the coast of French Algeria. On 3 July 1940, the British Royal Navy conducted an attack on French Navy ships after France had signed

armistices with Germany and Italy. The purpose was to prevent the French fleet from falling into Axis hands. During these bombardments, more than twelve hundred French sailors died. See Warfare History Network website, https://warfarehistorynetwork.com/2015/10/27/operation-catapult-the-attack-on-mers-el-kebir/.

45 Luneau, *Je vous écris de France*, 84: "Monsieur, Un jour en me promenant, j'aperçois une affiche, mais une affiche! Jugez par vous-même. Un soldat Boche portant un enfant français dans ses bras et cet enfant mangeait une énorme tartine de pain et dessous cette inscription: 'Faites confiance aux Allemands.' Ça alors! Une autre affiche. Dans le lointain, un amas de pierres, plus rapprochée une femme assise qui contemplait sans doute sa maison avec deux enfants auprès d'elle et dessus, il y avait ceci: 'C'est les Anglais qui ont fait cela.' Mais des mains françaises ont écrit ceci: 'Non, c'est Adolf.' Comme vous voyez, nous avons rétabli la vérité. Encore une autre affiche. Cette fois, c'est un marin français qui sombre en mer, et dans un effort avant de disparaître, il lève le drapeau français. Inscription allemande: 'N'oubliez pas Oran.' Mais Oran a disparu comme par enchantement. 'Oran' a été arraché ou effacé. Sur l'une de ces affiches, j'ai lu ce qui suit: 'N'oubliez pas… 1870, 1914–1918.'"

46 Luneau, *Je vous écris de France*, 26.

47 Ibid., 53.

48 Ibid., 68–9.

49 Lady Murdoch, preface to Morrow, *Abominable Epoch*, vi.

50 *Graduate Chorus Newsletter* 50, no. 3 (2013): 16, http://www.graduatewomenwa.org.au/sites/default/files/newsletters/GWWA%20Volume%2051%20Number%2003.pdf.

25

Robert Briffault's War Letters: A Divided Self under Fire

Emmanuel Roudaut

In May 1915, Robert Briffault (1873–1948) wrote the first of a long series of letters to his daughters, Muriel and Joan.[1] He had left them with their mother in Auckland, New Zealand, where he had abandoned his medical practice, and he was sailing for Europe to follow his son Lister, who had volunteered to join the air force on his eighteenth birthday. Travelling privately, Briffault would obtain a commission as a lieutenant in the RAMC (Royal Army Medical Corps) on his arrival in England and serve as a medical officer until the end of the war. I am still transcribing his war letters to Muriel and Joan, which were collected by the younger daughter, Joan, and donated to the British Library with other family papers in 1977.[2] Unfortunately, only one side of the correspondence has been kept, and the set does not include Briffault's letters to his wife, Anna Clarke. However, these letters provide a substantial amount of information, and I hope to publish an edition of their very idiosyncratic account of the war. In this chapter, I suggest a few possible lines of investigation and analysis, with all the caution required in the early stages of a project.

One may question the relevance of adding more personal correspondence from an Anzac soldier to the immense body of similar documents already published.[3] These letters, however, deserve attention on several counts. In the first place, if one excludes anthologies of texts written by unrelated individuals, there are few surviving collections of First World War letters spanning several years. *Letters from a Lost Generation: The War Letters of Vera*

Brittain and Four Friends, to take a famous example, does indeed span five years, from June 1913 to 1918, but Vera's four friends died on the Western Front.[4] She was the only one to go further afield during an assignment to a hospital in Malta as a voluntary nurse, whereas Briffault's letters cover a much wider range. In addition to his impressions of Honolulu, Vancouver, the Rocky Mountains, and New York, they include accounts of fighting in Gallipoli, the Somme, and Passchendaele as well as his views on the home front during his periods of leave in Alexandria and London. It is extremely rare to come across such a mass of evidence from a single witness. This breadth of experience enables him, for instance, to explain the evolution of German gunnery over two years and what it felt like to be on the receiving end (19 September 1917). We can also trace the shift from initial exhilaration to despondency, which led Briffault to frequent bouts of depression and nervous breakdowns.

There is nothing unusual in this growing disillusionment with the war, the most famous example being that of Siegfried Sassoon, who narrowly escaped being court-martialled for publishing his "Soldier's Declaration" in July 1917. However, it took a singular form in the case of this articulate middle-aged man born into a cosmopolitan family and presented as a "nomadic subject" by Luisa Passerini.[5] Briffault's mother, Margaret Stewart, was Scottish, and his father, Charles Frédéric Briffault, a French diplomat, had been a close collaborator of Prince Louis-Napoléon until the latter was proclaimed emperor as Napoléon III. Robert spent most of his youth between Britain and Italy. His letters also suggest that he had travelled widely around Europe and visited Egypt with his wife. Unlike Wilfrid Owen, John McCrae, and Vera Brittain's correspondents, he survived the war and became a prolific writer. Until his death in 1948, strong anti-British feelings would be a dominant feature of both his essays and novels, culminating in propagandist pamphlets published in France during the Second World War.[6] This chapter argues that his atypical case of anglophobia finds its origins in the trenches. In Briffault's letters, disenchantment with war is increasingly bound up with a strong distaste for the British way of conducting it. As I demonstrate later, this turns into a general indictment of Britain, which is to find an echo in his subsequent work.

WRITING AND RECEIVING LETTERS: PRACTICAL ASPECTS

The documentary value of these letters is inextricably linked with the specific nature of correspondence. It records (or conceals) events, but it also reflects the writer's mood and offers clues about the atmosphere on the front line or back home. According to the editors of Vera Brittain's correspondence, it conveys "the uncertainty, confusion, and almost unbearable suspense of wartime."[7] Above all, letters are not just a mirror but also an agent: they play an important psychological role. As her fiancé puts it in *Letters from a Lost Generation*, "[Vera Brittain's] letters help [him] to live."[8] It is also the case for Briffault, who, like many, makes it clear that a regular flow of correspondence is essential for his morale. "Write me lots," "Write me long letters," are recurrent pleas. This also applies to his family, which leads him to apologize to his daughters for not writing enough, although he sometimes admits that it is materially difficult for him to write, as in Gallipoli on 1 November 1915: "I am afraid my writing is very bad, but I am writing on my knees by the light of a candle stuck in a bottle & the wind is threatening to blow it all – so I can hardly see."

These allusions to the practical aspects of receiving and writing letters provide a useful glimpse of the mundane reality of warfare. Briffault often complains about an "erratic and unsatisfactory post" (Christmas Eve, 1916). Letters are lost, or frequently delayed, with a clear impact on his mood. Both sides resort to numbering their letters to keep count of those that have found their way. The sinking of mail boats seems to be the worst fear, as suggested by the self-pitying tone of the following extract:

> I have just this moment heard that a troopship, "the Caledonian," has been sunk near Lemnos – that she was carrying a very big mail. All your dear letters that I was so looking forward to have probably gone down with her! Oh! Well, it would be strange if I were vouchsafed a single little spark of joy – even that is taken away! There are some men going, I think, to Alexandria, so that I'll give them this, it will be posted in Egypt – with a fine Egyptian stamp & get to you quicker. Much love from Dad. (15 November 1915)

The likelihood of such incidents increased in 1917, when Germany launched into unrestricted naval warfare. Surprisingly, they are no longer referred to at this stage, as if Briffault had become inured to them. By then, even the attack of a train carrying troops on leave back to Britain hardly receives a casual mention (19 September 1917). The contrast is striking with the postal connection between Britain and the Western Front. Fussell describes it as "rapid and efficient" and quotes an officer who receives the London papers "only a day late, as we always do."[9] It is only when she arrives in Malta that Vera Brittain experiences delivery problems: "Malta, 20 October 1916. Just received my 1st mail since arriving here ... Oh! The glory of the mail! You who have never been further than France have no idea of it. I have just got 9 letters in all – ranging in dates from September 28 to Oct. 9."[10] For Briffault, however, it is "a treat" to receive the occasional *Auckland Weekly News*, and letters from home take at least a month to reach him. On the postal front, the plight of "colonials" was quite different from that of Englishmen. In addition to slower delivery, longer distances meant increased risks of loss or destruction.

There are occasional complaints about the compulsory writing in pencil and the poor quality of paper and envelopes (for instance on 18 July 1916), but the harsh reality of war becomes even more acute when he apologizes for the mud splashed on his letter by the explosion of a shell a few feet away. As already mentioned, letter writing is also impeded by bad weather conditions, especially during the very cold autumn of 1916. As Briffault puts it, "I cannot write either with my head or my hand ... I'll try and write more when I have thawed" (19 November 1916). For his family, the brevity of some letters may have been more telling, and worrying, than long and detailed ones.

Many letters are loosely structured as they are written at intervals, with interruptions caused by a "biff" (slang for fighting), the dressing of wounds, or emergency surgery. The occasional deterioration of spelling, syntax, and handwriting (at best very idiosyncratic) can be explained by the urge to write as much as possible during the rare moments grabbed, but it also reveals signs of exhaustion, or the onset of depression, especially when he is hospitalized in Alexandria or writes from the Somme. Even more than the content of these letters, it is their material presentation that conveys the sense of dislocation caused by the trauma of war.

The only letter that seems to have been intercepted by the censors was sent on 2 July 1916. Perhaps this incident reveals an increased level of nervousness on the part of the authorities in the wake of the disastrous fiasco of the Somme offensive. When told by his daughter, Briffault notes with amusement that it is about the only letter in which he has taken any precautions (11 October 1916). Indeed, most of his references to censorship are rather tongue-in-cheek. "I shall be shot at dawn if this gets into the hands of the censor," he writes on 26 August 1916. Like his son, Lister, he occasionally resorts to conundrums to indicate his whereabouts. It is interesting to note that Briffault, like others, used his personal correspondence to spread news or rumours censored in the press – for instance, to give details about Zeppelin raids on London that the press was not allowed to mention, mainly to avoid informing the Germans about the accuracy of their strikes. He also notes, on his way to Europe, the conspicuous presence of German sailors and pro-German literature in Honolulu, an American territory (2 June 1915). As we see later, the acrimony of his remarks increases as his disillusion with the conduct of war grows.

The diffusion of embarrassing news through private correspondence was obviously a concern for the authorities, but preserving this tenuous link with families and loved ones was essential to sustain morale, as suggested above. Obstructing the flow of communication would not only have been impractical but also counter-productive. Moreover, far from being subversive, most letters sent from the front line were bland in content and plodding in style. At least, that is suggested by Briffault's disparaging remarks when he himself has to check the letters of those under his command:

> It is very funny reading all the Tommies' letters, which I still have to do now and again. – their literary importance is pathetic. They are almost all identical, generally begin "I take up my pencil to write a few lines and let you know that I am in the pink hoping that you are the same." One love-letter began "My darling" and ended "I remain yours truly"! Goodbye for the present, dear, "I hope you are in the pink" Dad. (1 November 1915)

Vera Brittain's officer friends make similar remarks about the "prosaic and unimaginative" content of their men's letters.[11] Class consciousness is probably at work here. It will become more apparent with the arrival of conscripts, when Briffault, despite his communist sympathies, gives vent to his patrician contempt for the "English yokels" (30 July 1916) and "riff-raff" (8 September) who soil the beauty of the French countryside with their litter and the paraphernalia of war.

TRAVELOGUE, "PEN PICTURES," AND CONTRADICTIONS

Briffault was certainly aware of the documentary value of his letters. He is happy, for instance, to see an extract reproduced in his daughter's school magazine, even though he regrets that she has not corrected the punctuation (13 October 1917). There is clearly a didactic dimension to this correspondence and he jokingly admits he can be "a blooming schoolmaster" at times (26 September 1915). This caring father, frustrated by his absence, worries about his daughters' health; he also follows school results, gives advice, discusses literature (15 October 1916), and never loses an opportunity to draw historical parallels when he describes his travels or life on the front line. The early letters read like a travelogue punctuated by amusing anecdotes, and Briffault shares the general conviction that the war will not last long. More than Honolulu or Vancouver, it is clearly New York that makes the biggest impression on him and marks the beginning of a life-long fascination with the US. Even the chase of his ship by a U-boat on the way to Liverpool is presented as a light-hearted adventure. Briffault seeks to educate and entertain at the same time, especially when he writes to his younger daughter, Joan. And since his readers are teenagers living in the Antipodes, he provides details and explanations that would be taken for granted by an educated adult, especially by Europeans who have had first-hand experience of air raids on the home front. A good example of what he terms his "pen pictures from Picardy" is the "imaginary trip round the trenches" on which he takes Joan on 13 August 1916. It provides a detailed and vivid account of the labyrinthine world of the front line and the uncanny proximity between German and Allied troops.

Unsurprisingly, the tone of the "pen pictures from Picardy" is darker than his descriptions of the Pacific and North America. The account of his second journey to Gallipoli in September 1915 still reads like a history and geography lesson, full of mythological references, but the sobering experience of the front line has already put paid to his illusions, and he expresses his misgivings: for him, Gallipoli will end in disaster, like the expedition of Charles V against Algiers in 1541. However, even in the darkest days of the war, the word "picturesque" recurs in his descriptions of people and places. Musings over the beauty of the countryside are juxtaposed with reports of harrowing events. In some cases, nature almost serves as an antidote to the horrors of war, as in this letter dated 27 July 1916, reminiscent of John McRae's "In Flanders Fields" with its evocation of larks chirping above the battleground:

> My dear Muriel, …
> All our close fighting is mostly done with hand grenades, an awful weapon. The other night a party of our fellows got into the next Boche trench, did in the sentries and threw half a dozen hand-grenades down the steps of each dug-out. They inflict frightful wounds and when one goes down into the dug-out one finds nothing but corpses and horribly mutilated men. I went round giving the poor devils morphine. That is the usual mode of warfare now, it is called capturing a trench by surprise. …
>
> Later. It's a beautiful evening. I was standing in one of the oldest parts of the trench. The wild flowers hung in high festoons over the hedges, scarlet poppies, daisies, buttercups, veronicas. Above, wide clouds … in the pale blue sky. Two white butterflies were chasing each other, and a lark was chirping high up, above the hiss of passing shells. It was hard to think that all the madness of slaughter was real, that looking over the parapet one would see dead bodies turned black, purple. Nature was beautiful, spoke of other things than slaughter.
>
> Good night,
> Dad

Unfortunately, even the beauty of nature can be destroyed by warfare, as in the following spectacle of desolation (30 July 1916): "The banks are littered with the flotsam and jetsam of an English army, here a heap of broken, mud-covered rifles, sandbags filled with cartridges and rusted bombs, old boots, old tins, remains of webbing equipment, tin water bottles, steel hats, empty cartridge boxes." These images re-emerge in his interwar fiction: "But everywhere the detritus and garbage of murderous madness. It was one vast scrap heap."[12] Elegiac descriptions of the ravages inflicted on the landscape do not exclude an appreciation of the aesthetic quality of war itself. This is particularly true of air fights, and planes in general, which seem to evince a kind of fascination. Occasionally, even the chromatic effects of bombardments prompt an admiring comment:

> The firing in the middle of the night was a most magnificent sight, wherever you looked at there were tongues of flame of all sorts of different colours – red, blue, citron yellow flashing in every direction, and the whole sky was quivering with the lightning of guns amid which rose star shells sent up by the Boches to try and see what we were up to. (16 July 1916)

The pastoral dimension of war poetry and correspondence has often been studied, and Briffault's letters, like many, can be related to a long tradition of "bucolic interludes."[13] As for his fascination with plane fights, I have found at least one similar example in Vera Brittain's correspondence,[14] but this oscillation between attraction and repulsion is not confined to aesthetics. Briffault seems to be torn between contradictory attitudes, which probably reflects his mental state. He can fume against cowards and cases of self-inflicted wounds, and suggest that some soldiers ought to be shot. "The only way to do anything with them would be to adopt the German plan of having officers at the back as well as at the front and shoot a few – to encourage the others" (8 September 1916). But in the same letter, he envies the "lucky blighters," by which he means the soldiers who are repatriated because of their wounds, and confesses that he has just had to shake off the "fit," his term for a bout of depression. He had been down with dysentery in Gallipoli, but we understand that it was mainly for depression that he was hospitalized in Alexandria, and the after-effects were to continue sporadically at least until the end of the war. Significantly, when he is in hospital or on leave, he is always impatient to return to the front line despite his disgust for the war, as if the whirl of battle were the only place where he can temporarily forget his unquiet state. These contradictory impulses seem to have been experienced by many others, including Wilfrid Owen and Siegfried Sassoon. As his friend Robert Graves said of Sassoon, "he varied between happy warrior and bitter pacifist."[15] This also applies to George Sherston, Sassoon's alter ego in his autobiographical trilogy.

Like Sassoon's, Briffault's fiction derived from first-hand memories. One finds a similar ambivalence in *Europa in Limbo*, a novel set in the First World War and published in 1937. There is clearly an autobiographical dimension in the urge felt by the narrator and the main character, Julian Bern, to return to the front of a loathsome war and escape from England. An evocation of the solitude felt by the hero on leave in London is strikingly similar to some of Briffault's letters.[16] Descriptions of fighting in the Somme and at Passchendaele include details that seem to have been directly inspired by them, although these documents were no longer in his possession.[17] Other parallels include the fetid smell of German corpses ("Fetor Germanicus") embedded in the trench occupied by Julian. The "regular spraying of chlorine of lime over the offensive portion of the wall"[18] closely echoes a detail of the "imaginary trip round the trenches" written on 13 August 1916. Likewise, a comical altercation between British soldiers and a French peasant, based on the linguistic incompetence and arrogance of the former,[19] can be traced to similar anecdotes related by Briffault the polyglot to his daughters. Other characters, like the "horsey colonel,"[20] are clearly based on Briffault's personal experience (16 December 1916).

Political statements are relatively rare in these letters written to teenagers, but they foreshadow the numerous diatribes of the novel, a novel that Cyril Connolly dismissed as "a work of propaganda by a passionate Communist giving free rein to his rabid detestation of English imperialism and English democracy, 'just a bad cartoon' in spite of many interesting glimpses of Europe in wartime."[21] In this passage, written on 30 August 1917, Briffault is scathing about "the policy of lies and jingoism … carried out since the beginning of the war," but, at the same time, he condemns the prospect of a "patched up Peace negotiation," adding: "The object of the war (unavowed by England) is the crushing of militarism for ever, and the expulsion of autocracy from Europe. This is worth even some thousands of lives more." He is prepared to "to wait till next year & the Yanks. Nothing but a democratic army can fix up German autocracy. We are too much like them" (7 August 1917). This is clearly at variance with Sassoon's recently published, and widely commented, call for the curtailment of the war: "I am making this statement as an act of wilful defiance of military authority because I believe that the war is being deliberately prolonged by those who have the power to end it." Sassoon goes on to argue that, if they had been clearly stated, the initial purposes of the war "would now be attainable by negotiation."[22] His views of "the political errors and insincerities for which the fighting men are sacrificed" are as trenchant as those of Briffault, but he reaches the opposite conclusion. However, both men were noted for their bravery in action during a war that they regarded with loathing.

BUDDING ANGLOPHOBIA

The most distinctive feature of the correspondence is not its ambivalence about the war but the hatred that Briffault seems to develop for England as the conflict drags on. Arriving in June 1915, he still admires the beauty of the English landscape and seems pleased to guide fellow New Zealanders through the London of his youth. But the mood has already changed in September, when he returns from his first assignment in Gallipoli. In subsequent leaves, accounts of visits to the theatre or the cinema give way to contempt and sarcasm. The most ferocious remarks are for Wimbledon, presented as the epitome of English dullness and hypocrisy:

> This afternoon I went to Wimbledon ... Wimbledon is really a masterpiece which never ceases to amaze one and make one's hair stand on end. ... The dwelling-place of a smug people suffering from complete atrophy of the brain. There is not a discordant note anywhere, everything speaks of artless English suburbanites undefiled. It appears to be peopled by chambermaids and Sunday-school teaching shopboys. The girls are rosy-cheeked imbeciles. It is an unpatriotic place, because it makes one pro-German to look at it. Every flame of thought and living feeling is as much extinguished here as in an atmosphere of carbonic acid.
>
> Yesterday I saw the great Joffre and M. Briand as they were stepping into a motor-car from the French embassy; he looked magnificent, smilingly Olympian. (11 June 1916)

Impatience with civilian life on the home front was not confined to Briffault, and the troops often indulged in fantasies of violent retribution. They are voiced in some of Sassoon's verse. In "Fight to a Finish," for instance, he imagines returning troops on a Victory Parade turning their bayonets on the cheering crowds, the "Yellow-Pressmen" and "those Junkers" in Parliament.[23] "Blighters" features the audience of patriotic music halls being crushed to death by a tank.[24] Fussell lists other outbursts of hatred of civilian England experienced by soldiers returning from leave, concluding with this quotation from Philip Gibbs: "They hated the smiling women in the streets. They loathed the old men ... They desired that profiteers should die by poison-gas. They prayed God to get the Germans to send Zeppelins to England – to make the people know what war meant."[25]

Grievances against the "callous complacency with which the majority of those at home regard[ed] the continuation of agonies which they [did] not share and which they ha[d] not enough imagination to realise" were not always so violent,[26] but they were widespread. What was specific to Briffault, in the context of the First World War, was the polarization between an idealized France and a reviled Britain. Incipient in his letters from Gallipoli (1 November 1915), this polarization becomes obvious in the Somme. Anything or anyone French is put on a pedestal and contrasted with England and the English, as in his first letter after landing at Boulogne (5 July 1916): "It was so nice to tread the cobbles of heroic France that I felt better at once. ... What a contrast it all is with England – here the war is a reality, there is no picnicking and dilettantism." In *Europa in Limbo*, for instance, Paris in wartime is also contrasted with London.[27] A selective presentation enables both the letter writer and the novelist to build up an indictment of the English, especially in their conduct of war. Here is a telling example from a letter dated 11 November 1916: "It was rather curious to read the English and French communiqués the other day. The English said that owing to stormy weather we could do nothing. The same day the French took two villages and surrounded Chaulnes – in wind, mud & rain! We heard the din of their strafing here all day." The dithering of the British general staff is a favourite target, but not the only one. A more scathing (and more wide-ranging) comment follows a few weeks later: "There are 4 million British men in France spending 5 and a half millions a day very busy doing everything except fighting – as far as the war is concerned sitting on their bottoms" (27 December 1916). The point is made repeatedly. The soiling of the French countryside, conjured up in an imaginary idyllic past, is almost systematically blamed on British troops, "[d]irty, rugged louts, talking the thick speech of Yorkshire & Northumberland" (30 July 1916). On another occasion (3 November 1916), Briffault declares that he had "much rather be with the French than with the English. They are fighting splendidly."

However, if one looks for clues in the surrounding text, the most violent anti-English outbursts are often associated with moments of mental crisis. They could perhaps be read as signs of the dislocation brought about by war, especially in view of some inconsistencies. It is also interesting to note that he reverts to French expressions in moments of despondency, as if they triggered a desire to return to his French roots. The combination of the traumatic experience of war and Briffault's complex family history was probably instrumental in his evolution towards a radical form of anglophobia. According to Passerini, "Briffault's hatred for England possibly had psychological roots, and not only political ones, as Britain was the land of his austere mother, while France belonged to the mindless and charming father, whom he considered very similar to himself."[28] Significantly, in his first letter from the Somme, he admits that he "hardly know[s] France except Paris" (5 July 1916). Many years later, in 1938, he adds in despair: "I am denouncing the infamy of England all the time, but I write in English, and couldn't switch to another language and another culture."

Briffault's contrasted views of Britain and France also apply to their respective empires. This becomes apparent when he relates his journey from Alexandria to Britain in 1916:

> Well, the nice part of our trip is that we called at Algiers. Had two days there. Quite an unusual port of call, but the skipper had to get water & coal and, as he said to me "if we call at Malta or Gib they'll just say 'What the h**l do you want here!' and be as nasty as possible, whereas at a French port they can't do enough for you. The only objection, he said, is that it's just a bit sad calling at a French town because they are serious and take the war earnestly, whereas English towns are jolly because the fools of English are so frivolous & think they are having a picnic instead of a war." Very nice fellow that skipper. (28 May 1916)

After an enthusiastic ("It might be the most beautiful part of France") and probably idealized description of the Algerian landscape, he goes on to oppose the plight of native Arabs in Algiers and Egypt:

> The native Arabs are entirely different from those of Egypt. In the first place they mostly seem to be well off, one sees nothing of that population of beggars and ragamuffins which meets one everywhere on the Nile. I was not asked for "baksheesh" once. When I landed and went to a French café to take a "consommation" a little Arab boy came to clean my shoes (which needed it). When he had finished I told him I had no money – of his country, only rupees and piastras & all sorts of foreign coins. "Ça ne fait rien, Monsieur!" he answered and was going to walk away without being paid! The Arabs one sees appear all prosperous and clean.

There is clearly a bit of fantasy and self-delusion in this depiction of colonial Algeria, and Briffault's exhilaration will be echoed by his first impressions of France at war. But this contrasted vision of Algeria and Egypt heralds Briffault's indictment of British imperialism in the propaganda work he was to publish in occupied France during the Second World War. True, the destruction of Alexandria by the Royal Navy in 1882, dealt with at length in *L'Angleterre et l'Égypte* (1943), only gets a brief mention in the long letter (a "lecture" on the history of the city) dated 16 February 1916, but Briffault's most original contribution to the classic Egyptian chapter of anglophobia can be traced to the experience related in his letters. His denunciation of the brutal behaviour of Australian troops, "behaving like the worst savages," looting and brutalizing the local population,[29] closely reflects passages from his letters of 30 April and 28 May 1916, in which he also insists on the hatred between colonial and English troops: "People talk of cementing the colonies and mother country. The colonials are simply <u>detested</u>. The English men & officers hate them more than they do the Germans." In 1943, unsurprisingly, the propagandist amplifies this and establishes a causal link between the brutal behaviour of occupying troops and subsequent uprisings. However, it should also be said that Briffault was probably coerced into writing these *collaborationniste* publications as he and his American second wife, Herma Hoyt, spent the Second World War in Paris, where they were regarded as enemy aliens and arrested on several occasions. Moreover, since all revenue from copyrights in the United States and Britain was cut off, they lived in great deprivation, which was an incentive to accept hack work, however unpalatable this might have been. There is no suggestion in his writings of the 1930s that he drifted towards fascism. On the contrary, in a letter to Joan he comments that it was "the highest honour" ever paid to him to have one of his novels burned by the Nazis in 1938.[30]

CONCLUSION

This short chapter does not do justice to all the facets of a narrative that could prove a valuable addition to the literature of the First World War, despite its emotional bias, or perhaps because of it. Other aspects of documentary value include the insight into the privileged treatment of officers affected by depression and shell-shock. However, the specific interest of Briffault's letters lies in the clues they provide to the enigma of his evolution. This is a man who, decorated for gallantry, went far beyond the standard critique of the ineptness of military leaders. And the Marxist critique of a capitalist war, which is one of the main themes of *Europa in Limbo*, cannot alone explain the vicious tone of attacks specifically targeted on Britain. I make no claim to have solved the mystery at the present time, although I hope to have made some progress when the transcription is completed. But there is no doubt that the experience of the war played a crucial part in Briffault's drift towards the vitriolic anglophobia of his essays and novels. It certainly proved a watershed for him since he retired from medical practice after the war and devoted himself to writing, embarking on a bohemian life between France, the United States, and, paradoxically, Britain, where he would end his life. His first book, a work of anthropology,[31] was actually written in the trenches. Above all, the fulminating tone of his war letters would remain a constant feature of his subsequent writing.[32] As such, they provide essential reading for scholars seeking to explore the making of a rather strident interwar polemicist.

NOTES

1 Briffault's entry in *Who was Who* and a number of documents give 1876 as his year of birth, but other sources (e.g., Passerini) give 1873. Several elements suggest that the latter is correct. They include this passage from a letter dated 7 February 1939 ("I am sixty-six … and considerably weary"), a certificate of baptism in 1874 (British Library, Add MS 58442, f.1), and the following excerpt from a loosely autobiographical novel,

referring to the author's alter ego after joining up: "*Even though he had fraudulently curtailed his age by three years.*" See Robert Briffault, *Europa in Limbo* (New York: Charles Scribner's and Sons, 1937), 66.

2 British Library, Add MS 58440 for letters written before 9 April 1917; British Library, Add MS 58441 for subsequent ones. Presented in Arthur Searle, "Letters of Robert Briffault," *British Library Journal* 3 (1977): 169–76.

3 Early examples include Laurence Housman, *War Letters of Fallen English Englishmen* (London: Gollancz, 1930); and John Laffin, *Letters from the Front, 1914–1918* (London: Dent, 1973). For a more recent publication, see Glyn Harper, ed., *Letters from Gallipoli* (Auckland: Auckland University Press, 2011).

4 The same is true of Paul Fussell's path-breaking study, *The Great War and Modern Memory* (Oxford: Oxford University Press, 2013 [1975]), which derives exclusively from images of the trenches in France and Belgium.

5 Luisa Passerini, *Europe in Love, Love in Europe: Imagination and Politics in Britain between the Wars* (London: I.B. Tauris 1999), 149–87.

6 Robert Briffault, *La Démocratie: Instrument de la duperie anglaise* (Paris: Éditions du livre moderne 1941); Robert Briffault, *L'Angleterre et l'Egypte* (Paris: Centre d'études de l'agence inter-France, 1943); Robert Briffault, *La Fable anglaise* (Paris: Éditions Balzac, 1943); Robert Briffault, *L'Inde et l'Angleterre* (Corbeil: Imprimerie Crété 1943).

7 Alan Bishop and Mark Bostridge, *Letters from a Lost Generation* (London: Virago, 2008 [1998]), 1.

8 Bishop and Bostridge, "Roland to Vera, 9 May 1915," in *Letters from a Lost Generation*, 98.

9 Fussell, *Great War and Modern Memory*, 71.

10 Bishop and Bostridge, "Vera to Edward, 20 October 1916," in *Letters from a Lost Generation*, 279.

11 Bishop and Bostridge, "Roland to Vera, 29 April 1915," in *Letters from a Lost Generation*, 93.

12 Robert Briffault, *Europa in Limbo* (New York: Charles Scribner's and Sons, 1937), 290.

13 See, for instance, Fussell, *Great War and Modern Memory*, 251–92.

14 Bishop and Bostridge, "Roland to Vera, 29 April 1915," in *Letters from a Lost Generation*, 93.

15 Robert Graves, *Good-bye to All That: An Autobiography* (London: Jonathan Cape, 1929), 275, quoted in Fussell, *Great War in Modern Memory*, 107.

16 Briffault, *Europa in Limbo*, 139, 191.

17 Ibid., 290–3.
18 Ibid., 96.
19 Ibid., 86.
20 Ibid., 123.
21 *New Statesman and Nation*, 16 October 1936. Summarized in Passerini, *Europe in Love*, 172–3.
22 Siegfried Sassoon, "A Soldier's Declaration," *Bradford Pioneer*, 27 July 1917.
23 Siegfried Sassoon, *Counter-Attack and Other Poems* (London: Heinemann, 1918), 29.
24 Siegfried Sassoon, *The Old Huntsman and Other Poems* (London: Heinemann, [1917] 1918), 31.
25 Philip Gibbs, *Now It Can Be Told* (New York: Harper and Brothers, 1920), 143, quoted by Fussell, *Great War and Modern Memory*, 94.
26 Sassoon, "A Soldier's Declaration."
27 Briffault, *Europa in Limbo*, 225.
28 Passerini, *Europe in Love*, 180.
29 Briffault, *L'Angleterre et l'Égypte*, 101–3.
30 Passerini, *Europe in Love*, 178–9.
31 Robert Briffault, *The Making of Humanity* (London: Allen and Unwin, 1919).
32 Robert Briffault, *Reasons for Anger* (New York: Simon and Schuster, 1936).

Contributors

STÉPHANIE BELANGER is associate scientific director of the Canadian Institute for Military and Veteran Health Research, a unique consortium of forty-six Canadian universities dedicated to researching the health needs of military personnel, veterans, and their families. Dr Belanger is co-editor-in-chief of the *Journal of Military, Veteran and Family Health* (University of Toronto Press, founded in 2015). She is also co-founder of the Human Dimensions in Foreign Policy, Military Studies, and Security Studies series with McGill-Queen's University Press (founded in 2016). She is the co-editor of *War Memories: Commemoration, Recollections, and Writings on War* (MQUP, 2017); *Beyond the Line: Military and Veteran Health Research* (MQUP, 2013); *A New Coalition for a Challenging Battlefield* (CDA Press, 2012); *Shaping the Future* (CDA Press, 2011); as well as *Transforming Traditions: The Leadership of Women in the Canadian Navy* (CDA Press, 2010). She is also author of the monograph *Guerre, sacrifices et persécutions* (Paris: L'Harmattan, 2010). She co-chairs the CIMVHR annual Forums and Symposia, the bi-annual conferences on war memories (with Université de Rennes 2 and Paris VII), and the annual conferences on military ethics, and she partners with many other institutes to co-host workshops. She is a Board of Directors member of the North American chapter of the International Society for Military Ethics (ISME) and board chair of the Center for International and Defence Policy (CIDP). She was inducted as a member of the College of Young Scholars of the Royal Society of Canada in 2016. She is professor in the Department of French Language, Literature and Culture and the chair of the Master of Public Administration Programme at the Royal Military College of Canada, where her research focuses on war testimony, soldier identity, and moral injuries. She specializes in military ethics and just war theories. She completed her PhD degree at the University of Toronto in 2003 and her MPA degree at RMC in 2013. She has served in the Royal Canadian Navy as a reservist since 2004.

The leading scholar on the British painter Stanley Spencer, KEITH BELL, is professor emeritus of art history and visual culture in the Department of Art and Art History at the University of Saskatchewan. He is the author of *Stanley Spencer: A Complete Catalogue of the Paintings* (Phaidon Press) and curator of the Royal Academy of Arts, London, Stanley Spencer R.A. retrospective exhibition (282 works). He has published and spoken on Spencer on numerous occasions and has contributed several extensive catalogue essays on his work. In addition, he has curated numerous exhibitions on other subjects, including the work of the Scottish photographers Hill and Adamson, and historic Prairie photographs (*Plain Truth*), as well as other exhibitions on Prairie art. He has taught many classes on visual culture and historic and contemporary photography.

ANNA BRANACH-KALLAS is associate professor at Nicolaus Copernicus University in Toruń, Poland. Her research interests include the representation of trauma and war, postcolonialism, corporeality, health humanities, and comparative studies. She has published several books, including, most recently, *Comparing Grief in French, British and Canadian Great War Fiction (1977–2014)* (Brill-Rodopi, 2018), co-authored with Piotr Sadkowski. Her earlier monograph in Polish, *Uraz przetrwania* (*The Trauma of Survival: The (De)Construction of the Myth of the Great War in the Canadian Novel*) (NCU Press, 2014), was awarded a Pierre Savard Award by the International Council for Canadian Studies. She is the author of over eighty book chapters and articles, and has published in such academic journals as the *Journal of War and Culture Studies*, the *European Journal of English Studies*, the *Journal of Literature and Trauma Studies*, *Canadian Literature*, *Second Texts*, and *Studies in 20th- and 21st-Century Literature*. She has (co-)edited several essay collections, including *Re-Imagining the First World War: New Perspectives in Anglophone Literature and Culture* (Cambridge Scholars Publishing, 2015). She is currently head of the Institute of

Literary Studies at Nicolaus Copernicus University and has served as president of the Polish Association for Canadian Studies since 2016.

FLORENCE CABARET is a senior lecturer at the University of Rouen (Normandy, France) and a member of the interdisciplinary research group ERIAC. After a PhD dedicated to the status of fiction in Salman Rushdie's novels, she has mainly worked on novelists from the English-speaking Indian diaspora as well as on films produced by Indian diasporic artists in the UK, in the US, and in Canada, without forgetting television series staging characters from the British Asian and American Asian diasporas. She co-edited several collective works, such as *Mauvaises Langues !* (2013) with Nathalie Vienne-Guerrin; *Retranslating Children's Literature* (2014) with Virginie Douglas; and several issues of the online journal *TV/Series* (on revue.org) with Sylvaine Bataille and with Claire Cornillon. She is also the translator of the novels of Hanif Kureishi and of Chloé Hooper (published by Christian Bourgois in Paris).

NICOLE CLOAREC is a senior lecturer in English at the University of Rennes 1. Her research focuses on British- and English-speaking cinema and in particular on questions related to the cinematic apparatus, transmediality, adaptation, and the documentary. She is a member of the editorial board of *LISA e-journal* and *Film Journal*. Among her recent publications, she has co-edited *Social Class on British and American Screens* (McFarland, 2016); "The Specificities of Kitsch in the Cinema of English-Speaking Countries" (*LISA e-journal*, 2017); and "Actors Behind the Camera" (*Film Journal* 6, 2020); and co-written *Ian McEwan's Atonement and Joe Wright's Film Adaptation* (Ellipses, 2017).

SHEILA COLLINGWOOD-WHITTICK obtained her PhD (University of London, 1980) for her dissertation on the francophone literature of colonial Algeria in the decades preceding and up to the War of Independence. Following many years of living and teaching in Southeast Asia (resulting in further and broader reflection on the nature of colonialism), she switched the focus of her research to anglophone (post)colonial literatures on returning to Europe at the beginning of the 1990s. After an initial appointment at the University of Savoie, she occupied the post of senior lecturer in the Department of Anglophone Studies at the University of Grenoble III from

1996 until 2012. From 2002 to 2004, she was also appointed (concurrently) to teach postcolonial studies and literatures at Geneva University's Department of English. She is the author of more than forty published book chapters and scholarly articles, the editor of a collection of essays entitled *The Pain of Unbelonging: Alienation and Identity in Australasian Literature* (Rodopi, 2007), and the co-editor of a collection of essays entitled *Indigenous Peoples and Genetic Research: Towards an Understanding of the Issues* (Rodopi, 2012), to which she contributed two chapters. Since 1997, her research has centred on the (post)colonial literature and post-contact history of Australia, and she has published widely on these new foci of interest. Now retired from teaching, she continues to research, participate in conferences, and publish her work.

RAPHAËLLE COSTA DE BEAUREGARD is emeritus professor, Université de Toulouse Jean Jaurès. She is a specialist of English and American studies, in literature and visual arts, minature painting, and film studies. In 1993 she founded the SERCIA (Société d'Etudes et de Recherches sur le Cinéma Anglosaxon). Her publications include *Nicholas Hilliard et l'imaginaire élisabéthain* (CNRS, 1992); *Silent Elizabethans: The Language of Colour of Two Miniaturists* (CERRA, 2000); and several journal articles. She edited *Le Cinéma et ses objets-Objects in Film* (Poitiers, La Licorne, 1997), *Cinéma et Couleur: Film and Colour* (Michel Houdiard, 2009). Her present research focuses on phenomenology and film, mostly in early cinema.

CORINNE DAVID-IVES, who died in 2021, was a senior lecturer at Caen University (France). Her research work was in the field of Commonwealth studies, with a focus on New Zealand. She studied identity politics and the place of Indigenous peoples in the former British colonies of settlement. She was working on the strategies used nationally and internationally by Indigenous peoples to achieve recognition and empowerment. She published several studies in France and abroad on reconciliation politics and the representation of Indigenous minorities.

JEFFREY DEMSKY is associate professor of political science at San Bernardino Valley College. He specializes in American cultural representations of the Holocaust. Dr Demsky is an active scholar who has published his research in academic journals and presses located in the US, UK, France, Canada, Switzerland, the Netherlands, and Germany. He recently published *Irreverent Remembrance: Nazi and Holocaust Memorialization in Anglo-American Popular Culture, 1945–2020* (Palgrave Macmillan, 2021).

RENÉE DICKASON, who obtained her professorship in 2001, began her career over thirty years ago at Rennes 2 University, where she currently works, having taught for nine great and happy years at the University of Caen-Basse-Normandie. After enjoying fruitful and stimulating times, in cultural, linguistic, historical, and human terms, in the British Isles, she took the decisive step of pursuing studies on the "unfathomable and perfidious" Albion – a challenge in constant renewal as she chose a field of research related to contemporary history, more specifically immediate history, amid the historical realities inherited from the past and animated by present-day mutations in a world at war or at peace. She has favoured local immersive experiences and media approaches, mainly through visual and filmic prisms. Her observations and publications deal with social and political communication, the representation of the "real" in electoral periods and during conflicts, such as the "Troubles" in Northern Ireland and wars in the Gulf and in the Falklands. At the University of Caen-Basse-Normandie, she initiated (with the Maison de la Recherche en Sciences Humaines [CNRS], the Mémorial de Caen Museum, the Royal Military College of Canada [Kingston, Ontario], the International Committee of the Red Cross [Geneva], and researchers from various disciplinary and geographical horizons) a research network on war memories (WARMEM Project), which celebrated its tenth anniversary in 2020. Since her early childhood, Renée Dickason has travelled widely, untiringly seeking to discover new cultures and spaces; she likes to think of herself as an anthropologist of the living moment. Her objectives: to contribute to widening research in human sciences, transcending disciplinary frontiers so as to heighten awareness, to educate, to act, and to give a profoundly human and humanist meaning to ongoing and forthcoming academic and scientific missions. Her aspirations: to pursue and enrich reflections on war crimes, abuses, and human aggression that colour, through significant examples, the memorial construction of war phenomena in order to fight against barbarity and to reveal and confront atrocities committed in wartime, with the hope of establishing and maintaining peace, wherever it is imperilled.

Her professional curriculum vitae can be found at: https://perso.univ-rennes2.fr/renee.dickason.

GEORGES FOURNIER is professor in the Department of Foreign Languages at the University of Limoges. His main research interest lies in British-authored television. He has published many articles on the fictional representations of social and political issues. In 2019, he published the first book ever written on Peter Kosminsky, a prominent British filmmaker. He recently edited and co-wrote books on the BBC and is currently conducting research in factual programming.

MATTHEW GRAVES is professor of contemporary British history, culture, and society at Aix-Marseille University and a fellow of the Royal Geographical Society. His research interests lie in the fields of geohistory and critical geopolitics. He has published widely on issues of space, memory, and polity in nineteenth- to twenty-first-century British and Australian history.

ANITA JORGE is associate professor at the University of Toulouse II Jean-Jaurès, where she teaches British and American history and film studies. She holds a doctorate in British cultural history from the University of Lorraine and graduated from the École Normale Supérieure of Lyon. Her PhD dissertation examined the representation of the soundscape of Second World War-Britain in British official documentaries of the period, and her research interests revolve around documentary films, sound experimentation, propaganda, and electro-acoustic music.

ANDRÁS LÉNÁRT is senior lecturer at the Department of Hispanic Studies, University of Szeged, Hungary. He holds a PhD in contemporary history of the Hispanic world. His research interests include nineteenth- and twentieth-century history, international relations, cinema of Spain and Latin America, the relations between Latin America and the US, and the relations between universal history and cinema. Lénárt is the author of three books and the co-editor of several volumes. His book chapters, essays, and articles have appeared in Spanish, Hungarian, British, Italian, Mexican, German, and Czech journals; volumes of essays and books; and also on the website of the US Library of Congress, National Film Registry. He has been a lecturer at more than eighty international conferences in Spain, Portugal, Germany, the United Kingdom, France, Norway, Hungary, Serbia, and the Netherlands. He is currently the president of the International Federation for Latin American and Caribbean Studies (FIEALC) and is a member of various international research groups and associations, such as the Spanish Association for Contemporary History (AHC), the Association for Spanish and Portuguese Historical Studies, the Association for European Historians of Latin America (AHILA), and also of the General Assembly of the Hungarian Academy of Sciences.

DELPHINE LETORT is professor of film and American studies at the University of Le Mans (France). Her research focuses on the intertwining of history and politics, analyzing the power of cinema to shape memorial processes by producing or revising historical narratives and stereotypes. She has published *Du film noir au néo-noir: Mythes et stéréotypes de l'Amérique 1941–2008* (L'Harmattan, 2010) and *The Spike Lee Brand: A Study of Documentary Filmmaking* (SUNY, 2015). She has written many articles about film adaptations, documentary filmmaking, and African American cinema, and has co-edited several books (*L'Adaptation cinématographique: Premières pages, premiers plans*, 2014; *La Culture de l'engagement à l'écran*, 2015; *Social Class on British and American Screens: Essays on Cinema and Television*, 2016; *Women Activists and Civil Rights Figures in Auto/Biographical Literature and Film*, 2018) and two thematic issues for the CinémAction series (*Panorama mondial du film noir*, 2014; *Révoltes armées et terrorisme à l'écran*, 2019). She serves on the advisory editorial board of *Black Camera* (Bloomington, Indiana), as editor-in-chief for *Revue LISA/LISA e-journal*, and is the director of the 3L.AM research center – Langues, Littératures, Linguistique at the Universities of Le Mans and Angers (http://3lam.univ-lemans.fr/fr/index.html).

JOHN MAYNARD is a Worimi Aboriginal man from the Port Stephens region of New South Wales. He is currently chair of Aboriginal history at the University of Newcastle and director of the Purai Global Indigenous History Centre. He has held several major positions, including deputy chairperson of the Australian Institute of Aboriginal and Torres Strait Islander Studies (AIATSIS), and has served on numerous prominent organizations and committees. He was the recipient of the Aboriginal History (Australian National University)

Stanner Fellowship in 1996, was a New South Wales Premiers Indigenous History Fellow in 2003, an Australian Research Council Postdoctoral Fellow in 2004, and University of Newcastle Researcher of the Year in 2008 and 2012. And in 2014 he was elected a fellow of the prestigious Australian Social Sciences Academy. He has worked with and within many Aboriginal communities – urban, rural, and remote. Professor Maynard's publications have concentrated on the intersections of Aboriginal political and social history and the history of Australian race relations. He is the author of several books, including *Aboriginal Stars of the Turf*, *Fight for Liberty and Freedom*, *The Aboriginal Soccer Tribe*, *Aborigines and the Sport of Kings*, *True Light and Shade*, and *Living with the Locals*.

DOMINIQUE OTIGBAH is a former student of the School of Oriental and African Studies in London and holds two degrees in history. She is currently the Historical Research Fellow at the *Republic, a Journal of Nigerian and African Affairs*. Her current area of research is focused on exploring the historiography and narratives of the Nigerian Civil War, with special reference to the ethnic minority experience. Among her recent publications, she co-edited "Concerning Biafra and Nigerian National Futures" (2017), an issue published by the *Republic* on the Nigerian Civil War.

SYLVIE POMIÈS-MARÉCHAL is senior lecturer at the University of Orléans. Her research mainly deals with the impact of the First and Second World Wars on the social roles of women. Her works include studies not only on British women in the French Resistance (especially F section SOE agents) but also on British nurses who served on the Western Front. Some of these studies are based on the analysis of war diaries or narratives. She also translated Christine Morrow's *Abominable Epoch* into French (Christine Morrow, *Une abominable époque: Journal d'une Australienne en France 1940–1941* [Toulouse, Éditions Privat, 2008]).

TATIANA PROROKOVA-KONRAD is a postdoctoral researcher at the Department of English and American Studies, University of Vienna, Austria. She holds a PhD in American studies from the University of Marburg, Germany. She was a Visiting Researcher at the Forest History Society (2019), an Ebeling Fellow at the American Antiquarian Society (2018), and a Visiting Scholar at the University of South

Alabama, USA (2016). She is the author of *Docu-Fictions of War: U.S. Interventionism in Film and Literature* (University of Nebraska Press, 2019), the editor of *Cold War II: Hollywood's Renewed Obsession with Russia* (University Press of Mississippi, 2020) and *Transportation and the Culture of Climate Change: Accelerating Ride to Global Crisis* (West Virginia University Press, 2020), and is a co-editor of *Cultures of War in Graphic Novels: Violence, Trauma, and Memory* (Rutgers University Press, 2018).

MICHEL PRUM is professor emeritus at Université de Paris (formerly Université Paris Diderot). His field of research is the history of ideas in nineteenth-century Britain. His PhD dissertation and first publications were about British early Socialists, especially the so-called "Ricardian Socialists." When he was appointed professor in 1996 he extended his research to liberalism and then Darwinism. He created the Research Group on Eugenics and Racism (GRER) in 1998. The GRER seminar has been very active ever since, with some sixteen sessions every year as well as regular publications. He has edited or co-edited over thirty books, including *Racial, Ethnic and Homophobic Violence* (Routledge-Cavendish, 2007). He has co-organized international conferences in Africa (South Africa, Kenya, Senegal, Lesotho, and Botswana) on the questions of "race" and identity. He founded the Racism and Eugenics book series at l'Harmattan publishing house (Paris) in 2003 and has invited many authors to write on these issues (fifty-seven books so far). Since 2000 he has been coordinating the new French translation of Charles Darwin's Complete Works, published by Slatkine (Geneva) and Honoré Champion (Paris). He has published articles on Darwin's first translators, on racism, on the abolition of the slave trade, and on male feminism (John Stuart Mill and William Thompson). He hosted the Paris War Memories Conference in 2016 and was invited to the 2018 Kingston War Memories Conference as keynote speaker.

ELIZABETH RECHNIEWSKI is honorary senior lecturer at the University of Sydney in the School of Languages and Cultures. She has a long-standing research interest in the political uses of the national past and has published widely on remembrance of twentieth-century war in Australia, France, and New Caledonia, including on the commemoration of the role of Indigenous soldiers in these countries. She was commissioned (with Judith Keene) to write the official report for the French Ministry of Defence on war commemoration in Australia (in *La mémoire combattante, un regard international*, ed. Olivier Wieviorka et Antoine Prost [C2SD, Ministère de la Défense, 2009]). She was a chief investigator on the Australian Research Council project Seeking Meaning, Seeking Justice in a Post-Cold War World (2013–16), and in this context published articles on the impact of Cold War ideology on decolonization in French Cameroon and Madagascar. With Judith Keene, she published *Seeking Meaning, Seeking Justice in a Post-Cold War World* (Brill 2018). Dr Rechniewski's current projects include research into the tensions between France and Australia over control of the South Pacific in the late nineteenth and early twentieth centuries, and on the early period of the Anglo-French condominium in the New Hebrides (Vanuatu). She was awarded the title of Chevalier des palmes académiques by the French government in 2009 and the Ordre national du mérite in 2010.

EMMANUEL ROUDAUT is senior lecturer in British studies at Sciences Po Lille (France). Previous positions included the University of Valenciennes and Centrale Lille after an early career in secondary education (France and Turkey) and two years in the UK as a language assistant (Oundle School and the University of Manchester). His field of research is British social and cultural history (late nineteenth century and twentieth century). His doctorate (1997) examines the conflicting attitudes towards sports betting in British society between 1890 and 1961. He has published articles on this and other aspects of the history of leisure in Britain (horse racing, rambling and trespassing, foxhunting, the music hall, etc.). Other research interests include sexual politics (especially LGBT history), and he has almost completed the transcription of Robert Briffault's correspondence, with a view to publishing an edition.

NATHALIE SAUDO-WELBY is a former student of the École Normale Supérieure de Fontenay Saint-Cloud and holds the agrégation in English. She is a senior lecturer at the Université de Picardie in Amiens, France, where she teaches literature and translation. Her doctorate (2003) focuses on degeneration in British literature (1886–1913). She is accredited to supervise research in British literature. She has published over twenty articles on fin-de-siècle literature, women's writing, and women's perception of conflict. Her book, *New Woman Fiction, Le Courage de déplaire*, was published by Classiques Garnier in 2019.

MARZENA SOKOŁOWSKA-PARYŻ is associate professor at the Institute of English Studies, University of Warsaw, Poland, where she teaches courses on contemporary British and Commonwealth literature, with specific emphasis on war fiction and film in relation to history, memory, and national identity. She is the author of *Reimagining the War Memorial, Reinterpreting the Great War: The Formats of British Commemorative Fiction* (2012) and *The Myth of War in British and Polish Poetry, 1939–1945* (2002). Her work has appeared in edited volumes: *History of the Literary Cultures of East-Central Europe: Junctures and Disjunctures in the 19th and 20th Centuries* (2004); *Mnemosyne and Mars: Artistic and Cultural Representations of Twentieth-Century Europe at War* (2013); *Horrors of War: The Undead on the Battlefield* (2015); *North America, Europe and the Cultural Memory of the First World War* (2015); *Re-Imagining the First World War: New Perspectives in Anglophone Literature and Culture* (2015); *The Great War: From Memory to History* (2015); *The Long Aftermath: Historical and Cultural Legacies of Europe at War, 1936–1945* (2016); *Traumatic Memories of the Second World War and After* (2016); and *The English Canadian Novel in the Twenty-First Century: Interpretations* (2019). She has published in academic journals, including, among others, the *Journal of War and Culture Studies*; WLA: *War, Literature and the Arts*; ANGLISTIK: *International Journal of English Studies*; *Zeitschrift für Anglistik und Amerikanistik* (*A Quarterly of Language, Literature and Culture*); *and Violence against Women*. She has co-edited, together with Martin Löschnigg, *The Great War in Post-Memory Literature and Film* (2014) and *The Enemy in Contemporary Film* (2018). She is editor for the literature and culture issues of *Anglica: An International Journal for English Studies*.

PAUL STOCKER is a visiting fellow at the University of Northampton, United Kingdom. His work focuses on the history of right-wing extremism in the United Kingdom. His monograph, *Lost Imperium: Far Right Visions of the British Empire, c. 1920–1980* (Routledge, 2020), analyzes the far right's response to the decline and collapse of the British Empire during the twentieth century. His other major work is the book *English Uprising: Brexit and the Mainstreaming of the Far Right* (Melville House, 2017). *English Uprising* places the 2016 vote to leave the European Union within the historical context of anti-immigration politics in Britain. It argues that the increased mainstreaming of far-right ideas, particularly in England, which began at the turn of the twenty-first century, can help explain why the country voted to leave the European Union. Dr Stocker is also a senior fellow at the Centre for Analysis of the Radical Right.

VILASNEE TAMPOE-HAUTIN is professor at the University of La Réunion. The main thrust of her research is the study of cinema, both as an art and as an industry, in relation to questions of identity in Indian Ocean societies in colonial and postcolonial contexts (nineteenth/twentieth centuries). Her interest in cinema derives from her father, Robin Tampoe, and her grandfather, W.M.S. Tampoe, who were among those who laid the foundation stones of the Sinhala talkie during the formative period of South Indian and Sri Lankan cinemas (1935–65). Her mixed cultural heritage as well as her childhood during the throes of the Sinhala-Tamil ethnic conflict have enabled her to study cinema with a focus on ethno-politics, seen from within. Her more recent projects focus on the restoration, conservation, and accessing of film and non-film archives in South Asia, with a forthcoming publication, *Bringing Back a By-Gone Era: the Conservation of Sri Lanka's Cinema Heritage*. She is the author of a number of articles and books in French and in English, including two biographies of Sri Lankan filmmakers, *Robin Tampoe: Last of the Big Ones* (2008) and *Sumitra Peries: Poetess of Sinhala Cinema* (2011). In 2016, she published *Ethnicity, Politics and Cinema in Sri Lanka: From Global to 'Gama,' Casting a Celluloid Mould from Sinhala Cinema, 1900–1967*. A two-volume study of Sri Lankan cinema, a pioneering work on the topic published in 2011 by L'Harmattan, completes her bibliography: *Cinéma et Colonialisme: La genèse du cinéma au Sri Lanka (de 1896 à 1928)*; and *Cinéma et Conflits ethniques au Sri Lanka: Vers un cinéma cinghalais 'indigène' (de l'indépendance en 1948 à nos jours)*.

GILLES TEULIÉ is professor of British and South African studies at Aix-Marseille University. He has written extensively on South African history and the Victorian period. He has published a book on the Afrikaners and the Anglo-Boer War: *Les Afrikaners et la guerre anglo-Boer 1899–1902: Etude des cultures populaires et des mentalités en présence* (Montpellier University Press, 2000) as well as one on racial attitudes in Victorian South Africa: *La racialisation de l'Afrique du Sud dans l'imaginaire colonial* (Paris: L'Harmattan, 2015). He is the editor and co-editor of several collections of essays, among which are: *Religious Writings and War* (2006); *Victorian Representations of War* (2007); *War Sermons,* with Laurence Sterritt (2009); *Healing South African Wounds,* with Mélanie Joseph-Vilain (2000); *L'Afrique du Sud de Nouvelles identités?*, with Marie-Claude Barbier (2010); and *Spaces of History, History of Spaces*, with Matthew Graves (2017, https://journals.openedition.org/erea/5875). He is currently working on war memories, theology, and apartheid as well as the mediatization of European empires through early picture postcards.

STEPHEN J. WHITFIELD is emeritus professor of American studies at Brandeis University in Waltham, Massachusetts. He is the author of nine books, including *A Death in the Delta: The Story of Emmett Till*; *The Culture of the Cold War*; *In Search of American Jewish Culture*; and, in 2020, *Learning on the Left: Political Profiles of Brandeis University*. He served as a Fulbright visiting professor of American studies at the Hebrew University of Jerusalem and the Catholic University of Leuven and Louvain-la-Neuve. He has also served as a visiting professor at the University of Paris IV (Sorbonne) and the University of Munich.

VICTORIA YOUNG is professor of Modern Architectural History and chair of the Art History Department at the University of St Thomas in St Paul, Minnesota. She currently serves as the president of the international Society of Architectural Historians. Her research interests are sacred space and war museums. In 2014, Dr Young's book, *Saint John's Abbey Church: Marcel Breuer and the Creation of a Modern Sacred Space*, was published by the University of Minnesota Press. Most recently she has written on the role of concrete as a sacred material for the abbey church in the American Institute of Architects' journal, *Faith and Form*. This article compares Breuer's work to that of Voorsanger Architects in New York City and their design for a chapel of precast concrete panels in the National World War II Museum in New Orleans. This is a part of her current research on the museum's campus design. In spring 2018, Dr Young opened a digital and physical archive of the firm's work (https://voorsangerarchive.org/) and associated exhibition, *Preserving the Present: Creating the Voorsanger Architects Archive at the University of St Thomas*. Dr Young is the 2018 recipient of the Special Award from the American Institute of Architects Minnesota Chapter, given to an individual for exceptional contributions to AIA, the profession, or the quality of the built environment. In this same year the University of St Thomas faculty also named her as professor of the year.

Index